APPLIED ALGEBRA II

THOMAS J. McHALE
PAUL T. WITZKE
Milwaukee Area Technical College, Milwaukee, Wisconsin

ADDISON-WESLEY PUBLISHING COMPANY
Reading, Massachusetts
Menlo Park, California · London · Amsterdam · Don Mills, Ontario · Sydney

OTHER BOOKS OF INTEREST

APPLIED ALGEBRA I
INTRODUCTORY ALGEBRA - Programmed
INTRODUCTORY ALGEBRA
CALCULATION AND CALCULATORS

Reproduced by Addison-Wesley from camera-ready copy supplied by the authors.

Copyright © 1980 by Addison-Wesley Publishing Company, Inc. Philippines copyright 1980 by Addison-Wesley Publishing Company, Inc.

All rights reserved. No part of this publication may be reproduced, stored in a retrieval system, or transmitted, in any form or by any means, electronic, mechanical, photocopying, recording, or otherwise, without the prior written permission of the publisher. Printed in the United States of America. Published simultaneously in Canada.

ISBN 0-201-04775-6

20 21 22-DPC-02 01 00

PREFACE

APPLIED ALGEBRA I and APPLIED ALGEBRA II are designed to teach the algebraic concepts and skills needed in basic science, technology, pre-engineering, and mathematics itself. The word "applied" in the title does not mean that major emphasis is given to applied problems, though many basic types of applied problems are included. Rather, the word "applied" means that emphasis is given to topics that are broadly useful, including manipulations with formulas and problem-solving based on formulas. Written at an intermediate level, the texts presuppose one year of high school algebra or the equivalent. For students without that prerequisite or students with serious remedial problems, INTRODUCTORY ALGEBRA can be used as a preparatory text.

The content of APPLIED ALGEBRA II is presented in a programmed format with assignment self-tests with answers within each chapter and supplementary problems for each assignment at the end of each chapter. Answers for all supplementary problems are given in the back of the text.

A calculator is an integral part of the instruction in some chapters. It is used for powers and roots in Chapter 6, for powers of ten and common logarithms in Chapter 7, for powers of "e" and natural logarithms in Chapter 8, and for some determinants in Chapter 9.

APPLIED ALGEBRA II is accompanied by the book TESTS FOR APPLIED ALGEBRA II which contains a diagnostic test, thirty assignment tests, nine chapter tests, four multi-chapter tests (every two chapters except for Chapter 9), and a comprehensive test. Three parallel forms are provided for chapter tests, multi-chapter tests, and the comprehensive test. A full set of answer keys for the tests is included. Because of the large number of tests provided, various options are possible in using them. For example, an instructor can use the chapter tests alone, the multi-chapter tests alone, or some combination of the two types.

> Note: The test book is provided only to teachers. Copies of the tests for student use must be made by some copying process.

The following features make the instruction effective and efficient for students:

1. The instruction, which is based on a task analysis, contains examples of all types of problems that appear in the tests.

2. The full and flexible set of tests can be used as a teaching tool to identify learning difficulties which can be remedied by tutoring or class discussion.

3. Because of the programmed format and the full and flexible set of tests provided, the text is ideally suited for individualized instruction as well as more traditional methods.

The authors wish to thank Ms. Marylou Hansen and Ms. Arleen D'Amore who typed and proofread the camera-ready copy, Ms. Joan Fishler who prepared some of the graphs, Mr. Gail Davis who did the final proofreading, and Mr. Allan Christenson who prepared the Index.

HOW TO USE THE TEXT AND TESTS

This text and the tests available in TESTS FOR APPLIED ALGEBRA II can be used in various instructional strategies ranging from paced instruction with all students taking the same content to totally individualized instruction. The general procedure for using the materials is outlined below.

1. The diagnostic test can be administered either to simply get a measure of the entry skills of the students or as a basis for prescribing an individualized program.

2. Each chapter is covered in a number of assignments (see below). After the students have completed each assignment and the assignment self-test (in the text), the assignment test (from the test book) can be administered, corrected, and used as a basis for tutoring. The assignment tests are simply a teaching tool and need not be graded. The supplementary problems at the end of the chapter can be used at the instructor's discretion for students who need further practice.

   Note: Instead of using the assignment self-tests (in the text) as an integral part of the assignments, they can be used at the completion of a chapter as a chapter review exercise.

3. After the appropriate assignments are completed, either a chapter test or a multi-chapter test can be administered. Ordinarily, these tests should be graded. Parallel forms are provided to facilitate the test administration, including the retesting of students who do not achieve a satisfactory score.

4. After all desired chapters are completed in the manner above, the comprehensive test can be administered. Since the comprehensive test is a parallel form of the diagnostic test, the difference score can be used as a measure of each student's improvement.

ASSIGNMENTS FOR APPLIED ALGEBRA II

Ch. 1:	#1 (pp. 1-15)	Ch. 4:	#12 (pp. 169-184)	Ch. 7:	#21 (pp. 302-315)
	#2 (pp. 15-27)		#13 (pp. 184-198)		#22 (pp. 315-326)
	#3 (pp. 27-43)		#14 (pp. 198-211)		#23 (pp. 327-338)
	#4 (pp. 43-55)				#24 (pp. 338-348)
Ch. 2:	#5 (pp. 58-74)	Ch. 5:	#15 (pp. 215-229)	Ch. 8:	#25 (pp. 353-367)
	#6 (pp. 74-88)		#16 (pp. 230-244)		#26 (pp. 367-381)
	#7 (pp. 89-104)		#17 (pp. 244-257)		#27 (pp. 382-396)
Ch. 3:	#8 (pp. 109-124)	Ch. 6:	#18 (pp. 262-272)	Ch. 9:	#28 (pp. 401-411)
	#9 (pp. 124-137)		#19 (pp. 273-284)		#29 (pp. 411-423)
	#10 (pp. 137-152)		#20 (pp. 285-298)		#30 (pp. 423-431)
	#11 (pp. 153-165)				

C O N T E N T S

CHAPTER 1: POLYNOMIALS   (Pages 1-57)

- 1-1   The Meaning Of Powers   1
- 1-2   Multiplying Powers   3
- 1-3   Dividing Powers   5
- 1-4   Powers With "0" Exponents   7
- 1-5   Raising A Power To A Power   8
- 1-6   Polynomials In One Variable   10
- 1-7   Polynomials In More Than One Variable   12
  - Self-Test 1   15
- 1-8   Combining Like Terms   15
- 1-9   Simplifying Polynomials By Combining Like Terms   17
- 1-10   Adding Polynomials   19
- 1-11   Opposites Of Polynomials   21
- 1-12   Subtracting Polynomials   22
- 1-13   Multiplying And Squaring Monomials   24
- 1-14   Multiplying Other Polynomials By A Monomial   25
  - Self-Test 2   26
- 1-15   Multiplying Binomials   27
- 1-16   Multiplying The Sum And Difference Of Two Terms   30
- 1-17   Squaring Binomials   31
- 1-18   Factoring Monomials   33
- 1-19   Common Monomial Factors   35
- 1-20   Factoring The Difference Of Two Perfect Squares   37
- 1-21   Factoring Perfect-Square Trinomials   40
  - Self-Test 3   42
- 1-22   Factoring Trinomials Of The Form: $x^2 + bx + c$ 43
- 1-23   Factoring Trinomials Of The Form: $ax^2 + bx + c$ 47
- 1-24   Factoring Polynomials Completely   50
- 1-25   Dividing By Monomials   51
- 1-26   Dividing By Binomials   53
  - Self-Test 4   55
  - Supplementary Problems - Chapter 1   56

CHAPTER 2: LINEAR EQUATIONS   (Pages 58-108)

- 2-1   Identifying Linear Equations   58
- 2-2   Intercepts   60
- 2-3   Graphing Using Intercepts   62
- 2-4   Slope As A Signed Number   63
- 2-5   Slope As A Ratio   67
- 2-6   Slope-Intercept Form   71
  - Self-Test 5   73
- 2-7   Writing Linear Equations In Slope-Intercept Form   74
- 2-8   The Two-Point Formula For Slope   77
- 2-9   Finding The Equation Of A Line Given Its Slope And One Point   79
- 2-10   Finding The Equation Of A Line Given Two Points   80
- 2-11   Lines Through The Origin   81
- 2-12   Parallel Lines   84
- 2-13   Horizontal And Vertical Lines   85
  - Self-Test 6   88

2-14	Intercepts Of Linear Equations With Other Variables	89
2-15	Slope Of Linear Equations With Other Variables	92
2-16	Slope-Intercept Form Of Linear Equations With Other Variables	94
2-17	Lines Through The Origin With Other Variables	97
2-18	Estimating The Equations Of Graphed Lines	99
2-19	Slope And Axes With Different Scales	102
	Self-Test 7	104
	Supplementary Problems - Chapter 2	105

CHAPTER 3: SQUARE ROOT RADICALS (Pages 109-168)

3-1	Square Roots Of Positive Numbers	109
3-2	Formula Evaluations	111
3-3	Multiplying Radicals	113
3-4	Simplifying Radicals By Factoring Out Perfect-Square Numbers	116
3-5	Simplifying Radicals By Factoring Out Perfect-Square Powers	119
3-6	Dividing Radicals	122
	Self-Test 8	123
3-7	Simplifying Radicals Containing A Fraction	124
3-8	Simplifying Radicals Containing Indicated Squares	127
3-9	Squaring Radicals	129
3-10	Rationalizing Denominators	130
3-11	Rationalizing Numerators	132
3-12	Adding And Subtracting Radicals	134
	Self-Test 9	136
3-13	Radicals Containing Binomials	137
3-14	Operations With Radicals Containing Binomials	139
3-15	Squaring Terms Containing A Radical	143
3-16	Squaring Binomials Containing A Radical	146
3-17	Conjugates And Rationalizing	149
	Self-Test 10	152
3-18	Real And Imaginary Numbers	153
3-19	Adding And Subtracting Complex Numbers	155
3-20	Multiplying Complex Numbers And Conjugates	158
3-21	Dividing Complex Numbers	161
3-22	Imaginary Numbers And Complex Numbers As Solutions Of Equations	163
	Self-Test 11	165
	Supplementary Problems - Chapter 3	166

CHAPTER 4: RADICAL EQUATIONS AND FORMULAS (Pages 169-214)

4-1	The Squaring Principle For Radical Equations	169
4-2	Isolating Radicals In Radical Equations	173
4-3	Isolating Radicals And Solving Radical Equations	176
4-4	Formula Evaluations Requiring Equation-Solving	179
4-5	Extraneous Solutions	182
	Self-Test 12	183
4-6	The Squaring Principle And Radical Formulas	184
4-7	Isolating Radicals And Rearranging Formulas	190
4-8	Solving For Non-Radicand Variables	193
4-9	Radical Formulas And Squared Variables	196
	Self-Test 13	197

4-10	Simplifying Radical Solutions By Factoring Out Perfect Squares	198
4-11	The Equivalence Method And Formula Derivation	201
4-12	The Substitution Method And Formula Derivation	206
4-13	Formulas In Quadratic Form	208
	Self-Test 14	211
	Supplementary Problems - Chapter 4	212

CHAPTER 5: FUNCTIONS, NON-LINEAR GRAPHS, VARIATION   (Pages 215-261)

- 5-1   Functions   215
- 5-2   Graphs of Functions   218
- 5-3   Functional notation   219
- 5-4   Functions With More Than One Independent Variable   223
- 5-5   Linear Functions   224
- 5-6   Parabolas (Quadratic Functions)   225
-       Self-Test 15   229
- 5-7   Circles   230
- 5-8   Ellipses   232
- 5-9   Hyperbolas   234
- 5-10 Variation   235
- 5-11 Direct Variation   237
- 5-12 Inverse Variation   240
-       Self-Test 16   243
- 5-13 Direct Square Variation   244
- 5-14 Inverse Square Variation   248
- 5-15 Variation With More Than One Independent Variable   250
- 5-16 Joint Variation   252
- 5-17 Combined Variation   254
-       Self-Test 17   257
-       Supplementary Problems - Chapter 5   258

CHAPTER 6: EXPONENTS, POWERS, AND ROOTS   (Pages 262-301)

- 6-1   Powers   262
- 6-2   Multiplying Powers   264
- 6-3   Dividing Powers   266
- 6-4   Converting Division To Multiplication   268
- 6-5   Writing Expressions Without Negative Exponents   270
-       Self-Test 18   272
- 6-6   Raising A Power To A Power   273
- 6-7   Powers Of Monomials   274
- 6-8   Powers Of Fractions   276
- 6-9   Roots   278
- 6-10 Roots Of Powers   281
-       Self-Test 19   284
- 6-11 Powers With Fractional Exponents   285
- 6-12 Powers With Decimal Exponents   288
- 6-13 Finding Powers With A Calculator   290
- 6-14 Evaluating Formulas Containing Powers   293
- 6-15 Graphs Of Exponential Functions   295
-       Self-Test 20   298
-       Supplementary Problems - Chapter 6   299

CHAPTER 7: COMMON LOGARITHMS   (Pages 302-352)

- 7-1   Powers Of Ten   302
- 7-2   Powers Of Ten With Decimal Exponents   304
- 7-3   Graphs Of Base-10 Exponential Functions   306
- 7-4   Common Logarithms   308
- 7-5   Decimal Logarithms   311
- 7-6   Graph Of The Logarithmic Function   313
  Self-Test 21   314
- 7-7   Solving Power-Of-Ten And "Log" Equations   315
- 7-8   Evaluating "Log" Formulas   317
- 7-9   Evaluations Requiring Equation-Solving   321
- 7-10 Evaluations Requiring A Conversion To Power-Of-Ten Form   322
- 7-11 Rearranging "Log" Formulas   324
  Self-Test 22   326
- 7-12 Logarithmic Scales   327
- 7-13 Semi-Log And Log-Log Graphs   330
- 7-14 Laws Of Logarithms   332
- 7-15 Using The Laws Of Logarithms To Rearrange "Log" Formulas   334
  Self-Test 23   337
- 7-16 Exponential Equations   338
- 7-17 Combined Use Of The Laws Of Logarithms   341
- 7-18 Evaluating Exponential Formulas   343
  Self-Test 24   348
  Supplementary Problems - Chapter 7   349

CHAPTER 8: POWERS OF "e" AND NATURAL LOGARITHMS   (Pages 353-400)

- 8-1   Powers Of "e"   353
- 8-2   Graphing Functions Containing Powers Of "e"   355
- 8-3   Evaluating Formulas Containing Powers Of "e"   359
- 8-4   Graphing Formulas Containing Powers Of "e"   365
  Self-Test 25   366
- 8-5   Natural Logarithms   367
- 8-6   Solving Power-Of-"e" And "ln" Equations   370
- 8-7   Evaluating "ln" Formulas   372
- 8-8   Evaluations Requiring A Conversion To Power-Of-"e" Form   376
- 8-9   Evaluations Requiring A Conversion To "ln" Form   378
  Self-Test 26   381
- 8-10 Laws Of Natural Logarithms   382
- 8-11 Rearranging "ln" Formulas   384
- 8-12 Rearrangements Requiring A Conversion To Power-Of-"e" Form   386
- 8-13 Rearranging Formulas Containing Powers Of "e"   388
- 8-14 "exp" Notation For Power-Of-"e" Formulas   394
  Self-Test 27   396
  Supplementary Problems - Chapter 8   397

CHAPTER 9: SYSTEMS OF THREE EQUATIONS AND DETERMINANTS (Pages 401-433)

- 9-1 A Review Of Three Methods 401
- 9-2 Second Order Determinants 403
- 9-3 Using Determinants To Solve Systems Of Two Equations 405
- 9-4 Converting To Standard Form For The Determinant Method 408
  Self-Test 28 410
- 9-5 Systems Of Three Equations - The Addition Method 411
- 9-6 Converting To Standard Form For The Addition Method 414
- 9-7 Systems Of Three Equations - The Equivalence Method 416
- 9-8 Systems Of Three Formulas And Formula Derivation 419
  Self-Test 29 422
- 9-9 Third Order Determinants 423
- 9-10 Using Determinants To Solve Systems Of Three Equations 427
  Self-Test 30 431
  Supplementary Problems - Chapter 9 432

ANSWERS FOR SUPPLEMENTARY PROBLEMS (Pages 435-440)

INDEX (Pages 441-444)

Chapter 1 POLYNOMIALS

In this chapter, we will discuss powers, the basic laws of exponents, and the four basic operations with polynomials. Special products of binomials and various types of factoring polynomials are included.

1-1 THE MEANING OF POWERS

In this section, we will discuss the meaning of powers with positive whole-number exponents.

1. Expressions like 4^3 and 5^8 are in <u>base-exponent</u> form. The ordinary number at the left is called the <u>base</u>; the small number is called the <u>exponent</u>. That is:

    $\underset{\text{Base}}{\nearrow} 4^{3} \underset{\text{Exponent}}{\nwarrow} \qquad \underset{\text{Base}}{\nearrow} 5^{8} \underset{\text{Exponent}}{\nwarrow}$

    In 2^5:   a) The base is _____.   b) The exponent is _____.

2. A base-exponent expression is a short way of writing <u>a multiplication of identical factors</u>. For example:

    $5^3 = (5)(5)(5)$ (The exponent tells us to multiply <u>three</u> 5's.)

    $2^4 = (2)(2)(2)(2)$ (The exponent tells us to multiply <u>four</u> 2's.)

    Write each expression as a multiplication of identical factors.

    a) $8^3 = $ _____   b) $4^5 = $ _____

        a) 2   b) 5

3. In a base-exponent expression, the base can be a letter. Expressions of that type also stand for a multiplication of identical factors. That is:

    $x^2 = (x)(x)$ (The exponent tells us to multiply <u>two</u> x's.)

    $y^4 = (y)(y)(y)(y)$ (The exponent tells us to multiply <u>four</u> y's.)

    Write each expression as a multiplication of identical factors.

    a) $t^3 = $ _____   b) $d^6 = $ _____

        a) (8)(8)(8)
        b) (4)(4)(4)(4)(4)

        a) (t)(t)(t)
        b) (d)(d)(d)(d)(d)(d)

Polynomials

4. Any multiplication of identical factors can be written in base-exponent form. For example:

$(5)(5)(5) = 5^3$ (Since there are three 5's, the exponent is 3.)

$(x)(x)(x)(x) = x^4$ (Since there are four x's, the exponent is 4.)

Write each multiplication in base-exponent form.

a) (7)(7) = _____   b) (y)(y)(y)(y)(y)(y) = _____

a) 7^2 b) y^6

5. Any base-exponent expression is called a "power" of the base. For example:

6^2, 6^3, and 6^5 are called "powers of 6".

x^3, x^4, and x^7 are called "powers of x".

The following language is used to distinguish the powers of the same number or letter.

4^2 is called "4 to the second power".

4^3 is called "4 to the third power".

4^5 is called "4 to the fifth power".

Similarly: a) 2^4 is called "2 to the _____ power".

b) x^7 is called "x to the _____ power".

a) fourth
b) seventh

6. The words "squared" and "cubed" are frequently used for the second and third power of any base. That is:

5^2 is called "5 to the second power" or "5 squared".

x^3 is called "x to the third power" or "x cubed".

Write the base-exponent expression corresponding to these:

a) y squared = _____   b) 9 cubed = _____

a) y^2 b) 9^3

7. When the base is a number, we can convert the power to an ordinary number by performing the multiplication. That is:

$4^3 = (4)(4)(4) = 64$

a) $3^4 = (3)(3)(3)(3) =$ _____   b) $2^5 = (2)(2)(2)(2)(2) =$ _____

a) 81   b) 32

8. When the base is a letter, we can only convert the power to an ordinary number by substituting a number for the letter. That is:

If $y = 5$, $y^3 = 5^3 = (5)(5)(5) = 125$

if $t = 2$, $t^4 = 2^4 = (2)(2)(2)(2) =$ _____

Polynomials 3

9. Any power whose exponent is "1" is equal to its base. That is:

$$5^1 = 5 \qquad x^1 = x$$

The definition above fits the pattern for powers. That is:

$$5^3 = (5)(5)(5) \qquad x^3 = (x)(x)(x)$$
$$5^2 = (5)(5) \qquad x^2 = (x)(x)$$
$$5^1 = (5) \text{ or } 5 \qquad x^1 = (x) \text{ or } x$$

Using the definition above, complete these:

a) $7^1 = $ _____   b) $2^1 = $ _____   c) $y^1 = $ _____

16

---

10. By reversing the definition in the last frame, we can write any number or letter as a power. That is:

$$3 = 3^1 \qquad m = m^1 \qquad \text{a) } 8 = \underline{\qquad} \qquad \text{b) } t = \underline{\qquad}$$

a) 7
b) 2
c) y

---

a) 8^1 b) t^1

1-2 MULTIPLYING POWERS

In this section, we will discuss the law of exponents for multiplying powers with the same base.

11. Each expression below means "multiply 7^2 and 7^4".

$$7^2 \times 7^4 \qquad\qquad 7^2 \cdot 7^4 \qquad\qquad (7^2)(7^4)$$

To perform the multiplication, we can write each factor as a multiplication of 7's and then count the total number of 7's to get the exponent of the product. That is:

$$7^2 \cdot 7^4 = (7)(7) \cdot (7)(7)(7)(7) = 7^6$$

Count the total number of x's to write the product below as a power.

$$x^5 \cdot x^3 = (x)(x)(x)(x)(x) \cdot (x)(x)(x) = \underline{\qquad}$$

---

12. The two multiplications from the last frame are shown below.

$$7^2 \cdot 7^4 = 7^6 \qquad\qquad x^5 \cdot x^3 = x^8$$

In each, the exponent of the product is the sum of the exponents of the factors. That is:

$$7^2 \cdot 7^4 = 7^{2+4} = 7^6 \qquad x^5 \cdot x^3 = x^{5+3} = x^8$$

x^8

Continued on following page.

4   Polynomials

12. Continued

Therefore, we can use the following law of exponents to multiply powers with the same base.

$$a^m \cdot a^n = a^{m+n}$$

Using the law of exponents for multiplication, complete these:

a) $5^4 \cdot 5^3 = $ _____   b) $9^1 \cdot 9^2 = $ _____   c) $y^5 \cdot y^5 = $ _____

---

13. The law of exponents for multiplication applies <u>only if the powers have the same base</u>. For example, it does not apply to the multiplication below.

$$2^3 \cdot 5^2 = (2)(2)(2) \cdot (5)(5)$$

Note: The product cannot be written as a single power because all of the factors are not identical.

If it applies, use the law of exponents to write the products.

a) $3^3 \cdot 3^3 = $ _____   b) $5^3 \cdot 6^4 = $ _____   c) $x^2 \cdot y^6 = $ _____

a) 5^7
b) 9^3
c) y^{10}

---

14. To use the law of exponents for the multiplications below, we substituted 4^1 for 4 and x^1 for x.

$$4 \cdot 4^3 = 4^1 \cdot 4^3 = 4^4 \qquad x^5 \cdot x = x^5 \cdot x^1 = x^6$$

Use the same method to complete these:

a) $2 \cdot 2^5 = $ _____   b) $y \cdot y^4 = $ _____   c) $m^2 \cdot m = $ _____

a) 3^6
b) Does not apply
c) Does not apply

---

15. The law of exponents can also be used for multiplications containing more than two factors. For example:

$$x^3 \cdot x^2 \cdot x^5 \cdot x = x^{3+2+5+1} = x^{11}$$

Following the example, complete these:

a) $y^4 \cdot y^6 \cdot y^2 = $ _____   b) $m^3 \cdot m \cdot m^3 \cdot m^2 = $ _____

a) 2^6
b) y^5
c) m^3

---

16. When multiplying powers with different letters as the base, we write them side by side in alphabetical order. That is:

$$x^3 \cdot y^4 = x^3 y^4 \qquad b \cdot a^2 \cdot d^5 = a^2 b d^5$$

The multiplications below contain expressions like those above. Notice how we used the law of exponents twice in each.

$$(x^2 y^5)(x^3 y^4) = x^2 \cdot x^3 \cdot y^5 \cdot y^4 = x^5 y^9$$
$$(ab^4)(a^6 b) = a \cdot a^6 \cdot b^4 \cdot b = a^7 b^5$$

a) y^{12} b) m^9

Continued on following page.

16. Continued

    Following the examples, do these:

    a) $(m^4p^2)(m^7p^5) = $ _____   b) $(c^4d)(cd^9) = $ _____

---

17. Notice how we used the law of exponents three times in the multiplication below.

    $(xy^5z^6)(x^3y^4z) = x \cdot x^3 \cdot y^5 \cdot y^4 \cdot z^6 \cdot z = x^4y^9z^7$

    a) $m^{11}p^7$
    b) c^5d^{10}

    Following the example, do these:

    a) $(m^4pq^2)(mp^3q^6) = $ _____
    b) $(a^2b^4c)(a^2bc^5)(ab^3c^4) = $ _____

---

a) $m^5p^4q^8$
b) $a^5b^8c^{10}$

1-3 DIVIDING POWERS

In this section, we will discuss the law of exponents for dividing powers with the same base.

18. To perform the division below, we began by writing each term as a multiplication of 4's. Then we factored out $\left(\frac{4 \cdot 4 \cdot 4}{4 \cdot 4 \cdot 4}\right)$ which equals "1".

    $$\frac{4^5}{4^3} = \frac{4 \cdot 4 \cdot 4 \cdot 4 \cdot 4}{4 \cdot 4 \cdot 4} = \left(\frac{4 \cdot 4 \cdot 4}{4 \cdot 4 \cdot 4}\right)(4 \cdot 4) = (1)(4 \cdot 4) = 4 \cdot 4 = 4^2$$

    Write the quotient below as a power.

    $$\frac{x^4}{x^1} = \frac{x \cdot x \cdot x \cdot x}{x} = \left(\frac{x}{x}\right)(x \cdot x \cdot x) = (1)(x \cdot x \cdot x) = x \cdot x \cdot x = \underline{\qquad}$$

---

19. The two divisions from the last frame are shown below.

    $\dfrac{4^5}{4^3} = 4^2$ $\qquad$ $\dfrac{x^4}{x^1} = x^3$

    x^3

    In each, <u>the exponent of the quotient can be obtained by subtracting the exponent in the denominator from the exponent in the numerator.</u>

    $\dfrac{4^5}{4^3} = 4^{5-3} = 4^2$ $\qquad$ $\dfrac{x^4}{x^1} = x^{4-1} = x^3$

    Therefore, we can use the following law of exponents to divide powers with the same base.

    $$\boxed{\dfrac{a^m}{a^n} = a^{m-n}}$$

Continued on following page.

6   Polynomials

19. Continued

Using the law of exponents for division, complete these:

a) $\dfrac{5^6}{5^2} =$ _____   b) $\dfrac{9^5}{9^4} =$ _____   c) $\dfrac{t^8}{t^1} =$ _____

---

20. The law of exponents for division applies <u>only if the powers have the same base.</u> For example, it does not apply to the division below since we cannot factor out a fraction that equals "1".

$$\dfrac{6^4}{5^2} \qquad \dfrac{6 \cdot 6 \cdot 6 \cdot 6}{5 \cdot 5}$$

If it applies, use the law of exponents to write the quotient.

a) $\dfrac{2^5}{3^4} =$ _____   b) $\dfrac{x^2}{x^1} =$ _____   c) $\dfrac{m^8}{t^5} =$ _____

a) 5^4
b) 9^1 or 9
c) t^7

---

21. To use the law of exponents for the divisions below, we substituted 3^1 for 3 and y^1 for y.

$$\dfrac{3^3}{3} = \dfrac{3^3}{3^1} = 3^2 \qquad \dfrac{y^6}{y} = \dfrac{y^6}{y^1} = y^5$$

Use the same method to complete these:

a) $\dfrac{7^5}{7} =$ _____   b) $\dfrac{4^2}{4} =$ _____   c) $\dfrac{a^{10}}{a} =$ _____

a) Does not apply
b) x^1 or x
c) Does not apply

---

22. To simplify the expression below, we used the law of exponents twice.

$$\dfrac{x^5 y^4}{x^3 y} = \left(\dfrac{x^5}{x^3}\right)\left(\dfrac{y^4}{y}\right) = x^2 y^3$$

Following the example, simplify these.

a) $\dfrac{c^9 d^8}{c d^6} =$ _____   b) $\dfrac{p^7 q^2}{p^6 q} =$ _____

a) 7^4
b) 4^1 or 4
c) a^9

---

23. To simply the expression below, we used the law of exponents three times.

$$\dfrac{a^4 b^6 c^8}{a^3 b c^3} = \left(\dfrac{a^4}{a^3}\right)\left(\dfrac{b^6}{b}\right)\left(\dfrac{c^8}{c^3}\right) = a b^5 c^5$$

Following the example, simplify these:

a) $\dfrac{x^9 y^5 z^8}{x y^4 z^5} =$ _____   b) $\dfrac{m^2 p^{10} q^7}{m p^5 q} =$ _____

a) $c^8 d^2$
b) $p^1 q^1$ or pq

---

a) $x^8 y z^3$
b) $m p^5 q^6$

1-4 POWERS WITH "0" EXPONENTS

In this section, we will define the meaning of any power with a "0" exponent.

24. **Definition:** When the exponent of a power is "0", the power equals "+1". That is: $$5^0 = 1 \quad x^0 = 1 \quad \text{a) } 9^0 = \underline{\quad} \quad \text{b) } t^0 = \underline{\quad}$$	
25. The above definition of a power with a "0" exponent is consistent with the law of exponents for multiplication. To show that fact, we performed the same multiplications in two different ways below. 1. Substituting the definition of each power for the power. $$5^0 \cdot 5^2 = (1) \cdot (5)(5) = 5^2 \qquad x^3 \cdot x^0 = (x)(x)(x) \cdot (1) = x^3$$ 2. Using the law of exponents for multiplication. $$5^0 \cdot 5^2 = 5^{0+2} = 5^2 \qquad x^3 \cdot x^0 = x^{3+0} = x^3$$ Use the law of exponents to complete these: a) $3^0 \cdot 3^4 = \underline{\quad}$ b) $8^6 \cdot 8^0 = \underline{\quad}$ c) $m^0 \cdot m^5 = \underline{\quad}$	a) 1 b) 1
26. The definition of a power with a "0" exponent is also consistent with the law of exponents for division. To show that fact, we performed the same division in two different ways. 1. Using this fact: If any quantity is divided by "1", the quotient is the original quantity. $$\frac{3^5}{3^0} = \frac{3^5}{1} = 3^5 \qquad \frac{t^4}{t^0} = \frac{t^4}{1} = t^4$$ 2. Using the law of exponents for division. $$\frac{3^5}{3^0} = 3^{5-0} = 3^5 \qquad \frac{t^4}{t^0} = t^{4-0} = t^4$$ Use the law of exponents to complete these: a) $\frac{7^2}{7^0} = \underline{\quad}$ b) $\frac{2^1}{2^0} = \underline{\quad}$ c) $\frac{m^8}{m^0} = \underline{\quad}$	a) 3^4 b) 8^6 c) m^5
27. To show that the definition of a power with a "0" exponent is consistent with the law of exponents for division, we performed two more divisions in two different ways below. 1. Using this fact: If any quantity is divided by itself, the quotient is "1". $$\frac{4^3}{4^3} = 1 \qquad \frac{x^5}{x^5} = 1$$	a) 7^2 b) 2^1 or 2 c) m^8

Continued on following page.

8 Polynomials

27. Continued

    2. <u>Using the law of exponents for division.</u>

$$\frac{4^3}{4^3} = 4^{3-3} = 4^0 \text{ (or 1)} \qquad \frac{x^5}{x^5} = x^{5-5} = x^0 \text{ (or 1)}$$

Use the law of exponents to complete these:

a) $\frac{6^4}{6^3}$ = _____   b) $\frac{6^4}{6^4}$ = _____   c) $\frac{y^7}{y^6}$ = _____   d) $\frac{y^6}{y^6}$ = _____

---

28. Complete: a) $2 \cdot 2^0$ = _____   c) $x \cdot x^0$ = _____

                b) $\frac{2}{2^0}$ = _____   d) $\frac{x}{x^0}$ = _____

     a) 6^1 or 6
     b) 6^0 or 1
     c) y^1 or y
     d) y^0 or 1

---

29. Notice how we substituted "1" for x^0 below.

$$\frac{x^3 y^8}{x^3 y} = x^0 y^7 = 1(y^7) = y^7$$

Following the example, do these:

a) $\frac{p^{10} q^7}{p^8 q^7}$ = _____   b) $\frac{a^4 b^5 c^6}{a^3 b^5 c^5}$ = _____

     a) 2^1 or 2
     b) 2^1 or 2
     c) x^1 or x
     d) x^1 or x

---

     a) p^2 b) ac

1-5   RAISING A POWER TO A POWER

In this section, we will discuss the law of exponents for raising a power to a power.

30. In a base-exponent expression, the base can be a power. For example:

     In $(5^3)^2$: the base is 5^3; the exponent is "2".
     In $(y^2)^4$: the base is y^2; the exponent is "4".

Expressions like those above are also a short way of writing a multiplication of identical factors.

     $(5^3)^2 = 5^3 \cdot 5^3$ (There are <u>two</u> 5^3's.)
     $(y^2)^4 = y^2 \cdot y^2 \cdot y^2 \cdot y^2$ (There are <u>four</u> y^2's.)

Write each power as a multiplication of identical factors.

a) $(4^5)^3$ = _____

b) $(m^4)^5$ = _____

Polynomials 9

31. To raise a power to a power, we can convert to a multiplication. That is:

$$(3^5)^2 = 3^5 \cdot 3^5 = 3^{5+5} = 3^{10}$$

$$(y^3)^4 = y^3 \cdot y^3 \cdot y^3 \cdot y^3 = y^{3+3+3+3} = y^{12}$$

However, it is simpler to raise a power to a power <u>by multiplying the exponents</u>. That is:

$$(3^5)^2 = 3^{(5)(2)} = 3^{10} \qquad (y^3)^4 = y^{(3)(4)} = y^{12}$$

Therefore, we can use the following law of exponents to raise a power to a power.

$$\boxed{(a^m)^n = a^{mn}}$$

Using the law of exponents for raising a power to a power, complete these:

a) $(8^3)^2 = $ _____ b) $(p^2)^4 = $ _____ c) $(m^5)^6 = $ _____

a) $4^5 \cdot 4^5 \cdot 4^5$
b) $m^4 \cdot m^4 \cdot m^4 \cdot m^4 \cdot m^4$

32. Any power raised to the first power equals the original power. That is:

$$(2^4)^1 = 2^{(4)(1)} = 2^4 \qquad (y^2)^1 = y^{(2)(1)} = \underline{}$$

a) 8^6
b) p^8
c) m^{30}

33. Any power raised to the zero power equals "1". That is:

$$(8^3)^0 = 8^{(3)(0)} = 8^0 = 1 \qquad (m^5)^0 = m^{(5)(0)} = m^0 = \underline{}$$

y^2

34. To raise the expression below to a power, we converted to a multiplication.

$$(x^4y^5)^3 = (x^4y^5)(x^4y^5)(x^4y^5) = x^{12}y^{15}$$

However, it is simpler to multiply each exponent by 3. That is:

$$(x^4y^5)^3 = x^{(4)(3)}y^{(5)(3)} = x^{12}y^{15}$$

Using the shorter method, do these.

a) $(a^2b^5)^4 = $ _____ b) $(p^7q^8)^2 = $ _____

1

35. Notice how we substituted "x^1" for "x" in the example below.

$$(xy^4)^2 = (x^1y^4)^2 = x^2y^8$$

Following the example, do these:

a) $(c^5d)^3 = $ _____ b) $(pq)^5 = $ _____

a) a^8b^{20}
b) $p^{14}q^{16}$

10 Polynomials

36. The three basic laws of exponents are reviewed below.

$$a^m \cdot a^n = a^{m+n} \qquad \frac{a^m}{a^n} = a^{m-n} \qquad (a^m)^n = a^{mn}$$

Using the laws, complete these:

a) $7^4 \cdot 7^2 = $ _____   d) $x^5 \cdot x^3 = $ _____

b) $\frac{7^4}{7^2} = $ _____   e) $\frac{x^5}{x^3} = $ _____

c) $(7^4)^2 = $ _____   f) $(x^5)^3 = $ _____

a) $c^{15}d^3$
b) p^5q^5

37. Using the same laws, complete these:

a) $(xy^2)(x^2y) = $ _____   d) $(mp^2q)(m^7pq^6) = $ _____

b) $\frac{s^4t^5}{s^3t^5} = $ _____   e) $\frac{a^5b^7c^9}{ab^7c^9} = $ _____

c) $(cd^4)^4 = $ _____   f) $(xy)^{10} = $ _____

a) 7^6 d) x^8
b) 7^2 e) x^2
c) 7^8 f) x^{15}

a) x^3y^3 d) $m^8p^3q^7$
b) s e) a^4
c) c^4d^{16} f) $x^{10}y^{10}$

1-6 POLYNOMIALS IN ONE VARIABLE

In this section, we will define "polynomials in one variable" and introduce some terminology used in discussing them.

38. The terms below are called "<u>power-terms in one variable</u>". They contain a numerical coefficient and one power.

$3x^4 \qquad -2y^5 \qquad 6t^3 \qquad -9d^2$

To show that x^5 and $-y^3$ are power-terms, we can write their coefficients explicitly. That is:

x^5 can be written $1x^5$ $-y^3$ can be written _____

39. To show that $5x$ and $-7t$ are power-terms, we can write the exponent "1" of the power explicitly. That is:

$5x$ can be written $5x^1$ $-7t$ can be written _____

$-1y^3$

40. To show that "t" and "-p" are power-terms, we can write both the coefficient and the exponent of the letter explicitly. That is:

t can be written $1t^1$ -p can be written _____

$-7t^1$

Polynomials 11

41.	To show that numbers like 3 and -5 are power-terms, we can use the zero power of some letter. That is: $\quad$ 3 can be written $3x^0$, since $x^0 = 1$ $\quad$ -5 can be written $-5y^0$, since $y^0 =$ _____	$-1p^1$
42.	Any power-term in one variable is of the form ax^n where "a" is a signed whole number and "n" is a positive whole number or "0". Each term in the expressions below can be written explicitly as a power-term of that type. That is: $\quad 2x^3 - x^2 + x - 1$ can be written $2x^3 - 1x^2 + 1x^1 - 1x^0$ $\quad y^5 - 8y^3 - y + 5$ can be written _____	1
43.	However, when writing expressions containing power-terms, we do not usually write "1" and "-1" coefficients, "1" exponents, and zero powers. That is: $\quad 1t^3 - 1t^2 + 8t^1 + 3t^0$ is usually written $t^3 - t^2 + 8t + 3$ $\quad 2b^5 + 1b^4 - 1b^1 - 9b^0$ is usually written _____	$1y^5 - 8y^3 - 1y^1 + 5y^0$
44.	Any expression containing only power-terms in one variable is called a "<u>polynomial in one variable</u>". Some examples are shown. $\quad 3x \qquad y + 7 \qquad 2a^2 - 5a + 6 \qquad t^4 - 4t^3 + t^2 - 1$ A polynomial can contain one or more power-terms. For example: $\qquad 3x$ contains <u>one</u> power-term. $\qquad y + 7$ contains <u>two</u> power-terms. $\qquad t^4 \ldots$ contains _____ power-terms.	$2b^5 + b^4 - b - 9$
45.	... two, or three ...	four

12   Polynomials

45. Continued

State whether each expression is a monomial, binomial, or trinomial.

a) 8 _____   d) $5m^6 - 6m^5$ _____

b) $x^4 - 1$ _____   e) $-p$ _____

c) $2y^3 + y + 7$ _____   f) $a^4 - 3a^2 + 1$ _____

---

46. There are no special names for polynomials with more than three terms.

$2x^3 - x^2 + 4x - 6$ is called a polynomial with <u>four</u> terms.

$x^4 - 5x^3 + 2x^2 - x + 9$ is called a polynomial with _____ terms.

a) monomial
b) binomial
c) trinomial
d) binomial
e) monomial
f) trinomial

---

47. Polynomials in one variable are usually written <u>in</u> <u>descending</u> <u>order</u>. That is, they are written so that the power-terms are in order, with the largest exponent at the left. For example:

$3x^4 + 5x^3 + 6x^2 + 2x + 9$

To put each polynomial below in descending order, we simply rearrange the terms. That is:

$2 + 5y^4 + 3y^2$ is rearranged to $5y^4 + 3y^2 + 2$

$8b + b^2 + 1 + 4b^3$ is rearranged to _____

five

$4b^3 + b^2 + 8b + 1$

1-7 POLYNOMIALS IN MORE THAN ONE VARIABLE

In this section, we will define "polynomials in more than one variable" and introduce some terminology used in discussing them.

---

48. The terms below are called "<u>power-terms</u> <u>in</u> <u>more</u> <u>than</u> <u>one</u> <u>variable</u>". They contain a numerical coefficient and two or more powers.

$5x^2y^3$ $-7a^6b^5c^4$

To show that the expressions below are power-terms, we wrote "1" and "-1" coefficients or "1" exponents explicitly.

$c^8d^2 = 1c^8d^2$ $4ab^3 = 4a^1b^3$

$-x^5y^2z^4 = -1x^5y^2z^4$ $-9cx^2y = -9c^1x^2y^1$

Show that the expressions below are power-terms by writing both the coefficients and the "1" exponents explicitly.

a) $p^7q =$ _____   b) $-a^4mn =$ _____

49. Numbers like 2 and -7 can be written as power-terms in more than one variable by using some letters with "0" exponents. That is:

    $2 = 2x^0y^0$, since x^0 and $y^0 = 1$

    $-7 = -7a^0b^0c^0$, since a^0, b^0, and $c^0 =$ _____

    a) $1p^7q^1$
    b) $-1a^4m^1n^1$

50. Any power-term in two or three variables is of the form ax^my^n or $ax^my^nz^p$, where "a" is a signed whole number and the exponents are positive whole numbers or "0".

    Each term in the expression below can be written explicitly as a power term in "x" and "y". That is:

    $3x^3y + x^2y^5 - x - 4 = 3x^3y^1 + 1x^2y^5 - 1x^1y^0 - 4x^0y^0$

    Each term in the expression below can be written explicitly as a power term in "a", "b", and "c". That is:

    $ab^4c + a^3b^2 - 1 = 1a^1b^4c^1 + 1a^3b^2c^0 - 1a^0b^0c^0$

    However, when writing expressions like those above, we do not usually write "1" and "-1" coefficients, "1" exponents, and zero powers. That is:

    $1x^5y^1 - 3x^3y^0 - 1x^1y^0$ is written $x^5y - 3x^3 - x$

    $2c^1d^2t^3 - c^0d^1t^2 - 5c^4d^0t^0$ is written _____

    1

51. The "<u>degree</u>" of a power-term is the sum of the exponents of the variables. Some examples are shown.

Term	Degree	
$3x^4$	4	
$2a^3b^2$	5	(from 3 + 2)
$m^4p^7q^3$	14	(from 4 + 7 + 3)

    Write the degree of each term.

    a) t^6 _____   b) $4d^5p^2$ _____   c) $x^2y^3z^2$ _____

    $2cd^2t^3 - dt^2 - 5c^4$

52. Each term below contains one or more powers whose exponent is "1".

Term	Degree	
2t	1	
xy	2	(from 1 + 1)
$-4ab^4c$	6	(from 1 + 4 + 1)

    Write the degree of each term.

    a) $-8d^4t$ _____   b) $-x$ _____   c) pqr _____

    a) 6
    b) 7
    c) 7

Polynomials 13

14 Polynomials

53. The degree of any number-term is "0" because its variables have "0" exponents. That is: Since $4 = 4x^0$, the degree of 4 is 0. Since $-9 = -9a^0b^0$, the degree of -9 is _____	a) 5 b) 1 c) 3
54. Any expression containing only power-terms in more than one variable is called a "<u>polynomial in more than one variable</u>". Some examples are shown. $x^4y^2 - 5x^2y^3 + 1$ $5pq$ $a^4b^2c + abc^3$ State whether each polynomial is a monomial (one term), a binomial (two terms), or a trinomial (three terms). a) $p^4q^3 - p^2q$ _____ c) $x^2y^7z^5 + xyz^8$ _____ b) $7a^4b^3c$ _____ d) $m^4t - m^2 - 6$ _____	0
55. Polynomials in more than one variable are usually written <u>in descending order</u> for one of the variables. The polynomial below is written in descending order for "x". $x^4y - 2x^2y^3 + 1$ The polynomial below is written in descending order for "b". $ab^2c^5 + a^4bc - a^2$ Write the polynomial below so that it is in descending order for "m". $m^2t^5 - 3 - 5mt^4 + m^3t =$ _____	a) binomial b) monomial c) binomial d) trinomial
56. What is the degree of each term in the polynomial below? $7x^3y - x^2y^5 + xy - 1$ a) $7x^3y$ _____ b) $-x^2y^5$ _____ c) xy _____ d) -1 _____	$m^3t + m^2t^5 - 5mt^4 - 3$
	a) 4 b) 7 c) 2 d) 0

Polynomials 15

SELF-TEST 1 (pages 1-15)

Multiply.

1. $(y^3)(y)(y^5) = $ _____

2. $(rst)(rs^2t)(r^3st) = $ _____

Divide.

3. $\dfrac{a^6b^3}{a^2b} = $ _____

4. $\dfrac{h^4k^2p^3}{h^3k^2p} = $ _____

Complete: 5. $(x^3y^2)^3 = $ _____

6. $(dh)^6 = $ _____

Write each polynomial in descending order in "x".

7. $3x - x^3 + 7 - 2x^2$

8. $d^3x + 2d - 4d^2x^3 + dx^5$

State the "degree" of each term.

9. $3xy$ _____

10. $-5r^2st^3$ _____

State whether each of the following is a monomial, binomial, or trinomial.

11. $2w - 1$ _____

12. $3ab^2 + a^2b + 5$ _____

ANSWERS:
1. y^9
2. $r^5s^4t^3$
3. a^4b^2
4. hp^2
5. x^9y^6
6. d^6h^6
7. $-x^3 - 2x^2 + 3x + 7$
8. $dx^5 - 4d^2x^3 + d^3x + 2d$
9. Degree: 2
10. Degree: 6
11. Binomial
12. Trinomial

1-8 COMBINING LIKE TERMS

In this section, we will define "like" terms and show how they can be combined by addition and subtraction.

57. Two monomials with one power are "like" terms if they contain the same variable to the same power. The pairs below are "like" terms.

$5y^3$ and $4y^3$ $8t^5$ and $-6t^5$

Two monomials with one power are "unlike" terms if they contain the same variable to different powers or different variables to the same power. The pairs below are "unlike" terms.

$2x^2$ and $5x^3$ $-3t^5$ and $7y^5$

Which of the following are pairs of "like" terms? _____

a) $5m^4$ and $2m^4$

b) $-3p^5$ and $4p^2$

c) $-4b^2$ and $-2b^2$

d) $7y$ and $-4y^3$

Only (a) and (c)

16   Polynomials

58. Two monomials with more than one power are "like" terms if they contain exactly $\underline{the}$ $\underline{same}$ $\underline{variables}$ $\underline{to}$ $\underline{the}$ $\underline{same}$ $\underline{powers}$. Otherwise they are "$\underline{unlike}$" terms. Some examples are shown.

Like Terms	Unlike Terms
$3x^2y^3$ and $5x^2y^3$	$2p^4q^3$ and $5p^2q^3$
$2ab^4c^5$ and $-7ab^4c^5$	$-4c^2dt^5$ and $9c^2t^5$

Which of the following pairs are "like" terms? _____

a) $8xy^2$ and $2x^2y$ c) $3abc^3$ and $-9abc^3$

b) $-3m^4t^6$ and $4m^4t^6$ d) $5pq^2r^7$ and $7pr^7$

| Only (b) and (c)

---

59. When the terms in an addition are "like" terms, we can factor by the distributive principle to combine them. Doing so is the same as adding their coefficients. That is:

$$5y^3 + 2y^3 = (5 + 2)y^3 = 7y^3$$
$$-2ab^4 + 8ab^4 = [(-2) + 8]ab^4 = 6ab^4$$

Perform each addition by adding the coefficients.

a) $7m^4 + (-9m^4)$ = _____   b) $3pq^2r^8 + 2pq^2r^8$ = _____

---

60. When the coefficient of a term is "1", the "1" is not ordinarily written. However, it is helpful to insert the "1" when adding. For example:

$$5x^2 + x^2 = 5x^2 + 1x^2 = 6x^2$$
$$c^4d^6 + c^4d^6 = 1c^4d^6 + 1c^4d^6 = 2c^4d^6$$

Perform each addition by adding the coefficients.

a) $m^3 + 4m^3$ = _____   b) $x^2yz^4 + x^2yz^4$ = _____

| a) $-2m^4$
| b) $5pq^2r^8$

---

61. When the terms in a subtraction are "like" terms, we can also factor by the distributive principle to combine them. Doing so is the same as subtracting their coefficients. That is:

$$7x^3 - 4x^3 = (7 - 4)x^3 = 3x^3$$
$$3a^4b - 5a^4b = (3 - 5)a^4b = -2a^4b$$

Perform each subtraction by subtracting the coefficients.

a) $8p^3q^2t - 3p^3q^2t$ = _____   b) $7y^5 - 6y^5$ = _____

| a) $5m^3$
| b) $2x^2yz^4$

| a) $-5p^3q^2t$
| b) $1y^5$ or y^5

Polynomials 17

62. For each subtraction below, it is helpful to insert "1" as a coefficient. That is:

$$4x^7 - x^7 = 4x^7 - 1x^7 = 3x^7$$

$$b^4c^3 - 6b^4c^3 = 1b^4c^3 - 6b^4c^3 = -5b^4c^3$$

Do each subtraction.

a) $9mp^4t^3 - mp^4t^3 = $ _____   b) $t^5 - 2t^5 = $ _____

63. When the coefficients in a subtraction are identical, we get "0" as the difference. That is:

$$3y^2 - 3y^2 = (3 - 3)y^2 = 0y^2 = 0$$

$$a^8b^7 - a^8b^7 = (1 - 1)a^8b^7 = 0a^8b^7 = 0$$

Do each subtraction.

a) $10mt^7 - 10mt^7 = $ _____   b) $x^8 - x^8 = $ _____

a) $8mp^4t^3$

b) $-1t^5$ or $-t^5$

64. When the terms in an addition or subtraction are "unlike", they cannot be combined. For example, neither expression below can be simplified by combining terms.

$$2x^4 + 3y^7 \qquad 4b^3c^2 - bc$$

If possible, simplify each expression by combining like terms.

a) $3xy + y = $ _____   d) $2p^2 - x^2 = $ _____

b) $p^4 - 8p^4 = $ _____   e) $a^4b^3c - a^4b^3 = $ _____

c) $6d^2t + d^2t = $ _____   f) $mpq^3 - mpq^3 = $ _____

a) 0

b) 0

a) Not possible   b) $-7p^4$ c) $7d^2t$ d) Not possible   e) Not possible   f) 0

1-9 SIMPLIFYING POLYNOMIALS BY COMBINING LIKE TERMS

When a polynomial contains like terms, we can simplify it by combining the like terms. We will show the method in this section.

65. In the polynomial below, there are two pairs of like terms: ($4y^3$ and $5y^3$) and ($2y$ and $6y$). We simplified the polynomial by combining like terms.

$$4y^3 + 2y + 5y^3 + 6y = 9y^3 + 8y$$

Following the example, simplify these polynomials.

a) $8x^4 + 7x^2 + 2x^4 + 5x^2 = $

b) $2a^3b^2 + 5 + 3a^3b^2 + 1 = $

18   Polynomials

66. We simplified the polynomial below by combining like terms.

$$3x^2 - 5x - 7x^2 - 2x = -4x^2 - 7x$$

Following the example, simplify these polynomials.

a) $5t^4 + 3t^2 - 2t^4 - 9t^2 =$

b) $4bcy^4 - 3 - 6bcy^4 - 1 =$

a) $10x^4 + 12x^2$
b) $5a^3b^2 + 6$

67. When simplifying polynomials, it is helpful to write "1" and "-1" coefficients explicitly. Simplify these polynomials.

a) $x^3 - 5x + 3x^3 - x =$

b) $2a^5b^4 - ab^2 + a^5b^4 - ab^2 =$

a) $3t^4 - 6t^2$
b) $-2bcy^4 - 4$

68. When combining like terms to simplify a polynomial, we do not ordinarily write "1" and "-1" coefficients. Simplify these.

a) $2t^2 + t - t^2 - 2t =$

b) $5x^3y^2 - 7x^2y^3 - 4x^3y^2 + 6x^2y^3 =$

a) $4x^3 - 6x$
b) $3a^5b^4 - 2ab^2$

69. When combining like terms below, we got "$0x^2$" or "0". Notice that the "0" is not written in the final simplification.

$$3x^2 + x - 3x^2 - 5x = 0x^2 - 4x = 0 - 4x = -4x$$

Simplify these polynomials.

a) $5t^3 + 7 - 5t^3 - 2 =$

b) $p^2q + pq^3 + p^2q - pq^3 =$

a) $t^2 - t$
b) $x^3y^2 - x^2y^3$

70. There are three pairs of like terms in the polynomial below. Notice how we simplified it.

$$3x^2 - 5x + 1 + 2x^2 + 5x - 7 = 5x^2 + 0x - 6 = 5x^2 - 6$$

Simplify each polynomial.

a) $y^2 + 2y - 9 + y^2 - y - 3 =$

b) $at^4 - a^2t^2 + 3 - at^4 + 4a^2t^2 - 3 =$

a) 5
b) $2p^2q$

a) $2y^2 + y - 12$
b) $3a^2t^2$

Polynomials   19

71. There are only two pairs of like terms in the polynomial below. Notice how we wrote the terms in the simplified polynomial in descending order.

$$5a^3 + 2a^2 - a + a^2 - 3a + 8 = 5a^3 + 3a^2 - 4a + 8$$

Simplify this polynomial. Write the terms in descending order.

$$3t^5 + t^4 - 1 - t^5 - t^4 + 4t =$$

---

72. We simplified the polynomial below and wrote the terms in descending order for "x".

$$3xy^2 - 9 + 4x^2y - xy^2 - 1 = 4x^2y + 2xy^2 - 10$$

Simplify this polynomial. Write the terms in descending order for "b".

$$bc - 5b^3c^2 + 7b^2 + 2bc - b^3c^2 =$$

$2t^5 + 4t - 1$

$-6b^3c^2 + 7b^2 + 3bc$

1-10   ADDING POLYNOMIALS

In this section, we will discuss the procedure for adding polynomials both horizontally and vertically.

73. When writing an addition of polynomials horizontally, the polynomials are written in parentheses. For example:

Add $2x^2 + 5$ and $3x^2 - 1$ is written $(2x^2 + 5) + (3x^2 - 1)$

To perform the addition, we simply drop the parentheses and combine like terms. That is:

$$(2x^2 + 5) + (3x^2 - 1) = 2x^2 + 5 + 3x^2 - 1 = 5x^2 + 4$$

Following the example, do the following addition.

$$(ay^2 - 2by) + (2ay^2 - 3by) =$$

---

74. You should be able to perform an addition without rewriting the addition with the parentheses dropped. That is:

$$(y^4 - 2y^2 + 1) + (y^4 + 3y^2 - 9) = 2y^4 + y^2 - 8$$

Following the example, complete this addition.

$$(4ax^2 - bx - 1) + (ax^2 + bx - 3) =$$

$3ay^2 - 5by$

$5ax^2 - 4$

20 Polynomials

75. We added three polynomials below. Complete the other addition.

$(x^2 + 5) + (3x^2 - 7x + 2) + (2x - 10) = 4x^2 - 5x - 3$

$(by^3 - 2cy^2 - y) + (2by^3 - cy^2) + (2cy^2 + y) =$

---

76. Do each addition. Write each sum in descending order.

a) $(y^3 - 2) + (3y^4 - y^2 - 1) + (y^4 + 5) =$

b) $(4t^2 - 3t + 1) + (t^3 - 1) + (t^3 - 4t^2) =$

	$3by^3 - cy^2$

---

77. Do this addition. Write the sum in descending order for "x".

$(ax^2 - 3x) + (bx^3 + ax^2) + (bx^3 + 3x) =$

a) $4y^4 + y^3 - y^2 + 2$

b) $2t^3 - 3t$

---

78. To add polynomials vertically, we line up like terms in columns and then find the sum of the columns. An example is shown. Complete the other addition.

Add $2x^2 - 1$ and $5x^2 - 3$.   Add $y^3 + 4y$ and $-2y^3 - y$.

$$\begin{array}{r} 2x^2 - 1 \\ 5x^2 - 3 \\ \hline 7x^2 - 4 \end{array}$$

$2bx^3 + 2ax^2$

---

79. Following the example, complete the other addition.

Add $2x^2 + 5x - 1$, $3x + 5$,   Add $y + 6$, $y^2 - 6y + 1$,
and $x^3 - 4x^2 - 7$.   and $y^3 + 5y - 5$.

$$\begin{array}{r} 2x^2 + 5x - 1 \\ 3x + 5 \\ x^3 - 4x^2 - 7 \\ \hline x^3 - 2x^2 + 8x - 3 \end{array}$$

$$\begin{array}{r} y^3 + 4y \\ -2y^3 - y \\ \hline -y^3 + 3y \end{array}$$

---

80. Following the example, complete the other addition.

Add $bc + 1$, $ac^2 - 2bc - 1$,   Add $ax^4 - 7$, $ax^4 + 3bx^2 - 1$,
and $2ac^2 + bc - 3$.   and $-2ax^4 - bx^2 + 6$.

$$\begin{array}{r} bc + 1 \\ ac^2 - 2bc - 1 \\ 2ac^2 + bc - 3 \\ \hline 3ac^2 + 0bc - 3 \\ \text{or} \\ 3ac^2 - 3 \end{array}$$

$$\begin{array}{r} y + 6 \\ y^2 - 6y + 1 \\ y^3 + 5y - 5 \\ \hline y^3 + y^2 + 0y + 2 \\ \text{or} \\ y^3 + y^2 + 2 \end{array}$$

$$\begin{array}{r} ax^4 - 7 \\ ax^4 + 3bx^2 - 1 \\ -2ax^4 - bx^2 + 6 \\ \hline 0ax^4 + 2bx^2 - 2 \\ \text{or} \\ 2bx^2 - 2 \end{array}$$

1-11 OPPOSITES OF POLYNOMIALS

In this section, we will discuss the opposites of monomials and other polynomials.

---

81. Two monomials are a pair of opposites if their sum is "0". For example:

   Since $3x^2 + (-3x^2) = 0$: the opposite of $3x^2$ is $-3x^2$.
   the opposite of $-3x^2$ is $3x^2$.

   Since $ay^3 + (-ay^3) = 0$: a) the opposite of ay^3 is _____.
   b) the opposite of $-ay^3$ is _____.

---

82. To get the opposite of a monomial, we change the sign of its coefficient. That is:

   The opposite of $5t^4$ is $-5t^4$.

   The opposite of $-x^2y^3$ is x^2y^3.

   Write the opposite of each monomial.

   a) $9b^7$ _____   b) $-4cx^4$ _____   c) p^2qr^3 _____

   | a) $-ay^3$ |
   | b) ay^3 |

---

83. Two polynomials are a pair of opposites if each pair of like terms is a pair of opposites. For example:

   Since $(2x^2 + 5) + (-2x^2 - 5) = 0x^2 + 0 = 0$:

   the opposite of $2x^2 + 5$ is $-2x^2 - 5$.
   the opposite of $-2x^2 - 5$ is $2x^2 + 5$.

   Since $(y^3 - 2y - 7) + (-y^3 + 2y + 7) = 0y^3 + 0y + 0 = 0$:

   a) the opposite of $y^3 - 2y - 7$ is _____.
   b) the opposite of $-y^3 + 2y + 7$ is _____.

   | a) $-9b^7$ |
   | b) $4cx^4$ |
   | c) $-p^2qr^3$ |

---

84. To get the opposite of a polynomial, we change the sign of each term. That is:

   The opposite of $4t^3 - 5t^2 - 7$ is $-4t^3 + 5t^2 + 7$.

   The opposite of $ax^2 - 3bx + 1$ is $-ax^2 + 3bx - 1$.

   Write the opposite of each polynomial.

   a) $3cy^2 - 4$ _____   b) $x^3 - 7x + 1$ _____

   | a) $-y^3 + 2y + 7$ |
   | b) $y^3 - 2y - 7$ |

---

85. Write the opposite of each polynomial.

   a) $b^5 - b^4$ _____

   b) $-3ay^4 + 8by^2 - y$ _____

   | a) $-3cy^2 + 4$ |
   | b) $-x^3 + 7x - 1$ |

| a) $-b^5 + b^4$ |
| b) $3ay^4 - 8by^2 + y$ |

22   Polynomials

1-12   SUBTRACTING POLYNOMIALS

In this section, we will discuss the procedure for subtracting polynomials both horizontally and vertically.

86. To subtract polynomials, we add the opposite of the second polynomial. That is: $\quad\quad\quad\quad\quad\quad\quad\quad\quad\quad\quad$ Opposite $\quad\quad\quad\quad\quad\quad\quad\quad\quad\quad\quad\quad\downarrow$ $(3x^2 - 5) - (x^2 + 4) = (3x^2 - 5) + (-x^2 - 4) = 2x^2 - 9$ $(5y + 7) - (6y - 3) = (5y + 7) + (-6y + 3) = -y + 10$ Following the examples, complete these subtractions. a) $(y - 1) - (3y - 5) = (y - 1) + ($ $\quad\quad\quad$ $) =$ b) $(2t^4 + t^2) - (t^4 + t^2) = (2t^4 + t^2) + ($ $\quad\quad\quad$ $) =$	
87. To subtract below, we added the opposite of the second polynomial. $\quad\quad\quad\quad\quad\quad\quad\quad\quad\quad\quad$ Opposite $\quad\quad\quad\quad\quad\quad\quad\quad\quad\quad\quad\quad\downarrow$ $(7ax^4 - bx^2) - (ax^4 - 3bx^2) = (7ax^4 - bx^2) + (-ax^4 + 3bx^2) = 6ax^4 + 2bx^2$ Following the example, complete this subtraction. $(p^2q^3 + 1) - (4p^2q^3 - 3) = (p^2q^3 + 1) + ($ $\quad\quad\quad$ $) =$	a) $+ (-3y + 5) = -2y + 4$ b) $+ (-t^4 - t^2) = t^4$
88. Another subtraction of polynomials is shown below. Notice again that we added the opposite of the second polynomial. $\quad\quad\quad\quad\quad\quad\quad\quad\quad\quad\quad$ Opposite $\quad\quad\quad\quad\quad\quad\quad\quad\quad\quad\quad\quad\downarrow$ $(5y^3 - 2y^2 + 1) - (y^3 + 2y^2 + 1) = (5y^3 - 2y^2 + 1) + (-y^3 - 2y^2 - 1)$ $\quad\quad\quad\quad\quad\quad\quad\quad\quad\quad\quad\quad\quad\quad = 4y^3 - 4y^2$ Following the example, complete this subtraction. $(m^2 - m + 4) - (m^2 - 2m - 1) = (m^2 - m + 4) + ($ $\quad\quad\quad$ $)$ $=$	$+ (-4p^2q^3 + 3) = -3p^2q^3 + 4$
89. Perform each subtraction by adding the opposite of the second polynomial. Write each answer in descending order. a) $(2x^2 - 5x) - (x^4 + 3x) =$ b) $(6y - 1) - (-3y^3 + 6y - 1) =$	$+ (-m^2 + 2m + 1) = m + 5$
	a) $-x^4 + 2x^2 - 8x$ b) $3y^3$

90. Do this subtraction. Write the answer in descending order for "x".

$(4x^3y - 2xy) - (x^2y^2 + xy) =$

| | $4x^3y - x^2y^2 - 3xy$ |

91. To subtract polynomials vertically, we line up like terms in columns and then add the opposite of the bottom polynomial. An example is shown. Complete the other subtraction.

Subtract $2y - 7$ from $5y + 1$.

$\;\; 5y + 1$
$(-)\;\; 2y - 7$ becomes $\begin{array}{r} 5y + 1 \\ -2y + 7 \\ \hline 3y + 8 \end{array}$

Subtract $x^3 + x^2$ from $3x^3 - 2x^2$.

$\;\; 3x^3 - 2x^2$
$(-)\;\; x^3 + x^2$ becomes

92. Following the example, complete the other subtraction.

Subtract $2x^2 - 3x + 5$ from $7x^2 - x - 3$.

$\;\; 7x^2 - x - 3$
$(-)\;\; 2x^2 - 3x + 5$ becomes $\begin{array}{r} 7x^2 - x - 3 \\ -2x^2 + 3x - 5 \\ \hline 5x^2 + 2x - 8 \end{array}$

Subtract $t^4 - t^2 - 1$ from $t^4 + t^2 + 1$.

$\;\; t^4 + t^2 + 1$
$(-)\;\; t^4 - t^2 - 1$ becomes

| | $\begin{array}{r} 3x^3 - 2x^2 \\ -x^3 - x^2 \\ \hline 2x^3 - 3x^2 \end{array}$ |

93. Following the example, complete the other subtraction.

Subtract $2xy - 5$ from $x^2y + 4xy$.

$\;\; x^2y + 4xy$
$(-)\;\; 2xy - 5$ becomes $\begin{array}{r} x^2y + 4xy \\ -2xy + 5 \\ \hline x^2y + 2xy + 5 \end{array}$

Subtract $ay^2 + 3$ from $4ay^2 - by + 1$.

$\;\; 4ay^2 - by + 1$
$(-)\;\; ay^2 + 3$ becomes

| | $\begin{array}{r} t^4 + t^2 + 1 \\ -t^4 + t^2 + 1 \\ \hline 2t^2 + 2 \end{array}$ |

| | $\begin{array}{r} 4ay^2 - by + 1 \\ -ay^2 - 3 \\ \hline 3ay^2 - by - 2 \end{array}$ |

24   Polynomials

1-13 MULTIPLYING AND SQUARING MONOMIALS

In this section, we will discuss the procedure for multiplying and squaring monomials.

94. To perform the multiplications below, we multiplied the coefficients and used the law of exponents for multiplication. $(3x^2)(5x^4) = (3)(5)(x^2)(x^4) = 15x^6$ $(-2y^3)(y^2) = (-2)(y^3)(y^2) = -2y^5$ Using the same method, complete these: a) $(4t^3)(-3t^4) = $ _____ b) $(-m^5)(-5m^5) = $ _____	
95. We used the law of exponents more than once in the multiplication below. $(2ab^2)(5a^4b) = (2)(5)(a)(a^4)(b^2)(b) = 10a^5b^3$ Following the example, complete these: a) $(-5x^2y^2)(-3xy^7) = $ _____ b) $(a^3b^4c)(-6abc^3)(-a^4bc^5) = $ _____	a) $-12t^7$ b) $5m^{10}$
96. We used the law of exponents with only some variables in the multiplication below. $(apq^3)(pq^2r^5) = (a)(p)(p)(q^3)(q^2)(r^5) = ap^2q^5r^5$ Following the example, complete these: a) $(9x^3y^4)(7bx) = $ _____ b) $(-b^2c)(5ab^2)(c^4d) = $ _____	a) $15x^3y^9$ b) $6a^8b^6c^9$
97. To square a monomial, we multiply the monomial by itself. For example: $(3y)^2 = (3y)(3y) = 9y^2$ $(x^2y^3)^2 = (x^2y^3)(x^2y^3) = x^4y^6$ Following the examples, complete these. a) $(-4ab^5)^2 = (-4ab^5)(-4ab^5) = $ _____ b) $(p^2qr^4)^2 = (p^2qr^4)(p^2qr^4) = $ _____	a) $63bx^4y^4$ b) $-5ab^4c^5d$
	a) $16a^2b^{10}$ b) $p^4q^2r^8$

98. We used the same method to square the monomial below.

$$(4x^3y^4)^2 = (4x^3y^4)(4x^3y^4) = 16x^6y^8$$

As you can see, squaring a monomial is the same as <u>squaring</u> the <u>coefficient</u> <u>and</u> <u>doubling</u> <u>each</u> <u>exponent</u>. Using the shorter method, complete these:

a) $(2ab^3)^2 =$ _____ b) $(-3p^5q^7)^2 =$ _____

99. Use the shorter method to complete these.

a) $(xy)^2 =$ _____ d) $(5x^2y^3)^2 =$ _____

b) $(-6r^2)^2 =$ _____ e) $(-2mp^5q)^2 =$ _____

c) $(-t^{10})^2 =$ _____ f) $(c^3d^6f^9)^2 =$ _____

a) $4a^2b^6$
b) $9p^{10}q^{14}$

a) x^2y^2 b) $36r^4$ c) t^{20} d) $25x^4y^6$ e) $4m^2p^{10}q^2$ f) $c^6d^{12}f^{18}$

1-14 MULTIPLYING OTHER POLYNOMIALS BY A MONOMIAL

In this section, we will show how the distributive principle can be used to multiply other polynomials by a monomial.

100. To multiply $4y^3 + 5$ by y^2 below, we used the distributive principle. Notice that we multiplied each term in the binomial by y^2.

$$y^2(4y^3 + 5) = y^2(4y^3) + y^2(5)$$
$$= 4y^5 + 5y^2$$

Following the example, complete this multiplication.

$$3x(x^3 - 2x) = (\quad)(\quad) - (\quad)(\quad)$$
$$= \underline{\quad\quad} - \underline{\quad\quad}$$

101. To multiply $2x^4 - x^2 + 7$ by $5x$ below, we also used the distributive principle. Notice that we multiplied each term in the trinomial by $5x$.

$$5x(2x^4 - x^2 + 7) = 5x(2x^4) - 5x(x^2) + 5x(7)$$
$$= 10x^5 - 5x^3 + 35x$$

Following the example, complete this multiplication.

$$p^4(p^3 + 2p^2 - 3p) = (\quad)(\quad) + (\quad)(\quad) - (\quad)(\quad)$$
$$= \underline{\quad\quad} + \underline{\quad\quad} - \underline{\quad\quad}$$

$3x(x^3) - 3x(2x)$
$3x^4 - 6x^2$

$p^4(p^3) + p^4(2p^2) - p^4(3p)$
$p^7 + 2p^6 - 3p^5$

26   Polynomials

102. Complete these by multiplying each term in the other polynomial by the monomial.

  a) $a^3(2a + 5)$ = _____   b) $5t^2(t^3 - 2t)$ = _____

---

103. Use the same method to complete these.

  a) $2t(t^4 - 5t^2 - 8)$ = _____

  b) $3p^5(4p^3 - 6p^2 + p)$ = _____

a) $2a^4 + 5a^3$
b) $5t^5 - 10t^3$

---

104. Notice how we multiplied each term in the binomial below by xy^2.

$$xy^2(3x^3y - 2x) = xy^2(3x^3y) - xy^2(2x)$$
$$= 3x^4y^3 - 2x^2y^2$$

Complete these by multiplying each term in the other polynomial by the monomial.

  a) $5a^3b^2(3a^4b^2 + a^2b)$ = _____

  b) $pq(p^2q - pq^5 - 4)$ = _____

a) $2t^5 - 10t^3 - 16t$
b) $12p^8 - 18p^7 + 3p^6$

---

a) $15a^7b^4 + 5a^5b^3$
b) $p^3q^2 - p^2q^6 - 4pq$

SELF-TEST 2 (pages 15-27)

Do these additions. Write each sum in descending order for "x".

1. $(5x + 1) + (4x^2 - x - 3) + (x^2 + 4)$

2. Add $dx^2 - 1$, $kx^4 - 2dx^2 + 1$, and $dx^2 - rx$.

Do these subtractions. Write each sum in descending order for "t".

3. $(2t^3 - 3t^2 - t) - (t^3 - 3t^2 + 3)$

4. Subtract $bt^2 - 2at$ from $ht^4 + bt^2 - 1$.

Continued on following page.

SELF-TEST 2 (pages 15-27) - Continued

Do these multiplications.

5. $(-2xy^3)(6x^2y)$	6. $(-3ab)(5a^2b)(-2a^3b^2)$	7. $(-4tw^3)^2$

8. $4y(y^2 + 3y - 2)$	9. $2p^2w(3p^3w^3 - pw + 5)$

ANSWERS:
1. $5x^2 + 4x + 2$
2. $kx^4 - rx$
3. $t^3 - t - 3$
4. $ht^4 + 2at - 1$
5. $-12x^3y^4$
6. $30a^6b^4$
7. $16t^2w^6$
8. $4y^3 + 12y^2 - 8y$
9. $6p^5w^4 - 2p^3w^2 + 10p^2w$

1-15 MULTIPLYING BINOMIALS

In this section, we will discuss the procedure for multiplying binomials.

105. To multiply a binomial by a binomial, we can use the distributive principle three times. An example is shown.

$$(a + b)(c + d) = a(c + d) + b(c + d)$$
$$= ac + ad + bc + bd$$

Note: 1. First, we used the distributive principle to multiply (c + d) by both "a" and "b".

2. Then we used the distributive principle twice for the multiplications a(c + d) and b(c + d).

Following the example, complete this multiplication.

$$(p + q)(x + y) = (\quad)(x + y) + (\quad)(x + y)$$
$$= \underline{\quad} + \underline{\quad} + \underline{\quad} + \underline{\quad}$$

106. The two steps used to multiply (p + q) and (x + y) are shown below.

$$(p + q)(x + y) = p(x + y) + q(x + y)$$
$$= px + py + qx + qy$$

See next frame.

Continued on following page.

28 Polynomials

106. Continued

In the second step, there are four terms in the product. Using the FOIL method, we can skip the first step and write those four terms directly. We get:

$$\underbrace{(p + q)(x + y)}_{FOIL} = \underset{F}{px} + \underset{O}{py} + \underset{I}{qx} + \underset{L}{qy}$$

Note: 1. To get F (or "px"), we multiplied the first terms of the binomials.

2. To get O (or "py"), we multiplied the outside terms of the binomials.

3. To get I (or "qx"), we multiplied the inside terms of the binomials.

4. To get L (or "qy"), we multiplied the last terms of the binomials.

Using the FOIL method, write the four terms of the product below.

(c + d)(m + t) = _____ + _____ + _____ + _____

---

107. In the FOIL method, we multiply both terms in the second binomial:

1) first by the first term in the first binomial,

2) then by the second term in the first binomial.

Using the FOIL method, write each four-term product.

a) (a + 2b)(3x + y) = _____ + _____ + _____ + _____

b) (2m + p)(q + 4t) = _____ + _____ + _____ + _____

cm + ct + dm + dt

---

108. In the multiplication below, the second binomial is a difference. We used the FOIL method. Write the other product.

(c + d)(p − q) = cp − cq + dp − dq

(a + b)(x² − y) = _____ + _____ + _____ + _____

a) 3ax + ay + 6bx + 2by

b) 2mq + 8mt + pq + 4pt

---

109. In the multiplication below, the first binomial is a difference. We used the FOIL method. Write the other product.

(2a − b)(5c + 3d) = 10ac + 6ad − 5bc − 3bd

(p − 4q)(2x + 5y) =

ax² − ay + bx² − by

2px + 5py − 8qx − 20qy

110. In the multiplication below, both binomials are differences. We used the FOIL method. Write the other product.

$(c^2 - d^2)(p^3 - q^3) = c^2p^3 - c^2q^3 - d^2p^3 + d^2q^3$

$(2a - b^4)(x^3 - 3y) = $

---

111. Complete each multiplication.

a) $(3pq + 2p)(q^2 - 2p^2q) = $ _____

b) $(2x - 4xy)(y^4 - 3x^2) = $ _____

	$2ax^3 - 6ay - b^4x^3 + 3b^4y$

---

112. After multiplying two binomials, we can sometimes combine like terms in the product to get a trinomial. For example:

$(p + 3q)(2p - q) = 2p^2 - pq + 6pq - 3q^2$
$= 2p^2 + 5pq - 3q^2$

Do this multiplication and combine like terms in the product.

$(2x - 3y)(2x + y) = $ _____
$= $ _____

a) $3pq^3 - 6p^3q^2 + 2pq^2 - 4p^3q$

b) $2xy^4 - 6x^3 - 4xy^5 + 12x^3y$

---

113. Combine like terms in each product below.

a) $(4x + y)(2x + y) = $ _____
$= $ _____

b) $(a - 2b)(a - 3b) = $ _____
$= $ _____

$4x^2 + 2xy - 6xy - 3y^2$
$4x^2 - 4xy - 3y^2$

---

114. Combine like terms in each product below.

a) $(b^2 + c)(b^2 - 3c) = $ _____
$= $ _____

b) $(2x^2 - y^2)(x^2 - 4y^2) = $ _____
$= $ _____

a) $8x^2 + 4xy + 2xy + y^2$
$8x^2 + 6xy + y^2$

b) $a^2 - 3ab - 2ab + 6b^2$
$a^2 - 5ab + 6b^2$

---

a) $b^4 - 3b^2c + b^2c - 3c^2$
$b^4 - 2b^2c - 3c^2$

b) $2x^4 - 8x^2y^2 - x^2y^2 + 4y^4$
$2x^4 - 9x^2y^2 + 4y^4$

30 Polynomials

1-16 MULTIPLYING THE SUM AND DIFFERENCE OF TWO TERMS

In this section, we will discuss multiplications of the sum and difference of the same two terms.

115. In the multiplication below, both binomials contain an "x" and a "y". One binomial is a sum; the other binomial is a difference. Since $-xy + xy = 0$, the product simplifies to a binomial.

$$(x + y)(x - y) = x^2 - xy + xy - y^2$$
$$= x^2 - y^2$$

Following the example, complete this multiplication.

$(a + b^2)(a - b^2) = $ _____

$= $ _____

116. In the multiplication below, both binomials contain a "2p" and a "q". One is a sum; the other is a difference. Since $-2pq + 2pq = 0$, the product simplifies to a binomial.

$$(2p + q)(2p - q) = 4p^2 - 2pq + 2pq - q^2$$
$$= 4p^2 - q^2$$

Following the example, complete this multiplication.

$(x^2 + 3y)(x^2 - 3y) = $ _____

$= $ _____

[Answer to 115:]
$a^2 - ab^2 + ab^2 - b^4$
$a^2 - b^4$

117. Here is a multiplication performed earlier.

$$(x + y)(x - y) = x^2 - y^2$$

Notice these points about the product $x^2 - y^2$.

1) x^2 is the square of "x".
2) y^2 is the square of "y".
3) The two squares are <u>subtracted</u>.

Following the above pattern, complete these.

a) $(c + d)(c - d) = $ _____   b) $(a + b)(a - b) = $ _____

[Answer to 116:]
$x^4 - 3x^2y + 3x^2y - 9y^2$
$x^4 - 9y^2$

[Answer to 117:]
a) $c^2 - d^2$
b) $a^2 - b^2$

118. Here is another multiplication performed earlier.

$$(x^2 + 3y)(x^2 - 3y) = x^4 - 9y^2$$

Notice these points about the product $x^4 - 9y^2$.

    1) x^4 is the square of "x^2".

    2) $9y^2$ is the square of "3y".

    3) The two squares are subtracted.

Following the above pattern, write each product.

    a) $(2d + t^3)(2d - t^3) =$ _____   b) $(a^2 + b^4)(a^2 - b^4) =$ _____

119. Write each binomial product.

    a) $4d^2 - t^6$
    b) $a^4 - b^8$

    a) $(2x + 5y)(2x - 5y) =$ _____   c) $(p^3 + q^2)(p^3 - q^2) =$ _____

    b) $(3m + 4n)(3m - 4n) =$ _____   d) $(2a^2 + 3b^5)(2a^2 - 3b^5) =$ _____

a) $4x^2 - 25y^2$ b) $9m^2 - 16n^2$ c) $p^6 - q^4$ d) $4a^4 - 9b^{10}$

1-17 SQUARING BINOMIALS

In this section, we will show how the FOIL method can be used to square binomials. Then we will show a shortcut for the same operation.

120. To square a binomial, we multiply the binomial by itself. That is:

$$(a + b)^2 = (a + b)(a + b)$$
$$(m^2 - 2t)^2 = (m^2 - 2t)(m^2 - 2t)$$

Therefore, squaring a binomial is the same as multiplying two identical binomials. We can use the FOIL method to do so. For example:

$$(a + b)^2 = (a + b)(a + b) = a^2 + ab + ab + b^2$$
$$= a^2 + 2ab + b^2$$

Following the example, complete this squaring.

$$(m^2 - 2t)^2 = (m^2 - 2t)(m^2 - 2t) = \underline{\hspace{4cm}}$$
$$= \underline{\hspace{4cm}}$$

$m^4 - 2m^2t - 2m^2t + 4t^2$
$m^4 - 4m^2t + 4t^2$

121. If we use the FOIL method to square (x + 3y), we get:

$$(x + 3y)^2 = x^2 + 6xy + 9y^2$$

There is a shortcut that can be used to square (x + 3y). To see the shortcut, let's examine $x^2 + 6xy + 9y^2$.

1) The <u>first term</u> (x^2) is the square of "x".
2) The <u>second term</u> (6xy) is double the product of the two terms of the binomial. That is: 6xy = 2(x)(3y).
3) The <u>third term</u> ($9y^2$) is the square of "3y".

Let's use the shortcut to square (2p + q).

a) Squaring "2p", we get _____.
b) Doubling the product of "2p" and "q", we get _____.
c) Squaring "q", we get _____.
d) Therefore, $(2p + q)^2 =$ _____.

---

122. Let's use the shortcut to square ($t^2 + 4y$).

a) Squaring "t^2", we get _____.
b) Doubling the product of "t^2" and "4y", we get _____.
c) Squaring "4y", we get _____.
d) Therefore, $(t^2 + 4y)^2 =$ _____.

a) $4p^2$
b) $4pq$
c) q^2
d) $4p^2 + 4pq + q^2$

---

123. Use the shortcut for these. Be sure to <u>double</u> the product of the two terms to get the middle term of the trinomial.

a) $(x + y)^2 =$ _____
b) $(3b + 2c^3)^2 =$ _____

a) t^4
b) $8t^2y$
c) $16y^2$
d) $t^4 + 8t^2y + 16y^2$

---

124. If we use the FOIL method to square (5a - b), we get:

$$(5a - b)^2 = 25a^2 - 10ab + b^2$$

Notice that we can also use the shortcut to get $25a^2 - 10ab + b^2$. However, when using it, <u>we must remember to subtract the middle term</u>.

Let's use the shortcut to square ($p^5 - 2q$).

a) Squaring "p^5", we get _____.
b) Doubling the product of "p^5" and "2q", we get _____.
c) Squaring "2q", we get _____.
d) Therefore, $(p^5 - 2q)^2 =$ _____

a) $x^2 + 2xy + y^2$
b) $9b^2 + 12bc^3 + 4c^6$

Polynomials 33

Be sure to **subtract** the middle term of the

a) p^{10}
b) $4p^5q$
c) $4q^2$
d) $p^{10} - 4p^5q + 4q^2$

a) $c^2 - 2cd + d^2$
b) $9x^4 - 24x^2y + 16y^2$

$+ 4x^{12}$ c) $a^4b^4 + 2a^2b^2 + 1$ d) $x^4y^2 - 2x^3y^3 + x^2y^4$

ocess for monomials.

127. ...e reverse of multiplication. Therefore:

Since $(7)(x) = 7x$, $7x$ can be factored into $(7)(x)$

Since $(y)(y) = y^2$, y^2 can be factored into $(y)(y)$

Factor these: a) $by = ($ $)($ $)$ b) $p^2 = ($ $)($ $)$

a) (b)(y)
b) (p)(p)

128. We factored $3x^2$ and ab^2 in various ways below.

$3x^2 = (3)(x^2)$ $ab^2 = (a)(b^2)$
$3x^2 = (3x)(x)$ $ab^2 = (ab)(b)$
$3x^2 = (x)(3x)$ $ab^2 = (b)(ab)$

Write the missing factor in each blank.

a) $5t^2 = ($ $)(t^2)$ b) $x^2y = (x)($ $)$ c) $dp^2 = ($ $)(p)$

a) $(\underline{5})(t^2)$
b) $(x)(\underline{xy})$
c) $(\underline{dp})(p)$

129. We factored $10t$ and $16y^2$ in various ways below.

$10t = t(10)$ $16y^2 = 16(y^2)$
$10t = 5t(2)$ $16y^2 = 8(2y^2)$
$10t = 2(5t)$ $16y^2 = (4y)(4y)$

Write the missing factor in each blank.

a) $12m = ($ $)(6m)$ b) $8x^2 = ($ $)(4x)$ c) $20d^2 = (5)($ $)$

34   Polynomials

130. We factored x^4 and y^9 in various ways below.

$x^4 = (x)(x^3)$ $\quad\quad$ $y^9 = (y)(y^8)$
$x^4 = (x^2)(x^2)$ $\quad\quad$ $y^9 = (y^3)(y^6)$
$x^4 = (x^3)(x)$ $\quad\quad$ $y^9 = (y^7)(y^2)$

Write the missing factor in each blank.

a) $t^3 = (\quad)(t^2)$ $\quad$ b) $v^5 = (v^3)(\quad)$ $\quad$ c) $a^8 = (\quad)(a^2)$

a) $(\underline{2})(6m)$
b) $(\underline{2x})(4x)$
c) $5(\underline{4d^2})$

---

131. We factored ax^5 and $6t^{10}$ in various ways below.

$ax^5 = (a)(x^5)$ $\quad\quad$ $6t^{10} = (6)(t^{10})$
$ax^5 = (ax)(x^4)$ $\quad\quad$ $6t^{10} = (6t^3)(t^7)$
$ax^5 = (x^2)(ax^3)$ $\quad\quad$ $6t^{10} = (3t^5)(2t^5)$

Write the missing factor in each blank.

a) $7m^3 = (\quad)(m^2)$ $\quad$ b) $by^4 = (b)(\quad)$ $\quad$ c) $8v^6 = (\quad)(2v)$

a) $(\underline{t})(t^2)$
b) $(v^3)(\underline{v^2})$
c) $(\underline{a^6})(a^2)$

---

132. We factored x^2y^2 and c^5d^7 in various ways below.

$x^2y^2 = (x^2)(y^2)$ $\quad\quad$ $c^5d^7 = (cd^4)(c^4d^3)$
$x^2y^2 = (x)(xy^2)$ $\quad\quad$ $c^5d^7 = c^2(c^3d^7)$
$x^2y^2 = (xy)(xy)$ $\quad\quad$ $c^5d^7 = (c^5d^5)(d^2)$

Write the missing factor in each blank.

a) $p^2q^2 = (\quad)(p^2q)$ $\quad$ c) $xy^6 = (\quad)(y^2)$
b) $m^4y^4 = (my^3)(\quad)$ $\quad$ d) $v^9t^8 = (v^6)(\quad)$

a) $(\underline{7m})(m^2)$
b) $(b)(\underline{y^4})$
c) $(\underline{4v^5})(2v)$

---

133. We factored ax^3y^3 and $12p^7t^3$ in various ways below.

$ax^3y^3 = (a)(x^3y^3)$ $\quad\quad$ $12p^7t^3 = (6)(2p^7t^3)$
$ax^3y^3 = (ay^3)(x^3)$ $\quad\quad$ $12p^7t^3 = (4p^2t^2)(3p^5t)$
$ax^3y^3 = (axy^2)(x^2y)$ $\quad\quad$ $12p^7t^3 = (p^4t)(12p^3t^2)$

Write the missing factor in each blank.

a) $bdy^4 = (by)(\quad)$ $\quad$ c) $10p^5q^3 = (5pq)(\quad)$
b) $mx^6y^6 = (\quad)(x^6)$ $\quad$ d) $18a^4b = (\quad)(2a^3)$

a) $(\underline{q})(p^2q)$
b) $(my^3)(\underline{m^3y})$
c) $(\underline{xy^4})(y^2)$
d) $(v^6)(\underline{v^3t^8})$

---

a) $(by)(\underline{dy^3})$
b) $(\underline{my^6})(x^6)$
c) $(5pq)(\underline{2p^4q^2})$
d) $(\underline{9ab})(2a^3)$

134. Any term can be factored into itself and "1". For example:

$$3x = (3x)(1) \qquad ap^3q = (ap^3q)(1)$$
$$y^2 = (1)(y^2) \qquad 8v^6t^6 = (1)(8v^6t^6)$$

Write the missing factor in each blank.

a) $d^8 = (\quad)(d^8)$ b) $24xy^9 = (24xy^9)(\quad)$

a) $(\underline{1})(d^8)$ b) $(24xy^9)(\underline{1})$

1-19 COMMON MONOMIAL FACTORS

In this section, we will show how the distributive principle can be used to factor a common monomial factor out of polynomials containing two or more terms.

135. When factoring a common <u>numerical</u> factor out of a polynomial, <u>we always factor out the largest possible factor</u>. For example:

$$4x^2 + 12 = 4(x^2 + 3)$$
$$12t^2 - 18t - 6 = 6(2t^2 - 3t - 1)$$

Factor out the largest possible <u>numerical</u> factor.

a) $10y^3 - 5 = $ _____

b) $16m^2 + 4m - 8 = $ _____

136. We were able to factor a common first power out of the polynomial below.

$$x^3 - 3x^2 + x = x(x^2 - 3x + 1)$$

Factor out a first power.

a) $4t^2 - 7t = $ _____

b) $y^5 + 9y^3 - y = $ _____

a) $5(2y^3 - 1)$

b) $4(4m^2 + m - 2)$

137. When factoring a common power out of a polynomial, <u>we always factor out the largest possible power</u>. For example:

$$2y^6 - 7y^4 + y^2 = y^2(2y^4 - 7y^2 + 1)$$

Factor out the largest possible power.

a) $3x^5 + 7x^3 = $ _____

b) $a^8 + 4a^6 - a^4 = $ _____

a) $t(4t - 7)$

b) $y(y^4 + 9y^2 - 1)$

138. In the example below, we factored out the largest possible numerical factor and power.

$$10t^5 - 30t^2 = 10t^2(t^3 - 3)$$

Factor out the largest possible numerical factor and power.

a) $8m^9 + 4m^5 =$ _____

b) $6x^3 - 12x^2 - 9x =$ _____

a) $x^3(3x^2 + 7)$

b) $a^4(a^4 + 4a^2 - 1)$

---

139. To check a factoring, we multiply by the distributive principle. That is:

$$10x^3 - 5x^2 = 5x^2(2x - 1) \text{ is correct, since } 5x^2(2x - 1) = 10x^3 - 5x^2$$

Factor these and check your results by multiplying.

a) $7y^4 + 5y =$ _____

b) $18d^7 - 24d^5 - 12d^3 =$ _____

a) $4m^5(2m^4 + 1)$

b) $3x(2x^2 - 4x - 3)$

---

140. In the example below, we factored out the largest power of each variable.

$$3x^2y^2 - xy^3 = xy^2(3x - y)$$

Factor out the largest power of each variable. Check your results.

a) $c^4d + 7c^2d =$ _____

b) $2p^5q^3 - 5p^4q - p^3q^2 =$ _____

a) $y(7y^3 + 5)$

b) $6d^3(3d^4 - 4d^2 - 2)$

---

141. In the example below, we factored out the largest numerical factor and the largest powers of the variables.

$$12x^2y - 20x^3y = 4x^2y(3 - 5x)$$

Factor out the largest numerical factor and powers. Check your results.

a) $6p^4q^4 + 3p^2q^2 =$ _____

b) $8c^5d - 4c^4d^2 + 4c^3d^3 =$ _____

a) $c^2d(c^2 + 7)$

b) $p^3q(2p^2q^2 - 5p - q)$

---

142. Factor out the largest possible factor.

a) $10t^4 - 8t^3 =$ _____

b) $6m^2p + 9m^4p =$ _____

c) $2y^6 + 5y^5 - 3y^4 =$ _____

d) $12a^3x^4 - 6ax^3 + 18a^2x^2 =$ _____

a) $3p^2q^2(2p^2q^2 + 1)$

b) $4c^3d(2c^2 - cd + d^2)$

---

a) $2t^3(5t - 4)$ b) $3m^2p(2 + 3m^2)$ c) $y^4(2y^2 + 5y - 3)$ d) $6ax^2(2a^2x^2 - x + 3a)$

Polynomials 37

1-20 FACTORING THE DIFFERENCE OF TWO PERFECT SQUARES

In this section, we will discuss the procedure for factoring the difference of two perfect squares. Before doing so, we will discuss "perfect squares" and their square roots.

143. Any whole number whose square root is a whole number is a <u>perfect square</u>. For example:

$$36 \text{ is a perfect square, since } \sqrt{36} = 6$$

Which of the following are perfect squares? _____

    a) 1   b) 7   c) 25   d) 63   e) 81

144. When a power is squared, the exponent of its square is always an <u>even number</u>. For example:

$$(x)^2 = x^2 \qquad (y^3)^2 = y^6 \qquad (t^8)^2 = t^{16}$$

Therefore, any power whose exponent is <u>even</u> is a <u>perfect square</u>. To find the square root of a perfect-square power, we divide the exponent by 2 (cut it in half). That is:

$$\sqrt{m^2} = m^{\frac{2}{2}} = m^1 \text{ or } m \qquad \sqrt{b^{10}} = b^{\frac{10}{2}} = b^5$$

Find the square root of each perfect square below.

    a) $\sqrt{q^2}$ = _____   b) $\sqrt{d^6}$ = _____   c) $\sqrt{x^{20}}$ = _____

Answer: (a), (c), and (e)

145. Both $4x^2$ and $25y^8$ are perfect squares because their coefficients and powers are both perfect squares. To find the square root of each, we find the square root of the coefficient and the power. That is:

$$\sqrt{4x^2} = 2x \qquad \sqrt{25y^8} = 5y^4$$

Find the square root of each perfect square below.

    a) $\sqrt{9p^2}$ = _____   b) $\sqrt{16t^4}$ = _____   c) $\sqrt{100m^{12}}$ = _____

Answers:
a) q^1 or q
b) d^3
c) x^{10}

146. When monomials like those below are squared, the exponent of each power is always an even number. That is:

$$(xy^2)^2 = x^2y^4 \qquad (a^3b^6)^2 = a^6b^{12}$$

Therefore, any literal monomial in which the exponent of each power is even is a <u>perfect square</u>. To find its square root, we divide each exponent by 2. For example:

$$\sqrt{p^{10}q^4} = p^{\frac{10}{2}}q^{\frac{4}{2}} = p^5q^2$$

Answers:
a) 3p
b) $4t^2$
c) $10m^6$

Continued on following page.

38   Polynomials

146. Continued

   Find the square root of each perfect square below.

   a) $\sqrt{c^2 d^6}$ = _____   b) $\sqrt{x^8 y^{14}}$ = _____

147. The monomial $49m^4v^2$ is a perfect square because its numerical coefficient and both powers are perfect squares. To find its square root, we find the square root of the coefficient and both powers. That is:

$$\sqrt{49m^4v^2} = 7m^2v$$

   Find the square root of each perfect square below.

   a) $\sqrt{64x^2y^8}$ = _____   b) $\sqrt{81p^{10}q^8}$ = _____

a) cd^3
b) x^4y^7

148. In the multiplication below, the product is the difference of two perfect squares.

$$(x^2 + y)(x^2 - y) = x^4 - y^2$$

   By reversing the two sides above, we can factor $x^4 - y^2$. That is:

$$x^4 - y^2 = (x^2 + y)(x^2 - y)$$

   Notice these points about the factoring:

   1) "x^2" is the square root of "x^4".
   2) "y" is the square root of "y^2".
   3) One factor is a sum; the other factor is a difference.

   Following the pattern above, factor these:

   a) $c^2 - d^8$ = _____   b) $p^{10} - q^6$ = _____

a) $8xy^4$
b) $9p^5q^4$

149. In the multiplication below, the product is the difference of two perfect squares.

$$(3x + 2y^3)(3x - 2y^3) = 9x^2 - 4y^6$$

   By reversing the two sides, we can factor $9x^2 - 4y^6$. That is:

$$9x^2 - 4y^6 = (3x + 2y^3)(3x - 2y^3)$$

   Notice these points about the factoring:

   1) "$3x$" is the square root of "$9x^2$".
   2) "$2y^3$" is the square root of "$4y^6$".
   3) One factor is a sum; the other factor is a difference.

   Following the pattern above, factor these.

   a) $16t^4 - v^8$ = _____   b) $25c^{10} - 49d^2$ = _____

a) $(c + d^4)(c - d^4)$
b) $(p^5 + q^3)(p^5 - q^3)$

150. The binomial below is the difference of two perfect squares. We used the same pattern to factor it.

$$x^2y^2 - x^4y^6 = (xy + x^2y^3)(xy - x^2y^3)$$

Following the example, factor these.

a) $p^6q^4 - p^4q^2 = $ _____

b) $c^8d^8 - c^2d^2 = $ _____

| a) $(4t^2 + v^4)(4t^2 - v^4)$ |
| b) $(5c^5 + 7d)(5c^5 - 7d)$ |

---

151. The binomial below is also the difference of two perfect squares. We used the same pattern to factor it.

$$9a^2b^6 - 4a^4b^2 = (3ab^3 + 2a^2b)(3ab^3 - 2a^2b)$$

Following the example, factor these.

a) $x^6y^6 - 25x^2y^2 = $ _____

b) $36p^4q^2 - 16p^2q^4 = $ _____

| a) $(p^3q^2 + p^2q)(p^3q^2 - p^2q)$ |
| b) $(c^4d^4 + cd)(c^4d^4 - cd)$ |

---

152. The factoring pattern we have been using applies only to the difference of two perfect squares. It does not apply to the sum of two perfect squares. Therefore, it does not apply to either binomial below.

$$x^4 + 100 \qquad a^6 + b^4$$

Use the pattern to factor these if possible.

a) $t^6 - 9 = $ _____   c) $x^2 - y^2 = $ _____

b) $v^{12} + 25 = $ _____   d) $a^8 + b^6 = $ _____

| a) $(x^3y^3 + 5xy)(x^3y^3 - 5xy)$ |
| b) $(6p^2q + 4pq^2)(6p^2q - 4pq^2)$ |

---

153. The factoring pattern applies only when both terms are perfect squares. Therefore, it does not apply to the binomials below.

$$x^3 - 36 \qquad 7a^2b^2 - c^8 \qquad 4m^6 - a^5m^8$$

Use the pattern to factor these if possible.

a) $y^4 - 29 = $ _____

b) $c^6 - d^8 = $ _____

c) $p^4q^4 - 9p^2 = $ _____

d) $t^{10} - 10t^2v^6 = $ _____

| a) $(t^3 + 3)(t^3 - 3)$ |
| b) Does not apply |
| c) $(x + y)(x - y)$ |
| d) Does not apply |

| a) Does not apply |
| b) $(c^3 + d^4)(c^3 - d^4)$ |
| c) $(p^2q^2 + 3p)(p^2q^2 - 3p)$ |
| d) Does not apply |

40 Polynomials

1-21 FACTORING PERFECT-SQUARE TRINOMIALS

The square of a binomial is called a "perfect-square" trinomial. We will discuss the procedure for factoring perfect-square trinomials in this section.

154. We squared two binomials below. Each square is called a "perfect-square" trinomial.

$$(x + 5)^2 = x^2 + 10x + 25$$
$$(3a - 4b)^2 = 9a^2 - 24ab + 16b^2$$

Notice these three characteristics of the perfect-square trinomial.

1. The first and last terms are perfect squares.
2. Though the middle term may be positive or negative, the last term is always positive.
3. The middle term is double the product of the square roots of the first and last terms. That is:

$$10x = 2 \cdot \sqrt{x^2} \cdot \sqrt{25} = (2)(x)(5)$$
$$24ab = 2 \cdot \sqrt{9a^2} \cdot \sqrt{16b^2} = (2)(3a)(4b)$$

Since perfect-square trinomials have three characteristics, three tests are needed to identify one. The first test is this:

The first and last terms must be perfect squares.

In which trinomials below are the first and last terms perfect squares?

a) $y^2 - 8y + 16$ c) $a^2 - 2ab + ab^2$
b) $4c^2 + 12cd + 7d^2$ d) $9x^2y^2 + 6xy^2 + y^2$

155. The second test needed to identify a perfect-square trinomial is this:

Though the middle term may be positive or negative, the last term is always positive.

Which of the following cannot be perfect-square trinomials for the reason above? _____

a) $y^2 + 4y - 4$ c) $a^2 - 2ab - b^2$
b) $c^2 - 6cd + 9d^2$ d) $4p^4q^4 + 4p^3q^2 + p^2$

Only (a) and (d)

Both (a) and (c)

Polynomials 41

156. The third test needed to identify a perfect-square trinomial is this:

   <u>The middle term must be double the product of the square roots of the first and last terms.</u>

   The third test was applied to each example below.

   $x^2 + 14x + 49$ <u>is</u> a perfect-square trinomial, since:
   $$2 \cdot \sqrt{x^2} \cdot \sqrt{49} = 2(x)(7) = 14x$$

   $a^2 - 3ab + b^2$ <u>is not</u> a perfect-square trinomial, since:
   $$2 \cdot \sqrt{a^2} \cdot \sqrt{b^2} = 2(a)(b) = 2ab$$

   Use the above test to decide whether each trinomial below is a perfect square or not.

   a) $x^2 - 2x + 1$ _____   c) $c^4 - 7c^2d^2 + d^4$ _____

   b) $y^2 + 5y + 9$ _____   d) $p^2 + 2pq + q^2$ _____

157. The third test was also applied to each trinomial below.

   $9x^2 - 24xy + 16y^2$ <u>is</u> a perfect-square trinomial, since:
   $$2 \cdot \sqrt{9x^2} \cdot \sqrt{16y^2} = 2(3x)(4y) = 24xy$$

   $a^2b^2 + 4a^2b^3 + 4b^4$ <u>is not</u> a perfect-square trinomial, since:
   $$2 \cdot \sqrt{a^2b^2} \cdot \sqrt{4b^4} = 2(ab)(2b^2) = 4ab^3$$

   Use the above test to decide whether each trinomial below is a perfect square or not.

   a) $100t^2 - 10t + 1$ _____   b) $25x^4 + 70x^2y^3 + 49y^6$ _____

   Answers:
   a) Yes, since: $2 \cdot \sqrt{x^2} \cdot \sqrt{1} = 2x$
   b) No, since: $2 \cdot \sqrt{y^2} \cdot \sqrt{9} = 6y$
   c) No, since: $2 \cdot \sqrt{c^4} \cdot \sqrt{d^4} = 2c^2d^2$
   d) Yes, since: $2 \cdot \sqrt{p^2} \cdot \sqrt{q^2} = 2pq$

158. Using the three tests, identify the perfect-square trinomials below. _____

   a) $t^2 + 6t - 9$ c) $p^2 + 2pq + q^2$
   b) $x^2 - 12x + 36$ d) $c^4d^4 - 2c^3d^3 + d^6$

   Answers:
   a) No, since: $2 \cdot \sqrt{100t^2} \cdot \sqrt{1} = 20t$
   b) Yes, since: $2 \cdot \sqrt{25x^4} \cdot \sqrt{49y^6} = 70x^2y^3$

159. Which of the following are perfect-square trinomials? _____

   a) $4y^2 + 6y + 9$ c) $100R^2 - 20RT - T^2$
   b) $25x^2 - 20x + 4$ d) $9a^4b^4 + 9a^2b^2 + 1$

   Only (b) and (c)

   Only (b)

42 Polynomials

160. Any perfect-square trinomial is the square of a binomial. <u>The terms of the binomial are the square roots of the first and last terms of the trinomial.</u>
For example:
$$x^2 + 10x + 25 = (x + 5)^2$$
$$4p^2 + 12pq + 9q^2 = (2p + 3q)^2$$

Following the examples, complete these:

a) $m^2 + 16m + 64 = ($ $)^2$

b) $16d^4 + 16d^2h + 4h^2 = ($ $)^2$

---

161. When the middle term of the trinomial is negative, the binomial is a difference. For example:
$$y^2 - 6y + 9 = (y - 3)^2$$
$$25a^2 - 10ab + b^2 = (5a - b)^2$$

Following the examples, complete these.

a) $F^2 - 14F + 49 = ($ $)^2$

b) $9p^6q^6 - 30p^4q^4 + 25p^2q^2 = ($ $)^2$

a) $(m + 8)^2$

b) $(4d^2 + 2h)^2$

---

162. Complete these:

a) $t^2 + 2t + 1 = ($ $)^2$

b) $4v^2 - 40v + 100 = ($ $)^2$

c) $9x^2y^2 - 6xy + 1 = ($ $)^2$

d) $4a^4b^4 + 36a^3b^3 + 81a^2b^2 = ($ $)^2$

a) $(F - 7)^2$

b) $(3p^3q^3 - 5pq)^2$

---

a) $(t + 1)^2$ b) $(2v - 10)^2$ c) $(3xy - 1)^2$ d) $(2a^2b^2 + 9ab)^2$

<u>SELF-TEST 3</u> (<u>pages 27-43</u>)

In Problems 1-4, find each product.

1. $(d + p)(r - t)$

2. $(3s - 2w)(2s - w)$

3. $(a^2 + 2b^2)(a^2 - 2b^2)$

4. $(2h - 3k)^2$

Continued on following page.

SELF-TEST 3 (pages 27-43) - Continued	
In Problems 5-8, factor each expression.	
5. $12m^2r + 8mr^2$	6. $25x^4 - 16y^2$
7. $4G^2 + 12G + 9$	8. $9p^2 - 30pt + 25t^2$

ANSWERS:
1. $dr - dt + pr - pt$
2. $6s^2 - 7sw + 2w^2$
3. $a^4 - 4b^4$
4. $4h^2 - 12hk + 9k^2$
5. $4mr(3m + 2r)$
6. $(5x^2 + 4y)(5x^2 - 4y)$
7. $(2G + 3)^2$
8. $(3p - 5t)^2$

1-22 FACTORING TRINOMIALS OF THE FORM: $x^2 + bx + c$

In this section, we will discuss the procedure for factoring trinomials in which the coefficient of the first term is "1".

163. When multiplying binomials, some products can be simplified to a trinomial. Three examples are shown below.

#1 $(x + 2)(x + 3) = x^2 + 5x + 6$

#2 $(y - 3)(y - 4) = y^2 - 7y + 12$

#3 $(t + 2)(t - 5) = t^2 - 3t - 10$

Notice that each product is a trinomial of the form $x^2 + bx + c$. In each case, "b" is the sum and "c" is the product of the number-terms in the factors. That is:

For #1, $b = 2 + 3 = 5$ and $c = (2)(3) = 6$

For #2, $b = (-3) + (-4) = -7$ and $c = (-3)(-4) = 12$

For #3, $b = 2 + (-5) = -3$ and $c = (2)(-5) = -10$

Continued on following page.

44   Polynomials

163. Continued

By adding the number terms to get "b" and multiplying the number terms to get "c", find each product below.

a) $(m + 1)(m + 7)$ = _____

b) $(x - 6)(x - 5)$ = _____

c) $(y - 3)(y + 8)$ = _____

d) $(t + 1)(t - 10)$ = _____

---

164. To factor a trinomial of the form $x^2 + bx + c$, we use the facts from the last frame. An example is discussed below.

In the trinomial below, both "b" and "c" are positive.

$$x^2 + 6x + 8$$

Since the product of the numbers in the binomials must be 8 and their sum must be 6, both numbers must be positive. The possible pairs of positive factors for 8 are (1 and 8) and (2 and 4). The pair whose sum is 6 is (2 and 4). Therefore:

$$x^2 + 6x + 8 = (x + 2)(x + 4)$$

Using the same method, factor these.

a) $y^2 + 6y + 5$ = _____

b) $m^2 + 15m + 36$ = _____

a) $m^2 + 8m + 7$
b) $x^2 - 11x + 30$
c) $y^2 + 5y - 24$
d) $t^2 - 9t - 10$

---

165. In the trinomial below, "b" is negative and "c" is positive.

$$x^2 - 7x + 10$$

Since the product of the numbers in the binomials must be 10 and their sum must be -7, both numbers must be negative. The possible pairs of negative factors for 10 are (-1 and -10) and (-2 and -5). The pair whose sum is -7 is (-2 and -5). Therefore:

$$x^2 - 7x + 10 = (x - 2)(x - 5)$$

Using the same method, factor these.

a) $t^2 - 3t + 2$ = _____

b) $m^2 - 10m + 16$ = _____

a) $(y + 1)(y + 5)$
b) $(m + 3)(m + 12)$

---

a) $(t - 1)(t - 2)$

b) $(m - 2)(m - 8)$

Polynomials 45

166. In the trinomial below, "c" is negative. Let's factor the trinomial.

$$x^2 + 5x - 6$$

Since the product of the numbers in the binomials must be −6, <u>one number must be positive and the other negative</u>. The possible pairs of factors for −6 are (−1 and 6), (−6 and 1), (−2 and 3), and (−3 and 2). The pair whose sum is 5 is (−1 and 6). Therefore:

$$x^2 + 5x - 6 = (x - 1)(x + 6)$$

Using the same method, factor these.

a) $m^2 + 4m - 5 =$ _____

b) $y^2 - 5y - 14 =$ _____

167. To check the factoring of a trinomial, multiply the two binomial factors to see whether you obtain the original trinomial. For example:

$$y^2 + 7y - 18 = (y - 2)(y + 9) \text{ is correct, since:}$$

$$(y - 2)(y + 9) = y^2 + 7y - 18$$

Check each factoring. State whether it is "correct" or "incorrect".

a) $x^2 - 8x + 12 = (x - 2)(x - 6)$ _____

b) $t^2 - 5t - 14 = (t - 2)(t + 7)$ _____

a) $(m - 1)(m + 5)$

b) $(y + 2)(y - 7)$

168. Remember that "c" is the key to factoring a trinomial. That is:

1) If "c" is <u>positive</u>, the numbers in the binomials have <u>the same sign</u>.

  Note: If "b" is <u>positive</u>, both are <u>positive</u>.
  If "b" is <u>negative</u>, both are <u>negative</u>.

2) If "c" is <u>negative</u>, the numbers in the binomials have <u>different signs</u>.

Factor each trinomial below and check your results.

a) $x^2 + 9x + 20 =$ _____

b) $y^2 - 3y - 18 =$ _____

c) $t^2 - 9t + 8 =$ _____

d) $m^2 + 4m - 12 =$ _____

a) Correct

b) Incorrect, since:
  $(t - 2)(t + 7) =$
  $t^2 + 5t - 14$

a) $(x + 4)(x + 5)$

b) $(y + 3)(y - 6)$

c) $(t - 1)(t - 8)$

d) $(m - 2)(m + 6)$

46 Polynomials

169. We factored some trinomials containing two variables below. Notice that the same general method is used.

$$x^2 + 5xy + 6y^2 = (x + 2y)(x + 3y)$$
$$a^2 - 3ab - 28b^2 = (a + 4b)(a - 7b)$$

Factor these and check your results.

a) $p^2 - 7pq + 6q^2 =$ _____

b) $m^2 + 6mt - 16t^2 =$ _____

---

170. We factored two more trinomials containing two variables below.

$$x^2y^2 - 7xy + 12 = (xy - 3)(xy - 4)$$
$$a^2b^2 + 4ab - 32 = (ab - 4)(ab + 8)$$

Factor these and check your results.

a) $p^2q^2 + 11pq + 10 =$ _____

b) $m^2v^2 - 5mv - 14 =$ _____

a) $(p - q)(p - 6q)$

b) $(m - 2t)(m + 8t)$

---

171. We factored some trinomials containing higher powers below.

$$x^4 + 11x^2 + 30 = (x^2 + 5)(x^2 + 6)$$
$$p^6q^6 - 3p^3q^3 - 4 = (p^3q^3 + 1)(p^3q^3 - 4)$$

Factor these and check your results.

a) $y^6 - 10y^3 + 16 =$ _____

b) $a^4b^4 + a^2b^2 - 6 =$ _____

a) $(pq + 1)(pq + 10)$

b) $(mv + 2)(mv - 7)$

---

172. Some trinomials of the form $x^2 + bx + c$ <u>cannot</u> be factored into binomials. Two examples are given below.

$$x^2 + 4x + 6 \qquad p^2 - 4pq - 9q^2$$

Factor if possible.

a) $y^2 - 8y + 3 =$ _____

b) $a^2b^2 + 3ab - 10 =$ _____

a) $(y^3 - 2)(y^3 - 8)$

b) $(a^2b^2 + 3)(a^2b^2 - 2)$

---

a) Not possible

b) $(ab - 2)(ab + 5)$

Polynomials 47

1-23 FACTORING TRINOMIALS OF THE FORM: $ax^2 + bx + c$

In this section, we will discuss the procedure for factoring trinomials in which the coefficient of the first term is a number other than "1".

173. The steps needed to multiply $(4x + 3)$ and $(x + 2)$ are shown below.

$$(4x + 3)(x + 2) = 4x^2 + 8x + 3x + 6 = 4x^2 + 11x + 6$$

Notice these points about the trinomial product:

"$4x^2$" is the product of "$4x$" and "x", the letter terms.

"6" is the product of "3" and "2", the number terms.

Using the facts above, if we multiplied $(2y - 4)$ and $(3y + 1)$:

a) The first term of the trinomial product would be _____.

b) The last term of the trinomial product would be _____.

---

174. The facts in the last frame are used to factor the trinomial below. The steps are described.

$$3x^2 + 14x + 8$$

1) The only possible pair of letter terms is $(3x$ and $x)$.

2) The possible pairs of number terms are $(1$ and $8)$ and $(2$ and $4)$.

3) The possible pairs of binomial factors are:

   A: $(3x + 1)(x + 8)$

   B: $(3x + 8)(x + 1)$

   C: $(3x + 2)(x + 4)$

   D: $(3x + 4)(x + 2)$

4) Only one pair of binomials is correct. It is the pair that produces "$14x$" as the middle term of the trinomial product.

a) Which pair has "$14x$" as the middle term of its product? Pair _____

b) Therefore: $3x^2 + 14x + 8 = ($ $)($ $)$

a) $6y^2$, from $(2y)(3y)$

b) -4, from $(-4)(1)$

---

175. Let's factor the trinomial: $2y^2 - 13y + 15$

1) The only possible pair of letter terms is $(2y$ and $y)$.

2) The possible pairs of number terms are $(-1$ and $-15)$ and $(-3$ and $-5)$.

3) The possible pairs of binomial factors are:

   A: $(2y - 1)(y - 15)$

   B: $(2y - 15)(y - 1)$

   C: $(2y - 3)(y - 5)$

   D: $(2y - 5)(y - 3)$

a) Pair C

b) $(3x + 2)(x + 4)$

Continued on following page.

48   Polynomials

175. Continued

The correct pair of binomial factors is the one with "-13y" as the middle term of its product.

a) Which pair has "-13y" as the middle term of its product? Pair _____

b) Therefore: $2y^2 - 13y + 15 = ($ $)($ $)$

---

176. Let's factor the trinomial: $7t^2 - 2t - 5$

    1) The only possible pair of letter terms is (7t and t).

    2) The possible pairs of number terms are (1 and -5) and (-1 and 5).

    3) The possible pairs of binomial factors are:

        A: (7t + 1)(t - 5)

        B: (7t - 5)(t + 1)

        C: (7t - 1)(t + 5)

        D: (7t + 5)(t - 1)

a) Which pair has "-2t" as the middle term of its product? Pair _____

b) Therefore: $7t^2 - 2t - 5 = ($ $)($ $)$

a) Pair C

b) $(2y - 3)(y - 5)$

---

177. When the coefficient of the first term is a number other than "1", factoring is a process of "trial and error". Each possible pair of factors has to be checked by multiplying the binomial factors. That is:

$2y^2 - y - 1 = (2y + 1)(y - 1)$ is correct, since:

$(2y + 1)(y - 1) = 2y^2 - y - 1$

Factor these and check your results.

a) $5x^2 + 17x + 6 =$ _____

b) $7m^2 - 16m + 4 =$ _____

c) $2t^2 - 5t - 7 =$ _____

a) Pair D

b) $(7t + 5)(t - 1)$

---

178. Let's factor the trinomial: $6x^2 + 19x + 8$.

    1) The possible pairs of letter terms are (x and 6x) and (2x and 3x).

    2) The possible pairs of number terms are (1 and 8) and (2 and 4).

    3) The possible pairs of binomial factors are:

        A: (x + 1)(6x + 8)   E: (2x + 1)(3x + 8)

        B: (x + 8)(6x + 1)   F: (2x + 8)(3x + 1)

        C: (x + 2)(6x + 4)   G: (2x + 2)(3x + 4)

        D: (x + 4)(6x + 2)   H: (2x + 4)(3x + 2)

Continued on following page.

a) $(5x + 2)(x + 3)$

b) $(7m - 2)(m - 2)$

c) $(2t - 7)(t + 1)$

178. Continued

   a) Which pair of factors has a product whose middle term is "19x"?
      Pair _____

   b) Therefore: $6x^2 + 19x + 8 = ($ $)($ $)$

---

179. We factored some trinomials containing two variables below. Notice that the same general method is used.

   $$2x^2 + 7xy + 3y^2 = (2x + y)(x + 3y)$$
   $$8t^2 - 2tv - 15v^2 = (4t + 5v)(2t - 3v)$$

   Factor these and check your results.

   a) $3a^2 - 5ab + 2b^2 = $ _____

   b) $4p^2 + 8pq - 5q^2 = $ _____

| a) Pair E |
| b) $(2x + 1)(3x + 8)$ |

---

180. We factored two more trinomials containing two variables below.

   $$5x^2y^2 - 12xy + 7 = (5xy - 7)(xy - 1)$$
   $$6a^2b^2 - 7ab - 3 = (3ab + 1)(2ab - 3)$$

   Factor these and check your results.

   a) $7p^2q^2 + 9pq + 2 = $ _____

   b) $10t^2v^2 + 9tv - 9 = $ _____

| a) $(3a - 2b)(a - b)$ |
| b) $(2p + 5q)(2p - q)$ |

---

181. We factored two trinomials containing higher powers below.

   $$3x^4 + 8x^2 + 5 = (3x^2 + 5)(x^2 + 1)$$
   $$8a^6b^6 - 10a^3b^3 - 3 = (2a^3b^3 - 3)(4a^3b^3 + 1)$$

   Factor these and check your results.

   a) $2y^6 - 11y^3 + 5 = $ _____

   b) $6p^4q^4 + 7p^2q^2 - 10 = $ _____

| a) $(7pq + 2)(pq + 1)$ |
| b) $(5tv - 3)(2tv + 3)$ |

---

182. Some trinomials of the form "$ax^2 + bx + c$" <u>cannot</u> be factored into two trinomials. Two examples are given below.

   $$2x^2 + 6x + 3 \qquad\qquad 3t^2 - tv - 5v^2$$

   Factor if possible:

   a) $4y^2 + 13y + 3 = $ _____

   b) $5a^2b^2 + 7ab - 2 = $ _____

| a) $(2y^3 - 1)(y^3 - 5)$ |
| b) $(6p^2q^2 - 5)(p^2q^2 + 2)$ |

| a) $(4y + 1)(y + 3)$ |
| b) Not possible |

50 Polynomials

1-24 FACTORING POLYNOMIALS COMPLETELY

In this section, we will discuss what is meant by "factoring a polynomial completely".

183. When factoring a binomial, we always begin by looking for a common monomial factor. For example:

$$3x^2 - 12 = 3(x^2 - 4)$$
$$4y^3 + 8y = 4y(y^2 + 2)$$

To "factor completely", we then look to see whether the binomial can be factored further. It can if it is the difference of two perfect squares. Therefore:

$3(x^2 - 4)$ can be factored to $3(x + 2)(x - 2)$

$4y(y^2 + 2)$ cannot be factored further.

Factor each of these completely.

a) $7a^2 - 7b^2 = $ _____

b) $9pq^2 - 3pt^2 = $ _____

184. When factoring a trinomial, we should also begin by looking for a common monomial factor. For example:

$$x^3 + 5x^2 + 6x = x(x^2 + 5x + 6)$$
$$4y^2 - 6y - 8 = 2(2y^2 - 3y - 4)$$

To "factor completely", we then look to see whether the trinomial can be factored further. For example:

$x(x^2 + 5x + 6)$ can be factored to $x(x + 2)(x + 3)$

$2(2y^2 - 3y - 4)$ cannot be factored further.

Factor each of these completely.

a) $5a^2b^2 - 10ab + 15 = $ _____

b) $ad^4 - 4ad^2 - 5a = $ _____

Answers:
a) $7(a + b)(a - b)$
b) $3p(3q^2 - t^2)$

185. After factoring out a common monomial from a trinomial, the remaining trinomial can be a perfect square. To "factor completely", we must factor the perfect square. For example:

$$2x^2 - 12x + 18 = 2(x^2 - 6x + 9) = 2(x - 3)^2$$

Factor completely.

a) $6m^3 - 3m - 9 = $ _____

b) $bx^2 + 2bxy + by^2 = $ _____

Answers:
a) $5(a^2b^2 - 2ab + 3)$
b) $a(d^2 - 5)(d^2 + 1)$

186. "Factoring completely" means this: Whenever you get a factor that can still be factored, factor it.

Factor each of these completely.

    a) $2xy^2 - 50x =$ _____

    b) $4bp^2 - 9bq^2 =$ _____

    c) $3x^2 + 12xy - 36y^2 =$ _____

    d) $9a^2t + 12abt + 4b^2t =$ _____

a) $3(2m^3 - m - 3)$

b) $b(x + y)^2$

---

187. After factoring $(1 - 16t^4)$ below, we were able to factor $(1 - 4t^2)$ further.

$$1 - 16t^4 = (1 + 4t^2)(1 - 4t^2) = (1 + 4t^2)(1 + 2t)(1 - 2t)$$

Following the example, factor this one completely.

$x^4 - y^4 =$ _____

a) $2x(y + 5)(y - 5)$

b) $b(2p + 3q)(2p - 3q)$

c) $3(x + 6y)(x - 2y)$

d) $t(3a + 2b)^2$

---

188. After factoring the trinomial below, we were able to factor $(x^2 - 1)$ further.

$$x^4 - 3x^2 + 2 = (x^2 - 2)(x^2 - 1) = (x^2 - 2)(x + 1)(x - 1)$$

Following the example, factor this one completely.

$y^4 - 5y^2 + 4 =$ _____

$(x^2 + y^2)(x + y)(x - y)$

$(y + 2)(y - 2)(y + 1)(y - 1)$

1-25   DIVIDING BY MONOMIALS

In this section, we will discuss the procedure for dividing a monomial or another polynomial by a monomial.

---

189. To divide a monomial by a number, we divide the coefficient by the number. That is:

$$\frac{12x^3}{6} = 2x^3 \qquad \text{a)} \ \frac{-6y}{2} = \underline{} \qquad \text{b)} \ \frac{-20t^2}{-5} = \underline{}$$

---

190. To divide monomials containing powers of the same variable or variables, we divide the coefficients and use the law of exponents for division. For example:

$$\frac{10x^5}{5x^2} = \left(\frac{10}{5}\right)\left(\frac{x^5}{x^2}\right) = 2x^3$$

$$\frac{-8a^3b^6}{2ab^5} = \left(\frac{-8}{2}\right)\left(\frac{a^3}{a}\right)\left(\frac{b^6}{b^5}\right) = -4a^2b$$

a) $-3y$

b) $4t^2$

Continued on following page.

52   Polynomials

190. Continued

Using the same method, complete each division.

a) $\dfrac{12b^2t^7}{3bt^3} = $ _____   b) $\dfrac{-18m^5}{-6m^4} = $ _____

---

191. When the coefficient of the quotient is "1" or "-1", the coefficient is not ordinarily written. That is:

$\dfrac{4x^7}{4x^2} = 1x^5$ or x^5 $\dfrac{-8m^5v^8}{8m^3v^7} = $ _____

a) $4bt^4$

b) $3m$

---

192. When the coefficient of the denominator is "1" or "-1", it is helpful to write the "1" or "-1" explicitly. For example:

$\dfrac{5x^3}{x} = \dfrac{5x^3}{1x} = 5x^2$ $\dfrac{7a^4t^5}{-a^3t^2} = \dfrac{7a^4t^5}{-1a^3t^2} = $ _____

$-1m^2v$ or $-m^2v$

---

193. When the quotient contains a power whose exponent is zero, that power is not ordinarily written since it equals "1". For example:

$\dfrac{12b^3x^4}{4b^3x} = 3b^0x^3 = 3(1)x^3 = 3x^3$

Using the same method, complete these.

a) $\dfrac{10t^5}{-2t^5} = $ _____   b) $\dfrac{a^2b}{a^2b} = $ _____

$-7at^3$

---

194. To divide a binomial by a number, we divide each term by the number. That is:

$\dfrac{8x^2 + 6}{2} = \dfrac{8x^2}{2} + \dfrac{6}{2} = 4x^2 + 3$

$\dfrac{9ay^3 - 12}{3} = \dfrac{9ay^3}{3} - \dfrac{12}{3} = $ _____

a) -5, from $-5t^0$

b) 1, from a^0b^0

---

195. To divide a trinomial by a number, we divide each term by the number. That is:

$\dfrac{5x^2 - 10x + 15}{5} = \dfrac{5x^2}{5} - \dfrac{10x}{5} + \dfrac{15}{5} = x^2 - 2x + 3$

$\dfrac{8p^4 + 6p^2 - 12}{2} = \dfrac{8p^4}{2} + \dfrac{6p^2}{2} - \dfrac{12}{2} = $ _____

$3ay^3 - 4$

$4p^4 + 3p^2 - 6$

196. In each division below, we also divide each term in the numerator by the denominator. That is:

   a) $\dfrac{6y^2 + 4y}{2y} = \dfrac{6y^2}{2y} + \dfrac{4y}{2y} = $ _____

   b) $\dfrac{ax^6 - bx^5 - x^4}{x^2} = \dfrac{ax^6}{x^2} - \dfrac{bx^5}{x^2} - \dfrac{x^4}{x^2} = $ _____

197. Complete each division.

   a) $\dfrac{3t^4 + t}{t} = $ _____ b) $\dfrac{6m^5 - 9m^3}{3m^3} = $ _____

   a) $3y + 2$
   b) $ax^4 - bx^3 - x^2$

198. Complete: a) $\dfrac{2y^3 + 6y^2 - 4y}{2y} = $ _____

   b) $\dfrac{2p^5 + 4p^4 - 6p^3}{2p^3} = $ _____

   a) $3t^3 + 1$
   b) $2m^2 - 3$

199. Complete: a) $\dfrac{x^5y^2 - x^3y^3 + x^2y}{x^2y} = $ _____

   b) $\dfrac{p^4q^4 + p^3q^3 - p^2q^2}{p^2q^2} = $ _____

   a) $y^2 + 3y - 2$
   b) $p^2 + 2p - 3$

a) $x^3y - xy^2 + 1$ b) $p^2q^2 + pq - 1$

1-26 DIVIDING BY BINOMIALS

To divide a trinomial or larger polynomial by a binomial, we use a procedure similar to long division. We will discuss the method in this section.

200. The two basic steps needed to divide $x^2 + 6x + 8$ by $x + 2$ are shown below.

   1) Dividing the first term "x^2" of the dividend by the first term of the divisor.

   $$\begin{array}{r} x \\ x+2\overline{\smash{)}x^2 + 6x + 8} \\ \underline{x^2 + 2x} \\ 4x \end{array}$$

   ← Dividing x^2 by x: $\dfrac{x^2}{x} = x$
   ← Multiplying the divisor by $\underline{x}$
   ← Subtracting $x^2 + 2x$ from $x^2 + 6x$

   2) Bringing down the next term "8" of the dividend and dividing the first term "4x" of $4x + 8$ by the first term "x" of the divisor.

   $$\begin{array}{r} x + 4 \\ x+2\overline{\smash{)}x^2 + 6x + 8} \\ \underline{x^2 + 2x} \\ 4x + 8 \\ \underline{4x + 8} \\ 0 \end{array}$$

   ← Dividing $4x$ by x: $\dfrac{4x}{x} = 4$

   ← Multiplying the divisor by 4
   ← Subtracting $4x + 8$ from $4x + 8$.

Continued on following page.

54   Polynomials

200. Continued

Therefore, $x^2 + 6x + 8$ divided by $x + 2$ equals $x + 4$ with a "0" remainder. To check the division, multiply the divisor and quotient below.

$(x + 2)(x + 4) = $ _____

---

201. Using the steps from the last frame, divide $x^2 + 8x + 15$ by $x + 3$ at the right.

$x + 3 \overline{) x^2 + 8x + 15}$

Therefore: $\dfrac{x^2 + 8x + 15}{x + 3} = $ _____

$x^2 + 6x + 8$

---

202. We completed one division below. Notice how subtracting $-2x$ from $-5x$ is the same as adding $+2x$ and $-5x$. Complete the other division.

$\begin{array}{r} x - 3 \\ x - 2 \overline{) x^2 - 5x + 6} \\ \underline{x^2 - 2x} \\ -3x + 6 \\ \underline{-3x + 6} \\ 0 \end{array}$

$y - 6 \overline{) y^2 - 11y + 30}$

$\begin{array}{r} x + 5 \\ x + 3 \overline{) x^2 + 8x + 15} \\ \underline{x^2 + 3x} \\ 5x + 15 \\ \underline{5x + 15} \\ 0 \end{array}$

$\dfrac{x^2 + 8x + 15}{x + 3} = x + 5$

---

203. We completed one division below. Notice how subtracting $-5y$ from $-y$ is the same as adding $+5y$ and $-y$. Complete the other division.

$\begin{array}{r} y + 4 \\ y - 5 \overline{) y^2 - y - 20} \\ \underline{y^2 - 5y} \\ 4y - 20 \\ \underline{4y - 20} \\ 0 \end{array}$

$t - 3 \overline{) t^2 + t - 12}$

$\begin{array}{r} y - 5 \\ y - 6 \overline{) y^2 - 11y + 30} \\ \underline{y^2 - 6y} \\ -5y + 30 \\ \underline{-5y + 30} \\ 0 \end{array}$

---

204. Following the example, complete the other division.

$\begin{array}{r} 2x + 1 \\ 3x - 4 \overline{) 6x^2 - 5x - 4} \\ \underline{6x^2 - 8x} \\ 3x - 4 \\ \underline{3x - 4} \\ 0 \end{array}$

$4x - 1 \overline{) 8x^2 + 2x - 1}$

$\begin{array}{r} t + 4 \\ t - 3 \overline{) t^2 + t - 12} \\ \underline{t^2 - 3t} \\ 4t - 12 \\ \underline{4t - 12} \\ 0 \end{array}$

$\begin{array}{r} 2x + 1 \\ 4x - 1 \overline{) 8x^2 + 2x - 1} \\ \underline{8x^2 - 2x} \\ 4x - 1 \\ \underline{4x - 1} \\ 0 \end{array}$

205. There is a remainder of 44 in the division below. Notice how we wrote the remainder as the fraction $\frac{44}{m-5}$ when writing the quotient. Complete the other division.

$$\begin{array}{r} m + 8 \\ m - 5 \overline{\smash{)}m^2 + 3m + 4} \\ \underline{m^2 - 5m} \\ 8m + 4 \\ \underline{8m - 40} \\ 44 \end{array}$$

$$t - 2\overline{\smash{)}t^2 + 5t - 1}$$

$$\frac{m^2 + 3m + 4}{m - 5} = m + 8 + \frac{44}{m - 5} \qquad \frac{t^2 + 5t - 1}{t - 2} = \underline{\hspace{3cm}}$$

$$\begin{array}{r} t + 7 \\ t - 2 \overline{\smash{)}t^2 + 5t - 1} \\ \underline{t^2 - 2t} \\ 7t - 1 \\ \underline{7t - 14} \\ 13 \end{array}$$

$$\frac{t^2 + 5t - 1}{t - 2} = t + 7 + \frac{13}{t - 2}$$

SELF-TEST 4 (pages 43-55)

Factor each polynomial completely.

1. $t^2 - 3t - 18$
2. $a^2 - 10ab + 16b^2$
3. $6x^2 + 17x + 12$
4. $4r^2s^2 + rs - 5$
5. $6R^2 - 54$
6. $20d^2 - 10dk - 30k^2$

Do these divisions.

7. $\dfrac{18p^3r^4}{-6pr^4}$
8. $\dfrac{x^3y + xy^3}{xy}$
9. $\dfrac{9b^4 - 3b^3 + 6b^2}{3b^2}$
10. $w + 4\overline{\smash{)}w^2 - w - 20}$
11. $4m - 5\overline{\smash{)}8m^2 + 2m - 15}$

ANSWERS:
1. $(t + 3)(t - 6)$
2. $(a - 2b)(a - 8b)$
3. $(2x + 3)(3x + 4)$
4. $(rs - 1)(4rs + 5)$
5. $6(R + 3)(R - 3)$
6. $10(d + k)(2d - 3k)$
7. $-3p^2$
8. $x^2 + y^2$
9. $3b^2 - b + 2$
10. $w - 5$
11. $2m + 3$

Polynomials

SUPPLEMENTARY PROBLEMS - CHAPTER 1

Assignment 1

Multiply.

1. $t^2 \cdot t^3$
2. $m^4 \cdot m \cdot m^2$
3. $(a^2b)(a^3b^2)$
4. $(pr^5)(pr)$
5. $(x^3y^2)(xy^3)$
6. $(s^3v^2w)(svw^4)$
7. $(d^3hk)(dhk^2)$
8. $(a^2d)(a^3c)(cd^2)$

Divide.

9. $\dfrac{p^5}{p}$
10. $\dfrac{x^3y^4}{x^2y}$
11. $\dfrac{r^7t^4}{r^4t^3}$
12. $\dfrac{b^2h^3}{b^2h}$
13. $\dfrac{a^3d^4k^2}{ad^3k}$
14. $\dfrac{t^5vw^4}{t^2w^3}$
15. $\dfrac{h^2pr^3}{hpr}$
16. $\dfrac{m^6n^2s^3}{m^3n^2s}$

Complete.

17. $(xy^2)^3$
18. $(r^4t^3)^2$
19. $(a^2b^5)^0$
20. $(d^2h^3p)^4$

Write each polynomial in descending order for "y".

21. $4y - 3y^2 - 1 + 2y^3$
22. $a^3y^2 + ay + a^2y^3 + 2a$
23. $ky + ry^5 - 5t - by^3$

State the degree of each term.

24. $-3w$
25. $5xy$
26. $7w^3$
27. $-2r^2s$
28. $4pt^3w^2$

State whether each polynomial is a monomial, binomial, or trinomial.

29. $x^2 - 4$
30. $3a^2b$
31. $y^2 - y + 2$
32. $2t^3 + t$

Assignment 2

Add. Write each sum in descending order for "x".

1. $(x^2 + 2) + (x^2 - 1)$
2. $(x^3 - 2x - 3) + (2x^3 + x + 3)$
3. $(3x^4 - x^2 + 1) + (x^4 + 2x^2 - 2)$
4. $(x^2 + x - 7) + (3x + 2) + (5x^2 - 4x)$
5. $(ax^2 + b) + (rx - 2b) + (kx^4 - rx)$

Subtract. Write each difference in descending order for "y".

6. $(3y^2 + 2) - (y^2 - 2)$
7. $(y^3 - 4y + 3) - (2y^3 - 7y + 1)$
8. $(2y + 3) - (3y - y^2)$
9. $(2xy^4 - 1) - (xy^4 + 2)$
10. $(3by^2 + dy + 2) - (by^2 + 2)$
11. $(ky^3 - ty + w) - (ky^3 - 2ty - w)$

Multiply these monomials.

12. $(3rt^2)(-4r^3t)$
13. $(bx^2y)(axy)$
14. $(-dk)(5d^2k^3)(-2dk^2)$
15. $(6p^3s)(-3a^2s)(ap)$

Square these monomials.

16. $(4cd)^2$
17. $(-2x^3)^2$
18. $(s^2tw^4)^2$
19. $(-5h^3p)^2$

Multiply these monomials, binomials, and trinomials.

20. $4x(x^2 - 1)$
21. $a^2(2a + b)$
22. $my(3y^3 - 2m)$
23. $2t(t^2 - t + 5)$
24. $8d(cd^2 + dk + p)$
25. $xy(x^2 + xy + y^2)$
26. $r^3w(2r^2 + 3rw + 5)$
27. $3mv^2(m^2v - 2v^2 + m)$
28. $5h^2s^3(3hs^2 - h^2s - 2hs)$

Assignment 3
Multiply.

1. $(b + d)(r + t)$
2. $(x + 3)(y - 2)$
3. $(a - 4c)(3p + w)$
4. $(h^2 - p^2)(s - 2t)$
5. $(3x - 2y)(2x + y)$
6. $(5h + 2k)(4h - k)$
7. $(c^2 - d)(c^2 - 2d)$
8. $(2r^2 + t^2)(4r^2 + 3t^2)$
9. $(m + n)(m - n)$
10. $(5p - 3w)(5p + 3w)$
11. $(r^2 + t^2)(r^2 - t^2)$
12. $(xy - 2)(xy + 2)$
13. $(2s + v)^2$
14. $(3x - 4y)^2$
15. $(a^2 - 5b^2)^2$
16. $(5r^2 + 3w)^2$

Write the missing factor in each blank.

17. $8d^2 = (4d)(\quad)$
18. $15t^3 = (5t)(\quad)$
19. $hk^3p^2 = (hkp)(\quad)$
20. $12a^2b^4 = (2a^2b)(\quad)$

Factor each polynomial.

21. $9y^2 - 15$
22. $10t^2 + 6t$
23. $x^2y - 2xy + xy^2$
24. $12a^4b^2 + 4a^3b^2 - 8a^2b^4$
25. $2x^4 + 3x^3 - x^2$
26. $5d^2h - 15dh^2 - 10dh$
27. $R^2 - r^2$
28. $p^2 - 16t^2$
29. $1 - b^2c^2$
30. $9s^4 - t^4$
31. $25a^2d^4 - 64p^2$
32. $k^2 + 6k + 9$
33. $4y^2 - 20y + 25$
34. $m^2 + 4mn + 4n^2$
35. $9P^2 - 12P + 4$
36. $16d^2h^2 - 8dh + 1$

Assignment 4
Factor each polynomial completely.

1. $x^2 + 4x + 3$
2. $t^2 - 6t - 7$
3. $w^2 - 7w + 10$
4. $d^2 + d - 12$
5. $a^2 - ab - 2b^2$
6. $x^2 - 4xy - 12y^2$
7. $r^4 + 5r^2 + 4$
8. $p^2t^2 + 7pt + 6$
9. $3y^2 + 5y + 2$
10. $5h^2 + 9h - 2$
11. $6x^2 - 13x + 6$
12. $4t^2 - 4t - 15$
13. $2r^2 + 5rs - 3s^2$
14. $5a^2 + 11ab + 2b^2$
15. $3t^2w^2 - 7tw - 6$
16. $4x^2 + 5xy - 6y^2$
17. $2p^2 - 18$
18. $5y^2 + 25y + 30$
19. $8d^2 - 8k^2$
20. $hr^2 + 4hrs + 4hs^2$

Divide.

21. $\dfrac{15x^3}{5x}$
22. $\dfrac{-6h^2p}{6hp}$
23. $\dfrac{4a^3bc^2}{-a^2bc}$
24. $\dfrac{8t^3 - 6t}{2t}$
25. $\dfrac{3r^2s^3 + 2rs^2}{rs}$
26. $\dfrac{30y^2 - 12y + 24}{6}$
27. $\dfrac{9p^4w + 3p^3x - 6p^2}{3p^2}$
28. $\dfrac{m^5t^3 - m^4t^2 - m^3t}{m^2t}$
29. $x + 8 \overline{)x^2 + 14x + 48}$
30. $d + 3 \overline{)d^2 - 2d - 15}$
31. $a - 4 \overline{)a^2 - 10a + 24}$
32. $2y + 5 \overline{)6y^2 + 7y - 20}$
33. $5r - 6 \overline{)10r^2 + 3r - 18}$
34. $4t - 3 \overline{)20t^2 - 23t + 6}$

Chapter 2 LINEAR EQUATIONS

An equation whose graph is a straight line is called a "linear" equation. We will discuss linear equations and their graphs in this section. Intercepts, slope, and slope-intercept form are emphasized. Though introduced in a discussion of linear equations in "x" and "y", the concepts are extended to linear equations in other variables and linear formulas.

2-1 IDENTIFYING LINEAR EQUATIONS

An equation whose graph is a straight line is called a "linear equation". In this section, we will discuss a method for identifying linear equations in "x" and "y".

1. The degree of a term is the sum of the exponents of "x" and "y". For example:

Term	Degree	
$-3x$	1	
$2xy$	2	(from 1 + 1)
y^2	2	
x^2y	3	(from 2 + 1)

    Write the degree of each term.

    a) $-y$ _____ b) $5x^2$ _____ c) $-4xy$ _____ d) x^2y^2 _____

2. A term whose degree is "1" is called a "first-degree" term. All first-degree terms are called "linear" terms. Some examples are:

    $7x \qquad -4y \qquad -x \qquad y$

    All terms that are not first-degree are called "non-linear" terms. Some examples are:

    $x^2 \qquad -5xy \qquad 9y^3 \qquad xy$

    Which of the following are linear terms? _____

    a) $12x$ b) y^3 c) $-y$ d) $-9xy$ e) $6x^2$

3. Which of the following are non-linear terms? _____

    a) $-x^2$ b) $-5y$ c) $-4x^3$ d) x e) xy

Answers:
a) 1
b) 2
c) 2
d) 4

(a) and (c)

58

Linear Equations

4. Any equation in "x" and "y" in which each term is <u>linear</u> is called a "<u>linear</u>" equation. Its graph is a straight line. Each equation below is <u>linear</u>.

   $4x + 3y = 5 \qquad y = 2x - 7 \qquad 6y = x$

   Any equation in "x" and "y" in which one or more terms is <u>non-linear</u> is called a "<u>non-linear</u>" equation. Its graph is a <u>curve</u>. Each equation below is <u>non-linear</u>.

   $xy = 12 \qquad y = 2x^2 - 1 \qquad y - 5xy = 0$

   Which of the following are <u>linear</u> equations? _____

   a) $y - 4x = 7$ b) $7xy = 5$ c) $y - x^2 = 1$ d) $y = 5x$

   Answer: (a), (c), and (e)

5. Which of the following are <u>non-linear</u> equations? _____

   a) $y = 5x + 3$ b) $6xy = 1$ c) $3x - 4y = 0$ d) $y = 2x^2$

   Answer: (a) and (d)

6. To decide whether a fractional equation is linear or non-linear, we clear the fraction or fractions first. For example:

   $y = \dfrac{2x - 1}{5}$ is equivalent to $5y = 2x - 1$. It is linear.

   $y = \dfrac{7}{x}$ is equivalent to $xy = 7$. It is non-linear.

   Clear the fraction in each equation below and then decide whether it is linear or non-linear.

   a) $\dfrac{y}{x} = 10$ b) $x = \dfrac{3y + 5}{9}$ c) $x = \dfrac{1}{y}$

   Answer: (b) and (d)

   a) linear, since: $y = 10x$
   b) linear, since: $9x = 3y + 5$
   c) non-linear, since: $xy = 1$

7. Clear the fraction in each equation below and then decide whether it is linear or non-linear.

   a) $y = \dfrac{2x + 9}{y}$ b) $1 = \dfrac{x}{y - 2}$ c) $\dfrac{y + 3}{x} = y$

   a) non-linear, since: $y^2 = 2x + 9$
   b) linear, since: $y - 2 = x$
   c) non-linear, since: $y + 3 = xy$

60   Linear Equations

2-2 INTERCEPTS

The points where a graphed line crosses the axes are called intercepts. In this section, we will discuss "intercepts" and show how the coordinates of intercepts can be obtained directly from linear equations.

8. Both lines at the right intersect both axes.

    Point A and D are called <u>vertical</u> <u>intercepts</u> or <u>y-intercepts</u>. Their coordinates are:

        A (0, 4)   D (0, -3)

    Points B and C are called <u>horizontal</u> <u>intercepts</u> or <u>x-intercepts</u>. Write their coordinates below.

        B (,)   C (,)

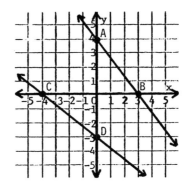

9. One of the coordinates of any intercept is always "0".

    Since any <u>horizontal</u> <u>intercept</u> (<u>x-intercept</u>) lies on the x-axis, its y-coordinate (or ordinate) is 0.

    Since any <u>vertical</u> <u>intercept</u> (<u>y-intercept</u>) lies on the y-axis, its x-coordinate (or abscissa) is 0.

    a) Which of the following points are x-intercepts? _____

        A (5, 0)   B (0, -1)   C (-2, 0)

    b) Which of the following points are y-intercepts? _____

        A (0, 8)   B (8, 0)   C (-9, 0)

B (3, 0)   C (-4, 0)

10. The linear equation below is graphed at the right. The coordinates of both intercepts are given.

    $$3x + 2y = 6$$

    We can get the coordinates of the intercepts directly from the equation in the following way.

    1) <u>Vertical</u> <u>intercept</u> (<u>y-intercept</u>)

        Since the x-coordinate is "0", we substitute "0" for "x" and solve for "y". We get:

            $3(0) + 2y = 6$
            $2y = 6$
            $y = 3$

        The coordinates of the vertical intercept are (0, 3).

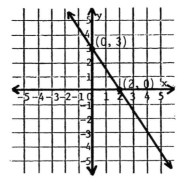

a) A and C

b) Only A

Continued on following page.

10. Continued

   2) Horizontal intercept (x-intercept)

   Since the y-coordinate is "0", we substitute "0" for "y" and solve for "x". We get:

   $$3x + 2(0) = 6$$
   $$3x = 6$$
   $$x = 2$$

   The coordinates of the horizontal intercept are (2, 0).

---

11. Here is another linear equation. $\boxed{3x + 4y = 24}$

   a) To find its vertical intercept, substitute "0" for "x" and solve for "y".

   The coordinates of the vertical intercept are (,).

   b) To find its horizontal intercept, substitute "0" for "y" and solve for "x".

   The coordinates of the horizontal intercept are (,).

---

12. By substituting "0" for "y", find the coordinates of the x-intercept for each equation below.

   a) $2x + 5y = 10$ b) $2y = 4x + 12$ c) $y = 5x - 6$

   a) (0, 6)   b) (8, 0)

---

13. By substituting "0" for "x", find the coordinates of the y-intercept for each equation below.

   a) $2x + 3y = 24$ b) $5y = x - 20$ c) $2y = 5x - 3$

   a) (5, 0)

   b) (-3, 0)

   c) ($\frac{6}{5}$, 0)

---

14. Find the coordinates of both intercepts for each equation below.

   a) $y = 2x - 3$ b) $2x - 3y = 1$

   a) (0, 8)

   b) (0, -4)

   c) (0, -$\frac{3}{2}$)

   x-intercept (,)   x-intercept (,)
   y-intercept (,)   y-intercept (,)

   a) x-intercept ($\frac{3}{2}$, 0)
      y-intercept (0, -3)
   b) x-intercept ($\frac{1}{2}$, 0)
      y-intercept (0, -$\frac{1}{3}$)

62 Linear Equations

2-3 GRAPHING USING INTERCEPTS

Since its graph is a straight line, we can graph a linear equation by plotting only two points. Frequently, the intercepts are used as the two points. In this section, we will graph linear equations by plotting their intercepts.

15. To graph $y = x + 2$ at the right, we plotted only the two points below.

    Point A: If $x = -4$, $y = -2$.

    Point B: If $x = 1$, $y = 3$.

    When graphing a line by plotting only two points, we always plot a third point as a check. If it lies on the line, the graph is probably correct.

    We plotted the point below on the same graph as a check.

    Point C: If $x = 3$, $y = 5$.

    a) Does Point C lie on the line? _____
    b) Is the graph probably correct? _____

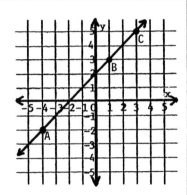

16. When graphing a linear equation, we frequently plot the intercepts as the two points.

    To graph $x + 2y = 4$, we plotted the two intercepts. Their coordinates are:

    (0, 2)   (4, 0)

    We also plotted (-4, 4) as a check.

    a) Does (-4, 4) lie on the line? _____
    b) Is the graph probably correct? _____

a) Yes   b) Yes

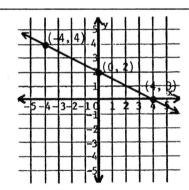

17. Find the intercepts for each equation below and use them to graph the equations at the right. (Plot a third point for each as a check.)

    $2x - y = 8$ $3y = 12 - 2x$

a) Yes   b) Yes

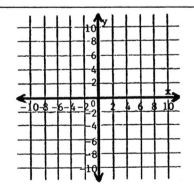

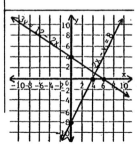

2-4 SLOPE AS A SIGNED NUMBER

The steepness of the rise or fall of a graphed line from left to right is called the "slope" of a line. In this section, we will show how the "slope" of a line can be expressed as a signed number.

18. The arrows at the right represent changes (increases or decreases) in x and y. The changes can be represented by signed numbers. That is:

    1) Any arrow to the right or upward is an increase. It is positive.

    2) Any arrow to the left or downward is a decrease. It is negative.

    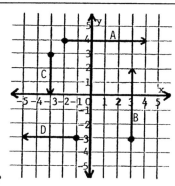

    What signed number represents each change?

    A: _____   B: _____   C: _____   D: _____

---

19. Points A (1, 2) and B (3, 5) are plotted on the line at the right. The changes in x and y from A to B are shown by arrows.

    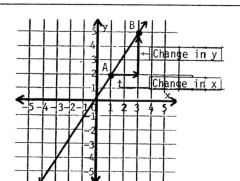

    a) The horizontal arrow shows an increase in x from 1 to 3. What signed number represents the change? _____

    b) The vertical arrow shows an increase in y from 2 to 5. What signed number represents the change? _____

    A: +6
    B: +5
    C: -3
    D: -4

---

20. The symbol "Δ" (pronounced "delta") is used as an abbreviation for the phrase "change in". Therefore:

    Δx means "change in x".
    Δy means "change in y".

    Points C (-4, 6) and D(6, -2) are plotted on the line at the right. Δx and Δy are the changes in x and y from C to D.

    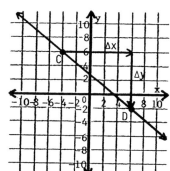

    a) Δx is an increase in x from -4 to 6. Therefore, Δx equals what signed number? _____

    b) Δy is a decrease in y from 6 to -2. Therefore, Δy equals what signed number? _____

    a) +2
    b) +3

---

a) +10   b) -8

64 Linear Equations

21. The "slope" of a line is a ratio of the "change in y" to the "change in x" from one point to another point on the line. The "slope" formula is:

$$\text{Slope} = \frac{\Delta y}{\Delta x} = \frac{\text{increase or decrease in y}}{\text{increase in x}}$$

Let's use the changes from P to Q to compute the slope of the line at the right.

a) Since x increases from 1 to 4, Δx =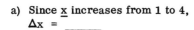

b) Since y increases from 1 to 5, Δy = _____

c) Therefore, the slope = $\frac{\Delta y}{\Delta x}$ = _____

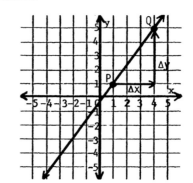

22. Let's use the changes from S to T to compute the slope of the line at the right.

a) Since x increases from −4 to 3, Δx = 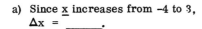.

b) Since y decreases from 2 to −2, Δy = _____.

c) Therefore, the slope = $\frac{\Delta y}{\Delta x}$ = _____

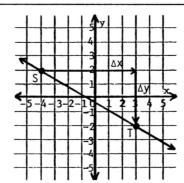

a) +3

b) +4

c) $\frac{+4}{+3}$ or $\frac{4}{3}$

23. Slope is a ratio or fraction. When computing a slope, the fraction should always be reduced to lowest terms.

a) If $\Delta x = 6$ and $\Delta y = 4$, the slope is _____.

b) If $\Delta x = 8$ and $\Delta y = -10$, the slope is _____.

a) +7

b) −4

c) $\frac{-4}{+7}$ or $-\frac{4}{7}$

24. When computing a slope, sometimes the ratio (or fraction) reduces to a whole number.

a) If $\Delta x = 4$ and $\Delta y = 12$, the slope is _____.

b) If $\Delta x = 1$ and $\Delta y = -2$, the slope is _____.

a) $\frac{2}{3}$ (from $\frac{4}{6}$)

b) $-\frac{5}{4}$ (from $\frac{-10}{8}$)

a) 3 (from $\frac{12}{4}$)

b) −2 (from $\frac{-2}{1}$)

Linear Equations 65

25. No matter which pair of points we choose to compute the slope of a line, we always get the same value for the slope. As an example, we graphed the changes from A to B and from C to D on the line at the right.

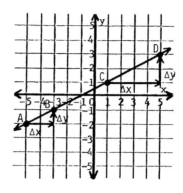

a) For A and B, Δx = 2 and Δy = 1, Therefore, the slope = _____

b) For C and D, Δx = 4 and Δy = 2. Therefore, the slope = _____

c) Did we get the same value for the slope with each pair? _____

---

26. We graphed the changes from P to Q and from S to T on the line at the right.

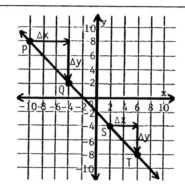

a) From P to Q, Δx = 6 and Δy = -6. Therefore, the slope = _____

b) From S to T, Δx = 4 and Δy = -4. Therefore, the slope = _____

c) Did we get the same value for the slope with each pair? _____

a) $\frac{1}{2}$

b) $\frac{1}{2}$ (from $\frac{2}{4}$)

c) Yes

---

27. Since we read graphs from left to right, we compute the slope of a line from a point on the left to a point on the right. Therefore:

1) Δx is <u>always</u> positive since it always represents an increase.

2) Δy may be <u>positive</u> or <u>negative</u> since it can represent an increase or a decrease.

a) -1 (from $\frac{-6}{6}$)

b) -1 (from $\frac{-4}{4}$)

c) Yes

Therefore, the <u>sign</u> of the slope is determined by Δy.

If Δy is <u>positive</u>, the slope is <u>positive</u>.
If Δy is <u>negative</u>, the slope is <u>negative</u>.

On the graph at the right, we have shown the changes from E to F on line #1 and from P to Q on line #2.

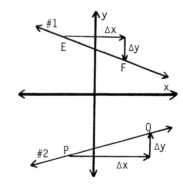

On line #1, F is <u>lower</u> than E. Therefore:

a) Δy is _____ (positive/negative).

b) The slope is _____ (positive/negative).

On line #2, Q is <u>higher</u> than P. Therefore:

c) Δy is _____ (positive/negative).

d) The slope is _____ (positive/negative).

66   Linear Equations

28.   The sign of the slope tells us whether a line rises or falls from left to right.

If its slope is <u>positive</u>, the line <u>rises</u>.
If its slope is <u>negative</u>, the line <u>falls</u>.

On the graph at the right, we have drawn four lines and labeled them #1, #2, #3, and #4.

a) Which lines have a positive slope? _____

b) Which lines have a negative slope? _____

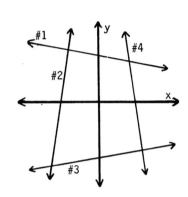

a) negative
b) negative
c) positive
d) positive

---

29.   The <u>sign</u> of the slope tells us whether the line <u>rises</u> or <u>falls</u> from left to right. The <u>absolute value</u> of the slope tells us <u>how steep the rise or fall is</u>.

We graphed two lines at the right. The rise of line #1 is steeper than the rise of line #2.

a) Using A and C, the slope of line #1 is _____.

b) Using A and B, the slope of line #2 is _____.

c) The steeper the rise of a line, the _____ (larger/smaller) is the absolute value of its slope.

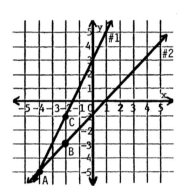

a) lines #2 and #3
b) lines #1 and #4

---

30.   We graphed two lines at the right. The fall of line #1 is steeper than the fall of line #2.

a) Using P and R, the slope of line #1 is _____.

b) Using P and Q, the slope of line #2 is _____.

c) The steeper the fall of a line, the _____ (larger/smaller) is the absolute value of its slope.

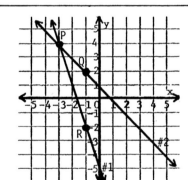

a) 2 (from $\frac{4}{2}$)
b) 1 (from $\frac{2}{2}$)
c) larger

---

31.   The slope of line #1 is "5" and the slope of line #2 is "3".

a) Do both lines <u>rise</u> or <u>fall</u> from left to right? _____

b) The rise of which line (#1 or #2) is steeper? _____

a) -3 (from $\frac{-6}{2}$)
b) -1 (from $\frac{-2}{2}$)
c) larger

Linear Equations   67

32. The slope of line #1 is $-\frac{3}{4}$ and the slope of line #2 is $-\frac{5}{4}$.

    a) Do both lines <u>rise</u> or <u>fall</u> from left to right? _____

    b) The fall of which line (#1 or #2) is steeper? _____

a) rise

b) #1, since 5 is larger than 3

---

33.   a) If line #1 has a slope of $\frac{7}{5}$ and line #2 has a slope of $\frac{1}{2}$, which line has the steeper rise? _____

    b) If line #1 has a slope of -1 and line #2 has a slope of $-\frac{4}{3}$, which line has the steeper fall? _____

a) fall

b) #2, since $\frac{5}{4}$ is larger than $\frac{3}{4}$

---

a) #1, since $\frac{7}{5}$ is larger than $\frac{1}{2}$ b) #2, since $\frac{4}{3}$ is larger than 1

2-5 SLOPE AS A RATIO

In this section, we will show that the slope of a line is a ratio that represents various pairs of changes in <u>x</u> and <u>y</u>.

34. The slope of a line represents a ratio of changes in <u>x</u> and <u>y</u>. We can see that fact by examining the graphs below.

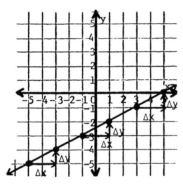 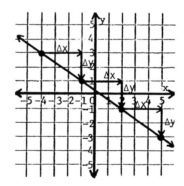

Each pair of arrows on the graph at the left shows that the slope is $\frac{1}{2}$.

    A slope of $\frac{1}{2}$ means: For any 2-unit <u>increase</u> in <u>x</u>, there is a 1-unit <u>increase</u> in <u>y</u>.

Each pair of arrows on the graph at the right shows that the slope is $-\frac{2}{3}$.

    A slope of $-\frac{2}{3}$ means: For any 3-unit <u>increase</u> in <u>x</u>, there is a ____-unit <u>decrease</u> in <u>y</u>.

<u>2</u>-unit

68   Linear Equations

35. Any fractional slope states an obvious ratio or pair of corresponding changes in x and y.

   a) A slope of $\frac{6}{5}$ means: For any 5-unit <u>increase</u> in <u>x</u>,
      there is a ____-unit _____
      (increase/decrease) in <u>y</u>.

   b) A slope of $-\frac{1}{4}$ means: For any 4-unit <u>increase</u> in <u>x</u>,
      there is a ____-unit _____
      (increase/decrease) in <u>y</u>.

36. When the slope is a whole number, the slope is still a ratio.
   For example:

   A slope of 2 means $\frac{2}{1}$.   A slope of -3 means $\frac{-3}{1}$.

   Therefore, a whole-number slope also states an obvious pair of corresponding changes in x and y.

   a) A slope of 2 or $\frac{2}{1}$ means: For any 1-unit <u>increase</u> in <u>x</u>,
      there is a ____-unit <u>increase</u> in <u>y</u>.

   b) A slope of -3 or $\frac{-3}{1}$ means: For any 1-unit <u>increase</u> in <u>x</u>,
      there is a ____-unit <u>decrease</u> in <u>y</u>.

   Answers:
   a) <u>6</u>-unit <u>increase</u>
   b) <u>1</u>-unit <u>decrease</u>

37. a) A slope of "1" or $\frac{1}{1}$ means: For any 1-unit <u>increase</u> in <u>x</u>,
      there is a ____-unit _____
      (increase/decrease) in <u>y</u>.

   b) A slope of "-1" or $\frac{-1}{1}$ means: For any 1-unit <u>increase</u> in <u>x</u>,
      there is a ____-unit _____
      (increase/decrease) in <u>y</u>.

   Answers:
   a) <u>2</u>-unit increase
   b) <u>3</u>-unit decrease

38. Sometimes a slope is stated as a decimal number rather than a fraction.
   For example:

   Instead of $\frac{7}{5}$, we use 1.4.   Instead of $-\frac{1}{4}$, we use -0.25.

   Decimal-number slopes also stand for a ratio. That is:

   A slope of 1.4 means $\frac{1.4}{1}$.   A slope of -0.25 means $\frac{-0.25}{1}$.

   Therefore, a decimal-number slope also states an obvious pair of corresponding changes in x and y.

   a) A slope of 1.4 or $\frac{1.4}{1}$ means: For any 1-unit <u>increase</u> in <u>x</u>,
      there is a ____-unit <u>increase</u> in <u>y</u>.

   b) A slope of -0.25 or $\frac{-0.25}{1}$ means: For any 1-unit <u>increase</u> in <u>x</u>,
      there is a ____-unit <u>decrease</u> in <u>y</u>.

   Answers:
   a) <u>1</u>-unit increase
   b) <u>1</u>-unit decrease

Linear Equations 69

39. A fractional slope can also be thought of as a ratio whose denominator is "1". That is:

$$\text{A slope of } \frac{3}{2} = \frac{\frac{3}{2}}{1}. \qquad \text{A slope of } -\frac{2}{5} = \frac{-\frac{2}{5}}{1}.$$

Therefore, a fractional slope also stands for a pair of changes in which the increase in x is "1".

a) A slope of $\frac{3}{2}$ or $\frac{\frac{3}{2}}{1}$ means: For any 1-unit <u>increase</u> in <u>x</u>, there is a ____-unit ____ (increase/decrease) in <u>y</u>.

b) A slope of $-\frac{2}{5}$ or $\frac{-\frac{2}{5}}{1}$ means: For any 1-unit <u>increase</u> in <u>x</u>, there is a ____-unit ____ (increase/decrease) in <u>y</u>.

a) 1.4-unit
b) 0.25-unit

40. Though a slope is always reported in lowest terms, it represents various pairs of changes in <u>x</u> and <u>y</u>. For example:

A slope of $\frac{1}{2}$ stands for any of the following equivalent $\frac{\Delta y}{\Delta x}$ ratios:

$$\frac{1}{2} \qquad \frac{2}{4} \qquad \frac{3}{6} \qquad \frac{5}{10} \qquad \frac{10}{20} \qquad \frac{50}{100}$$

Therefore, a slope of $\frac{1}{2}$ represents all of the following pairs of changes.

If <u>x</u> increases 2 units, <u>y</u> increases 1 unit.
If <u>x</u> increases 4 units, <u>y</u> increases 2 units.
If <u>x</u> increases 20 units, <u>y</u> increases ____ units.

a) $\frac{3}{2}$-unit <u>increase</u>
b) $\frac{2}{5}$-unit <u>decrease</u>

41. A slope of -2 stands for any of the following equivalent $\frac{\Delta y}{\Delta x}$ ratios:

$$\frac{-2}{1} \qquad \frac{-4}{2} \qquad \frac{-6}{3} \qquad \frac{-10}{5} \qquad \frac{-20}{10} \qquad \frac{-50}{25}$$

Therefore, a slope of -2 represents all of the following pairs of changes.

If <u>x</u> increases 1 unit, <u>y</u> decreases 2 units.
a) If <u>x</u> increases 2 units, <u>y</u> decreases ____ units.
b) If <u>x</u> increases 10 units, <u>y</u> decreases ____ units.

10 units

a) 4 units
b) 20 units

70   Linear Equations

42. A slope of $\frac{4}{3}$ stands for any $\frac{\Delta y}{\Delta x}$ ratio that reduces to $\frac{4}{3}$. To find the Δy corresponding to a specific Δx, we can substitute in the proportion below.

$$\boxed{\frac{4}{3} = \frac{\Delta y}{\Delta x}}$$

That is: If $\Delta x = 6$, $\frac{4}{3} = \frac{\Delta y}{6}$, and $\Delta y = 6\left(\frac{4}{3}\right) = 8$

If $\Delta x = 15$, $\frac{4}{3} = \frac{\Delta y}{15}$, and $\Delta y = 15\left(\frac{4}{3}\right) = $ _____

---

43. To find the Δy corresponding to a specific Δx for a line whose slope is $-\frac{1}{4}$, we can substitute in the proportion below.

$$\boxed{-\frac{1}{4} = \frac{\Delta y}{\Delta x}}$$

That is: If $\Delta x = 8$, $-\frac{1}{4} = \frac{\Delta y}{8}$, and $\Delta y = 8\left(-\frac{1}{4}\right) = -2$

If $\Delta x = 40$, $-\frac{1}{4} = \frac{\Delta y}{40}$, and $\Delta y = 40\left(-\frac{1}{4}\right) = $ _____

20

---

44. By setting up a proportion, find the corresponding change in "y" for each of these:

a) If the slope is $\frac{1}{3}$ and x increases 12 units,

y _____ (increases/decreases) _____ units.

b) If the slope is $-\frac{7}{5}$ and x increases 15 units,

y _____ (increases/decreases) _____ units.

-10

---

45. To find the Δy corresponding to a specific Δx for a line whose slope is 4, we can substitute in the following:

$$\boxed{4 = \frac{\Delta y}{\Delta x}}$$

That is: If $\Delta x = 2$, $4 = \frac{\Delta y}{2}$, and $\Delta y = 2(4) = 8$

If $\Delta x = 6$, $4 = \frac{\Delta y}{6}$, and $\Delta y = 6(4) = $ _____

a) increases 4 units

b) decreases 21 units

---

46. To find the Δy corresponding to a specific Δx whose slope is -1.5, we can substitute in the following:

$$\boxed{-1.5 = \frac{\Delta y}{\Delta x}}$$

That is: If $\Delta x = 2$, $-1.5 = \frac{\Delta y}{2}$, and $\Delta y = 2(-1.5) = -3$

If $\Delta x = 5$, $-1.5 = \frac{\Delta y}{5}$, and $\Delta y = 5(-1.5) = $ _____

24

-7.5

47. Using the same method, find the corresponding change in "y" for each of these:

    a) If the slope of a line is -3 and $\underline{x}$ increases 4 units,
   $\underline{y}$ _____ (increases/decreases) _____ units.

    b) If the slope of a line is 1.25 and $\underline{x}$ increases 10 units,
   $\underline{y}$ _____ (increases/decreases) _____ units.

48. When Δx is +1, the corresponding change in $\underline{y}$ is <u>the slope itself</u>. For example:

    If the slope is $\frac{3}{4}$, $\frac{3}{4} = \frac{\Delta y}{1}$, and $\Delta y = 1\left(\frac{3}{4}\right) = \frac{3}{4}$

    If the slope is -2, $-2 = \frac{\Delta y}{1}$, and $\Delta y = 1(-2) = -2$

Using the fact above, complete these:

    a) If the slope is $-\frac{5}{2}$, what happens to $\underline{y}$ if $\underline{x}$ increases 1 unit?

    b) If the slope is 6.4, what happens to $\underline{y}$ if $\underline{x}$ increases 1 unit?

a) decreases 12 units

b) increases 12.5 units

a) y decreases $\frac{5}{2}$ units   b) y increases 6.4 units

2-6 SLOPE-INTERCEPT FORM

In this section, we will discuss the slope-intercept form of linear equations.

49. Any linear equation can be written in the form below where "y" is "solved for" and "m" and "b" are numerical constants.

$$y = mx + b$$

"m" and "b" can be either positive or negative whole numbers, fractions, or decimal numbers. For example:

    In $y = 3x - 5$, $m = 3$ and $b = -5$

    a) In $y = -\frac{2}{3}x + \frac{1}{3}$, $m =$ _____ and $b =$ _____

    b) In $y = -2.5x - 1.5$, $m =$ _____ and $b =$ _____

a) $m = -\frac{2}{3}$, $b = \frac{1}{3}$

b) $m = -2.5$, $b = -1.5$

72   Linear Equations

50. $\boxed{y = mx + b}$ is called the "slope-intercept" form of a linear equation because:

"m" is the <u>slope</u> of the line.
"b" is the <u>vertical intercept</u> of the line.

To show the facts above, we graphed $\boxed{y = 2x - 3}$ below. The coordinates of points A and B are given.

1) In $y = 2x - 3$, $m = 2$. We can show that "2" is the <u>slope</u> of the line by using the Δx and Δy from point A to point B.

$$\text{slope} = \frac{\Delta y}{\Delta x} = \frac{4}{2} = 2$$

2) In $y = 2x - 3$, $b = -3$. We can see that -3 is the ordinate of the vertical intercept by examining the graph.

Note: Point A is the vertical intercept. Its coordinates are (0, -3)

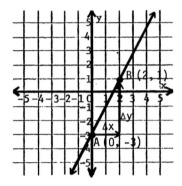

51. The equation $\boxed{y = -\frac{2}{3}x + 3}$ is graphed at the right. The coordinates of points C and D are given.

a) In $y = -\frac{2}{3}x + 3$, $m = -\frac{2}{3}$. We can show that $-\frac{2}{3}$ is the <u>slope</u> of the line by using the Δx and Δy from point C to point D.

$$\text{slope} = \frac{\Delta y}{\Delta x} = \underline{}$$

b) In $y = -\frac{2}{3}x + 3$, $b = 3$. We can see that 3 is the ordinate of the <u>y-intercept</u> by examining the graph.

Point C is the vertical intercept. Its coordinates are (,   )

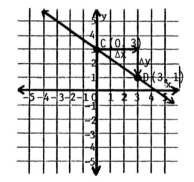

52. The symbol "m" is used for the "slope" of a line. Therefore, the slope formula is usually written:

$$\boxed{m = \frac{\Delta y}{\Delta x}}$$

Using the formula, complete these:

a) If $\Delta x = 5$ and $\Delta y = 3$, $m = \underline{}$

b) If $\Delta x = 1$ and $\Delta y = -4$, $m = \underline{}$

a) $-\frac{2}{3}$

b) (0, 3)

Linear Equations 73

53. The constant "m" is not explicitly written in the equations below.

$$y = x - 2 \qquad y = x + 5$$

However, since $x = 1x$ and $-x = -1x$:

a) In $y = x - 2$, $m =$ _____   b) In $y = -x + 5$, $m =$ _____

a) $\frac{3}{5}$
b) -4

---

54. In $y = mx + b$, "b" is the <u>ordinate</u> (or y-coordinate) of the <u>vertical intercept</u> (or y-intercept). Therefore:

For $y = x - 2$, the coordinates of the y-intercept are $(0, -2)$

For $y = -3x + 9$, the coordinates of the y-intercept are (,)

a) +1
b) −1

---

55. Write the slope-intercept form of the linear equations with the following slopes and y-intercepts.

	Slope	y-intercept	Equation
a)	4	$(0, -1)$	_____
b)	$-\frac{5}{2}$	$(0, 4)$	_____
c)	-1	$(0, -\frac{1}{2})$	_____

$(0, 9)$

a) $y = 4x - 1$ b) $y = -\frac{5}{2}x + 4$ c) $y = -x - \frac{1}{2}$

SELF-TEST 5 (pages 58-74)

1. Which of the following are linear equations? _____

   a) $y = 3x$ b) $y = 2x^2 - 5$ c) $y = \frac{5}{x}$ d) $\frac{4x}{y-1} = 7$ e) $\frac{y}{x+2} = 6x$

2. Find the coordinates of the x-intercept of $\boxed{5x - 4y = 20}$. _____

3. After finding the intercepts for $\boxed{2x + 3y = 6}$, use them to graph the equation.

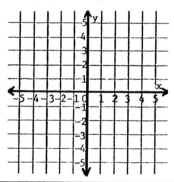

Find the slope of:

4. Line #1 _____   5. Line #2 _____

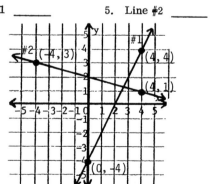

Continued on following page.

74   Linear Equations

SELF-TEST 5 (pages 58-74) - Continued

Four lines have these slopes: #1: -5   #2: 2   #3: $-\frac{9}{5}$ #4: $\frac{3}{4}$

6. Which line or lines fall from left to right? _____
7. Which line is steepest? _____

8. The slope of a line is $-\frac{2}{3}$.
   If $\Delta x = 12$, $\Delta y = $ _____

9. The slope of a line is 3. If x increases 6 units, then y increases _____ units.

The equation of a line is $\boxed{y = 5x - 8}$.

10. Find its slope. _____
11. Find the coordinates of its y-intercept. _____

12. Find the equation of the line whose slope is -4 and whose y-intercept is (0, 9).   _____

ANSWERS:
1. (a) and (d)
2. (4, 0)
3.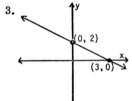
4. 2
5. $-\frac{1}{4}$
6. #1 and #3
7. #1
8. $\Delta y = -8$
9. 18 units
10. 5
11. (0, -8)
12. $y = -4x + 9$

2-7 WRITING LINEAR EQUATIONS IN SLOPE-INTERCEPT FORM

Linear equations are not always written in slope-intercept form. However, linear equations in other forms are frequently rearranged to slope-intercept form so that their slopes and y-intercepts are obvious. We will discuss that rearrangement process in this section.

56. Only linear equations of the form $\boxed{y = mx + b}$ are in slope-intercept form. Which of the following linear equations are in slope-intercept form?

   a) $x + 5y = 9$ b) $y = 2x - 7$ c) $1 = x - y$ d) $y = -\frac{3}{2}x + 6$

57. To put each equation below in slope-intercept form, we commuted the terms on the right side to get the x-term first.

   $y = 3 + 4x$ $y = 5 - 2x$
   $y = 4x + 3$ $y = -2x + 5$

   Following the examples, put each equation in slope-intercept form.

   a) $y = 1 + x$ b) $y = 4 - 7x$ c) $y = 1 - x$

   _____   _____   _____

Only (b) and (d)

Linear Equations 75

58. To put the equation below in slope-intercept form, we solved for "y". Put the other equation in slope-intercept form.

$$y + 3 = x \qquad\qquad y - 5 = 4x$$
$$y + 3 + (-3) = x + (-3)$$
$$y = x - 3$$

a) $y = x + 1$
b) $y = -7x + 4$
c) $y = -x + 1$

59. To put the equation below in slope-intercept form, we solved for "y". Notice how we wrote the x-term first on the right side. Put the other equation in slope-intercept form.

$$y - 2x = 7 \qquad\qquad x + y = 10$$
$$y - 2x + 2x = 2x + 7$$
$$y = 2x + 7$$

$y = 4x + 5$

60. To get the opposite of a binomial, we replace each term by its opposite. That is:

The opposite of $-9x + 5$ is $9x - 5$.

The opposite of $x - 1$ is $-x + 1$.

To put each equation below in slope-intercept form, we used the opposing principle. That is, we replaced each side by its opposite.

$$-y = -2x + 3 \qquad\qquad -y = x - 9$$
$$y = 2x - 3 \qquad\qquad y = -x + 9$$

Use the opposing principle to put each equation below in slope-intercept form.

a) $-y = -8x + 1$ b) $-y = 4x - 3$ c) $-y = -x + 7$

$y = -x + 10$

61. Notice how we used the opposing principle to put each equation below in slope-intercept form.

$$3x - y = 10 \qquad\qquad x = 5 - y$$
$$(-3x) + 3x - y = (-3x) + 10 \qquad x + (-5) = (-5) + 5 - y$$
$$-y = -3x + 10 \qquad\qquad x - 5 = -y$$
$$y = 3x - 10 \qquad\qquad y = -x + 5$$

Following the examples, put each equation below in slope-intercept form.

a) $x - y = 4$ b) $8 - y = 7x$

a) $y = 8x - 1$
b) $y = -4x + 3$
c) $y = x - 7$

76 Linear Equations

62. To put the equation below in slope-intercept form, we divided x + 3 by 5. Put the other equation in slope-intercept form.

$$5y = x + 3 \qquad\qquad 4y = -5x + 1$$
$$y = \frac{x + 3}{5}$$
$$y = \frac{1}{5}x + \frac{3}{5}$$

a) y = x − 4
b) y = −7x + 8

---

63. Notice how we reduced to lowest terms below. Put the other equation in slope-intercept form.

$$3y = x + 6 \qquad\qquad 2y = -6x + 1$$
$$y = \frac{x + 6}{3}$$
$$y = \frac{x}{3} + \frac{6}{3}$$
$$y = \frac{1}{3}x + 2$$

$y = -\frac{5}{4}x + \frac{1}{4}$

---

64. Following the example, put the other equation in slope-intercept form.

$$4x + 3y = 12 \qquad\qquad x + 5y = 3$$
$$3y = -4x + 12$$
$$y = \frac{-4x + 12}{3}$$
$$y = -\frac{4}{3}x + 4$$

$y = -3x + \frac{1}{2}$

---

65. Notice how we used the oppositing principle below. Put the other equation in slope-intercept form.

$$x - 4y = 7 \qquad\qquad 5x - 3y = 15$$
$$-4y = 7 - x$$
$$4y = x - 7$$
$$y = \frac{x - 7}{4}$$
$$y = \frac{1}{4}x - \frac{7}{4}$$

$y = -\frac{1}{5}x + \frac{3}{5}$

$y = \frac{5}{3}x - 5$

66. After putting an equation in slope-intercept form, we can easily identify its slope and y-intercept. For example:

Since $\boxed{2x + y = 5}$ is equivalent to $\boxed{y = -2x + 5}$:

Its slope is -2. Its y-intercept is (0, 5)

Since $\boxed{5x - 2y = 6}$ is equivalent to $\boxed{y = \frac{5}{2}x - 3}$

a) Its slope is _____. b) Its y-intercept is _____.

a) $\frac{5}{2}$ b) (0, -3)

2-8 THE TWO-POINT FORMULA FOR SLOPE

If we know the coordinates of two points on a line, there is a formula that can be used to find its slope. We will discuss that formula in this section.

67. If a line passes through the points (2, -1) and (4, 3), we can use the following steps to find its slope:

1) Plot the points on a graph.
2) Draw a line through them.
3) Draw arrows representing Δx and Δy.
4) Use Δx and Δy to compute the slope.

We did the first three steps on the graph at the right. Compute the slope.

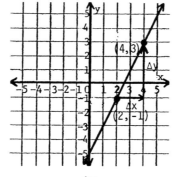

$m = \frac{\Delta y}{\Delta x} = $ _____

68. In the last frame, we found that the slope of the line through (2, -1) and (4, 3) is 2. We can find the same slope without graphing by using the formula below.

m = 2

$\boxed{m = \frac{y_2 - y_1}{x_2 - x_1}}$ where: (x_1, y_1) are the coordinates of one point.
(x_2, y_2) are the coordinates of the other point.
$y_2 - y_1$ is the change in y
$x_2 - x_1$ is the change in x

When using the above formula, we can use either (2, -1) or (4, 3) for (x_2, y_2).

Using (4, 3) as (x_2, y_2), we get: $m = \frac{3 - (-1)}{4 - 2} = \frac{4}{2} = 2$

Using (2, -1) as (x_2, y_2), we get: $m = \frac{(-1) - 3}{2 - 4} = \frac{-4}{-2} = 2$

Did we get the correct value for the slope using both methods? _____

78   Linear Equations

69. Let's use the same formula to find the slope of the line through (-2, 1) and (3, -3). To show that we can use either point as (x_2, y_2), we will do it both ways.

   a) Using (3, -3) as (x_2, y_2), we get: $m = \dfrac{y_2 - y_1}{x_2 - x_1} =$ _____

   b) Using (-2, 1) as (x_2, y_2), we get: $m = \dfrac{y_2 - y_1}{x_2 - x_1} =$ _____

   Yes

70. Let's use the same formula to find the slope of the line through (-4, 0) and (0, 2).

   Note: When using the formula, it is helpful to sketch the two points as we have done to avoid gross errors.

   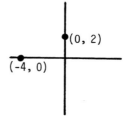

   Using either (-4, 0) or (0, 2) as (x_2, y_2): $m = \dfrac{y_2 - y_1}{x_2 - x_1} =$ _____

   a) $\dfrac{(-3) - 1}{3 - (-2)} = \dfrac{-4}{5} = -\dfrac{4}{5}$

   b) $\dfrac{1 - (-3)}{(-2) - 3} = \dfrac{4}{-5} = -\dfrac{4}{5}$

71. Use the formula to find the slope of the line through each pair of points below. (Sketch the points first.)

   a) (5, -7) and (7, 1)   b) (1, 0) and (-5, 4)

   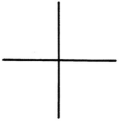   

   $m = \dfrac{y_2 - y_1}{x_2 - x_1} =$ _____ $m = \dfrac{y_2 - y_1}{x_2 - x_1} =$ _____

   $m = \dfrac{1}{2}$, from:

   $\dfrac{2 - 0}{0 - (-4)} = \dfrac{2}{4}$

   or

   $\dfrac{0 - 2}{(-4) - 0} = \dfrac{-2}{-4}$

72. Find the slope of the line through each pair of points. (Sketch the points first.)

   a) (-8, 0) and (0, 20)   b) (17, -11) and (-12, 18)

   $m =$ _____ $m =$ _____

   a) $m = 4$

   b) $m = -\dfrac{2}{3}$

   a) $m = \dfrac{5}{2}$ b) $m = -1$

2-9 FINDING THE EQUATION OF A LINE GIVEN ITS SLOPE AND ONE POINT

In this section, we will discuss the procedure for finding the equation of a line when its slope and one point are given.

73. The slope of the line at the right is $\frac{1}{2}$. It passes through the point (5, 4). From the graph, it looks as if the y-intercept is about $(0, \frac{3}{2})$. Therefore, it looks as if the equation of the line is:

$$y = \frac{1}{2}x + \frac{3}{2}$$

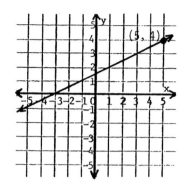

Knowing that $m = \frac{1}{2}$ and that the line passes through (5, 4), we can find **b** in $y = mx + b$ algebraically. The steps are:

1) Substitute $\frac{1}{2}$ for **m** in $y = mx + b$. $\qquad y = \frac{1}{2}x + b$

2) Using (5, 4) substitute 5 for **x** and 4 for **y** in the above equation and solve for **b**.

$4 = \frac{1}{2}(5) + b$

$4 = \frac{5}{2} + b$

$b = 4 - \frac{5}{2} = \frac{8}{2} - \frac{5}{2} = \frac{3}{2}$

3) Using $\frac{1}{2}$ for **m** and $\frac{3}{2}$ for **b**, write the equation in slope-intercept form. $\qquad y = \frac{1}{2}x + \frac{3}{2}$

Does this last equation make sense in terms of the above graph? _____

	Yes

74. Let's use the same steps to find the equation of a line whose slope is 2 if the line passes through (-1, -3).

1) Substitute 2 for **m** in $y = mx + b$. $\qquad y = 2x + b$

2) Using (-1, -3), substitute -1 for **x** and -3 for **y** in that equation and solve for **b**.

$-3 = 2(-1) + b$

$-3 = (-2) + b$

$b = -1$

3) Using 2 for **m** and -1 for **b**, write the equation in slope-intercept form. _____

$y = 2x - 1$

80 Linear Equations

75. The slope of a line is $\frac{2}{3}$ and the line passes through (9, 1).

a) Find b by substituting the values of (9, 1) into: $y = \frac{2}{3}x + b$

b) The equation of the line is: y = _____

76. The slope of a line is -3 and the line passes through (4, 0).

a) Find b by substituting the values of (4, 0) into: $y = -3x + b$

b) The equation of the line is: y = _____

a) b = -5   b) $y = \frac{2}{3}x - 5$

a) b = 12   b) y = -3x + 12

2-10 FINDING THE EQUATION OF A LINE GIVEN TWO POINTS

In this section, we will discuss the procedure for finding the equation of a line when the coordinates of two points are given.

77. To find the equation of the line through (3, 5) and (-3, 1), we use the steps below.

1) Draw a rough sketch of the line as we have done at the right.

2) Using the formula below, find m.

$$m = \frac{y_2 - y_1}{x_2 - x_1} = \frac{5 - 1}{3 - (-3)} = \frac{4}{6} = \frac{2}{3}$$

3) Substitute that value for m in y = mx + b and then use the coordinates of either point to solve for b.

Using (3, 5)

$y = \frac{2}{3}x + b$

$5 = \frac{2}{3}(3) + b$

$5 = 2 + b$

$b = 3$

Using (-3, 1)

$y = \frac{2}{3}x + b$

$1 = \frac{2}{3}(-3) + b$

$1 = (-2) + b$

$b = 3$

4) Using $\frac{2}{3}$ for m and 3 for b, write the equation of the line. y = _____

78. Let's find the equation of the line through (-2, -5) and (2, -1). (Draw a sketch first.)

a) Use the formula to find <u>m</u>.

b) Substitute that value for <u>m</u> in y = mx + b and substitute the coordinates of one point to find <u>b</u>.

c) The equation of the line is: y = _____

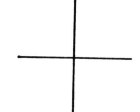

$y = \frac{2}{3}x + 3$

79. Find the equation of the line passing through each pair of points.

a) (0, -4) and (1, 0)   b) (-4, 3) and (2, -1)

a) m = 1

b) b = -3

c) y = 1x - 3
   or
   y = x - 3

a) y = 4x - 4   b) $y = -\frac{2}{3}x + \frac{1}{3}$

2-11   LINES THROUGH THE ORIGIN

In this section, we will discuss the slope-intercept form of equations whose graphed lines pass through the origin.

80. The line at the right passes through the origin. Therefore, both its x-intercept and its y-intercept are (0, 0).

Using the changes from point P to point Q, we can compute its slope.

$m = \frac{\Delta y}{\Delta x} = \frac{2}{1} = 2$

Since m = 2 and b = 0, the slope-intercept form of the equation of the line is:

y = 2x + 0
or   y = 2x

There is only one constant in y = 2x. Does it represent the slope or y-intercept of the line? _____

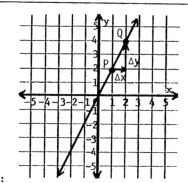

82 Linear Equations

81. The general slope-intercept form of linear equations is $y = mx + b$. However, the vertical intercept of all lines through the origin is $(0, 0)$. Since $b = 0$ for all lines through the origin, their slope-intercept form is:

$$\boxed{y = mx}$$

Which equations below graph as lines through the origin? _____

a) $y = -5x$ b) $y = 3x - 1$ c) $y = -4x + 3$ d) $y = 10x$

the slope

---

82. We can graph a linear equation by plotting only two points. If we know that its line passes through the origin, we can use the origin and one other point.

For example, to graph $\boxed{y = 4x}$, we can use $(0, 0)$ and find some other point by substituting for x.

If $x = 2$, $y = 8$

Plotting $(0, 0)$ and $(2, 8)$, we graphed $y = 4x$ at the right. Use the same method to graph each equation below on the same axes.

a) $\boxed{y = -2x}$ b) $\boxed{y = \frac{1}{2}x}$

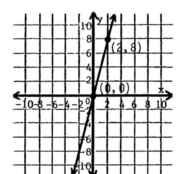

(a) and (d)

---

83. If we know the slope of a line through the origin, we can write its equation. That is:

If $m = \frac{3}{4}$, $y = \frac{3}{4}x$ If $m = -\frac{5}{2}$, $y = $ _____

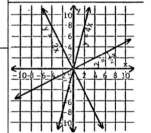

---

84. Knowing that $x = 1x$ and $-x = -1x$, complete these:

a) For $y = x$, $m = $ _____   b) For $y = -x$, $m = $ _____

$y = -\frac{5}{2}x$

---

85. Line #1 at the right passes through $(0, 0)$ and $(8, 6)$. We computed its slope below.

$$m = \frac{\Delta y}{\Delta x} = \frac{6}{8} = \frac{3}{4}$$

Line #2 passes through $(0, 0)$ and $(-4, 6)$. Use the same method to find the slope.

$$m = \frac{\Delta y}{\Delta x} = $$ _____

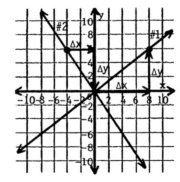

a) +1
b) -1

86. We know that the line at the right passes through (0, 0) and (2, 10). Let's use the two-point formula to find its slope.

$$m = \frac{y_2 - y_1}{x_2 - x_1} = \frac{10 - 0}{2 - 0} = \frac{10}{2} = 5$$

or

$$m = \frac{y_2 - y_1}{x_2 - x_1} = \frac{0 - 10}{0 - 2} = \frac{-10}{-2} = 5$$

The equation of the line is: y = _____

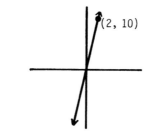

$m = \dfrac{-6}{4} = -\dfrac{3}{2}$

---

87. To find the equation of a line through the origin and one other known point, we only have to find its slope. We can use the two-point formula to do so.

Find the equation of the line passing through the origin and each point below. (Make a sketch.)

a) (3, 9) b) (-10, 8) c) (-6, -7)

y = 5x

---

88. We can find the slope of y - 5x = 0 by putting it in slope-intercept form. We get: y = 5x. Therefore, m = 5.

Find the slope of each line below by putting it in slope-intercept form.

a) 4x + 3y = 0   m = _____   b) x - y = 0   m = _____

a) y = 3x

b) y = $-\dfrac{4}{5}$x

c) y = $\dfrac{7}{6}$x

---

a) m = $-\dfrac{4}{3}$ b) m = 1

84 Linear Equations

2-12 PARALLEL LINES

In this section, we will discuss the equations of parallel lines. We will show that parallel lines are lines with <u>the same slope but different y-intercepts</u>.

---

89. When linear equations have <u>the same slope but different y-intercepts</u> their graphed lines are parallel. As an example, we graphed the two equations below at the right.

    #1 $y = 3x + 1$
    #2 $y = 3x - 4$

    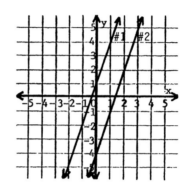

    Therefore, when linear equations are written in slope-intercept form, we can tell immediately whether their graphed lines are parallel or not. They are <u>parallel</u> if their slopes are the <u>same</u>; they are <u>not</u> parallel if their slopes are different.

    The graphs of which equations below are parallel? _____

    a) $y = \frac{1}{2}x$ b) $y = 2x + 5$ c) $y = \frac{1}{2}x - 3$

---

90. To decide whether the graphs of $4y - 3x = 8$ and $3x - 4y = 20$ are parallel or not, we must put them in slope-intercept form so that we can compare their slopes.

    a) Put $4y - 3x = 8$ in slope-intercept form.

    b) Put $3x - 4y = 20$ in slope-intercept form.

    c) Are the graphed lines parallel? _____

(a) and (c), because $m = \frac{1}{2}$ for each

---

91. Which of the following lines are parallel? _____

    a) $y = x$ c) $y = 2x - 3$ e) $y - x = 9$

    b) $y = 5x$ d) $3y = 3x + 5$ f) $6y = x$

a) $y = \frac{3}{4}x + 2$

b) $y = \frac{3}{4}x - 5$

c) Yes, since: $m = \frac{3}{4}$ for each

(a), (d), and (e), since $m = 1$ for each

2-13 HORIZONTAL AND VERTICAL LINES

In this section, we will discuss the equations and slopes of horizontal and vertical lines on the coordinate system.

92. Two horizontal lines are shown at the right.

    For line #1, the y-coordinate is 2 for every value of x. Therefore, the equation of the line is $y = 2$.

    $\boxed{y = 2}$ means: For every x-value, $y = 2$.

    For line #2, the y-coordinate is -3 for every value of x. Therefore, the equation of the line is $y = -3$.

    $\boxed{y = -3}$ means: For every x-value, $y = $ _____

    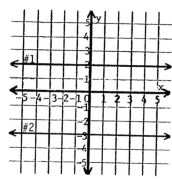

93. Four horizontal lines are drawn at the right. Following the examples from the last frame, write the equation of each line.

    #1 _____
    #2 _____
    #3 _____
    #4 _____

    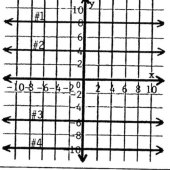

    $y = -3$

94. The x-axis is also a horizontal line on the coordinate system.

    a) On the x-axis, for every x-value, $y = $ _____.

    b) Therefore, the equation of the x-axis is _____.

    #1 $y = 8$ #3 $y = -6$
    #2 $y = 4$ #4 $y = -10$

95. Two vertical lines are shown at the right.

    For line #1, the x-coordinate is 4 for every value of y. Therefore, the equation of the line is $x = 4$.

    $\boxed{x = 4}$ means: For every y-value, $x = 4$.

    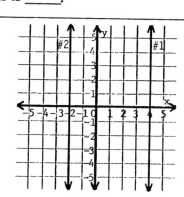

    a) 0   b) $y = 0$

Continued on following page.

86   Linear Equations

95. Continued

For line #2, the x-coordinate is -2 for every value of y. Therefore, the equation of the line is x = -2.

$\boxed{x = -2}$ means: For every y-value, x = _____.

---

96. Four vertical lines are drawn at the right. Following the examples from the last frame, write the equation of each line.

#1 _____
#2 _____
#3 _____
#4 _____

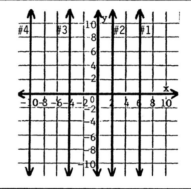

x = -2

---

97. The y-axis is also a vertical line on the coordinate system.

a) On the y-axis, for every y-value, x = _____.

b) Therefore, the equation of the y-axis is _____.

#1 x = 6   #3 x = -4
#2 x = 2   #4 x = -10

---

98. Four lines are drawn at the right. Write the equation of each line.

#1 _____
#2 _____
#3 _____
#4 _____

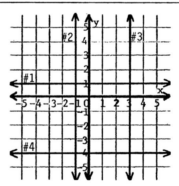

a) 0   b) x = 0

---

99. Slope is a measure of how fast a line rises or falls from left to right. Since a horizontal line does not rise or fall, it seems that its slope should be "0". Let's confirm that fact by examining the line at the right.

a) From point C to point D, Δx = _____

b) From point C to point D, Δy = _____

c) Therefore: $m = \dfrac{\Delta y}{\Delta x}$ = _____

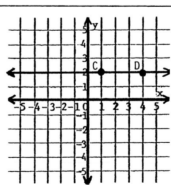

#1 y = 1   #3 x = 3
#2 x = -1   #4 y = -4

100. Let's examine the slope of the vertical line at the right. From point P to point Q:

$$\Delta y = 2$$

$$\Delta x = 0 \text{ (it does not change)}$$

$$m = \frac{\Delta y}{\Delta x} = \frac{2}{0}$$

But division by "0" is undefined. Therefore, the slope of the line is undefined.

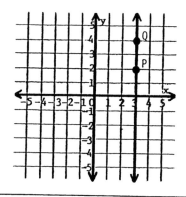

a) 3

b) 0

c) $\frac{0}{3} = 0$

---

101. In the last two frames, we saw these facts:

1) The slope of any horizontal line is "0".

2) The slope of any vertical line is undefined.

What is the slope of each line below?

a) y = 8   m = _____   c) y = -5   m = _____

b) x = -9 m = _____   d) x = 15   m = _____

---

102. Since the x-axis is a horizontal line:

a) its slope is _____

b) its equation is _____

Since the y-axis is a vertical line:

c) its slope is _____

d) its equation is _____

a) 0

b) undefined

c) 0

d) undefined

---

a) 0

b) y = 0

c) undefined

d) x = 0

SELF-TEST 6 (pages 74-88)

For the linear equation $\boxed{3x - 5y = 10}$:

1. Its slope-intercept form is _____
2. Its slope is _____.
3. The coordinates of its y-intercept are _____.

4. Find the slope of the line through (−4, 2) and (2, −6). _____

5. Find the equation of the line whose slope is 4 and which passes through (3, 5).

6. Find the equation of the line through (−8, 5) and (4, 2).

7. Find the equation of the line through the origin and (−2, −10).

8. Find the equation of the line whose slope is $-\frac{5}{2}$ and which passes through the origin.

9. Which equations below have graphs passing through the origin? _____

   a) $y = 8x$ b) $y = -\frac{2}{5}x$ c) $y = 4x - 3$ d) $x + y = 1$ e) $2x + y = 0$

10. Find the slope of $2x - 6y = 0$. _____

11. Which of the following lines are parallel? _____

    a) $y = 4x$ b) $4y = x$ c) $2y = 8x - 5$ d) $4x + y = 0$ e) $4x - y = 3$

12. Find the equation of the line parallel to the x-axis and 12 units below it.

13. Find the equation of the vertical line through (8, 5).

14. What is the slope of the line $y = 20$? $m = $ _____

ANSWERS:
1. $y = \frac{3}{5}x - 2$
2. $\frac{3}{5}$
3. (0, −2)
4. $-\frac{4}{3}$
5. $y = 4x - 7$
6. $y = -\frac{1}{4}x + 3$
7. $y = 5x$
8. $y = -\frac{5}{2}x$
9. a, b, e
10. $\frac{1}{3}$
11. a, c, e
12. $y = -12$
13. $x = 8$
14. 0

Linear Equations

2-14 INTERCEPTS OF LINEAR EQUATIONS WITH OTHER VARIABLES

In this section, we will discuss the intercepts of linear equations with variables other than "x" and "y". We will show that the intercepts depend on the choice of axes for the variables.

---

103. Terms containing variables other than "x" and "y" are either linear or non-linear.

    Each term below is <u>linear</u> because it is <u>a first-degree term</u>.

    $5a \qquad -t \qquad \frac{2}{3}q \qquad -1.4d$

    Each term below is <u>non-linear</u> because it is <u>higher than first-degree</u>.

    $a^2 \qquad -8pq \qquad -\frac{5}{4}m^3 \qquad 0.6RV$

    Which of the following are linear terms? _____

    a) $\frac{3}{2}d$ b) $0.5FH$ c) $2v^2$ d) $-4w$ e) $1.8t^3$

---

104. A two-variable equation is <u>linear</u> only if each term containing a variable is linear. Which of the following equations are linear? _____

    a) $2m + 3t = 6$ b) $5cd = 7$ c) $V = S^2 - 4$ d) $p = -2.8q$

    Answer: Both (a) and (d)

---

105. When graphing an equation containing variables other than <u>x</u> and <u>y</u>, <u>either variable can be scaled on either axis</u>. For example, when graphing $4m + 5v = 20$, either <u>m</u> or <u>v</u> can be scaled on either axis.

    a) If we are told to plot <u>m</u> as <u>x</u>, m is scaled on the _____ (horizontal/vertical) axis.

    b) If we are told to plot <u>m</u> as <u>y</u>, m is scaled on the _____ (horizontal/vertical) axis.

    Answer: Both (a) and (d)

---

106. When graphing $3d - 2t = 12$, either <u>d</u> or <u>t</u> can be scaled on either axis.

    a) If we are told to plot <u>d</u> as the abscissa, d is scaled on the _____ (horizontal/vertical) axis.

    b) If we are told to plot <u>d</u> as the ordinate, d is scaled on the _____ (horizontal/vertical) axis.

    Answer:
    a) horizontal
    b) vertical

Answer:
a) horizontal
b) vertical

Answer at top of 104 box: Both (a) and (d)

90   Linear Equations

107. When an equation with other variables is graphed, the coordinates of the intercepts depend upon which variable is scaled on which axis. For example, we graphed the equation below in two different ways. On the left, p is scaled on the horizontal axis; on the right, p is scaled on the vertical axis.

$$p + 2q = 8$$

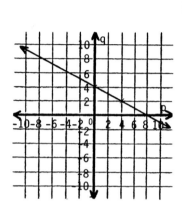 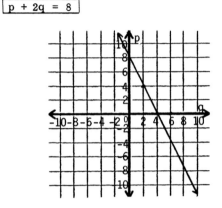

On the left: a) the coordinates of the horizontal (or p) intercept are (,   ).
b) The coordinates of the vertical (or q) intercept are (,   ).

On the right: c) The coordinates of the horizontal (or q) intercept are (,   ).
d) The coordinates of the vertical (or p) intercept are (,   ).

108. Let's find the coordinates of the intercepts for the equation below when S is plotted as the abscissa (or x). As an aid, a sketch of the axes is given at the right.

$$2S + 3T = 12$$

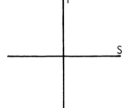

a) (8, 0)
b) (0, 4)
c) (4, 0)
d) (0, 8)

a) On the horizontal axis, T = 0. Therefore, to find the horizontal (or S) intercept, we substitute "0" for T and solve for S.

b) On the vertical axis, S = 0. Therefore, to find the vertical (or T) intercept, we substitute "0" for S and solve for T.

The horizontal intercept is (,   ).

The vertical intercept is (,   ).

a) (6, 0)
b) (0, 4)

109. Let's find the intercepts for $\boxed{2p + 5q = 20}$ when q is plotted as the abscissa. As an aid, a sketch of the axes is shown at the right.

   a) To find the horizontal intercept, substitute "0" for p. The coordinates are (,   ).

   b) To find the vertical intercept, substitute "0" for q. The coordinates are (,   ).

---

110. Draw your own sketch of the axes in this frame and the next.

   If R is plotted as the abscissa for $\boxed{3R = 4W - 36}$:

   a) The R-intercept is (,   ).

   b) The W-intercept is (,   ).

   a) (4, 0)
   b) (0, 10)

---

111. If m is plotted as the abscissa for $\boxed{5m = 4v + 40}$:

   a) The horizontal intercept is (,   ).

   b) The vertical intercept is (,   ).

   a) (-12, 0)
   b) (0, 9)

---

112. Let's graph the following formula by the intercept method. (Note: Since only positive values make sense for E and I, only the first quadrant is used on the graph.)

   $\boxed{E + 4I = 8}$

   If E is plotted as the abscissa:

   a) The E-intercept is (,   ).

   b) The I-intercept is (,   ).

   c) Plot the two intercepts and draw a line connecting them.

   a) (8, 0)
   b) (0, -10)

---

a) (8, 0)   b) (0, 2)   c)

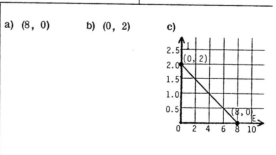

2-15 SLOPE OF LINEAR EQUATIONS WITH OTHER VARIABLES

In this section, we will discuss the slope of linear equations with variables other than "x" and "y". We will show that the slope depends on the choice of axes for the variables.

113. With variables other than "x" and "y", the slope of a line is still the ratio of a corresponding vertical change and horizontal increase. As examples, two lines are graphed below.

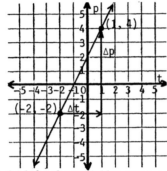 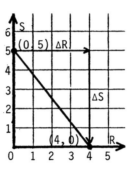

At the left, the variables are $\underline{t}$ and $\underline{p}$. $m = \dfrac{\Delta p}{\Delta t} = \dfrac{6}{3} = 2$

At the right, the variables are $\underline{R}$ and $\underline{S}$. $m = \dfrac{\Delta S}{\Delta R} = $ _____

114. With variables other than "x" and "y", the slope depends on the choice of axes for the variables. For example, we used intercepts to graph the following equation two different ways below.

$$\boxed{3h - 4s = 12}$$

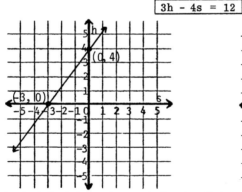

At the left, $\underline{h}$ is on the vertical axis. Using the coordinates of the intercepts:

$$m = \dfrac{\Delta h}{\Delta s} = \dfrac{4}{3}$$

At the right, $\underline{s}$ is on the vertical axis. Using the coordinates of the intercepts:

$$m = \dfrac{\Delta s}{\Delta h} = $$ _____

$-\dfrac{5}{4}$

115. We used intercepts to graph the following formula two different ways below. (Note: Only the first quadrant is used because only positive values of K and T are needed.)

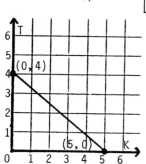

Using the coordinates of the intercepts, complete these:

a) At the left, $m = \dfrac{\Delta T}{\Delta K} =$ _____ b) At the right, $m = \dfrac{\Delta K}{\Delta T} =$ _____

$\dfrac{3}{4}$

116. Using the fact that $d = 8$ when $t = 4$, we graphed the formula below in two different ways.

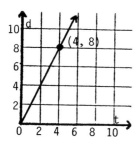
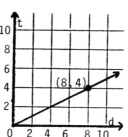

Using the origin $(0, 0)$ and the given point, complete these:

a) At the left, $m = \dfrac{\Delta d}{\Delta t} =$ _____ b) At the right, $m = \dfrac{\Delta t}{\Delta d} =$ _____

a) $-\dfrac{4}{5}$

b) $-\dfrac{5}{4}$

117. Two points are given on each line at the right. The variables are p and q.

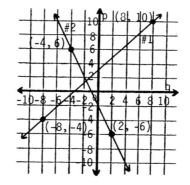

Using the two-point formula, we computed the slope of line #1 two ways below.

$m = \dfrac{p_2 - p_1}{q_2 - q_1} = \dfrac{10 - (-4)}{8 - (-8)} = \dfrac{14}{16} = \dfrac{7}{8}$

$m = \dfrac{p_2 - p_1}{q_2 - q_1} = \dfrac{(-4) - 10}{(-8) - 8} = \dfrac{-14}{-16} = \dfrac{7}{8}$

Use the two-point formula to compute the slope of line #2 below.

$m = \dfrac{p_2 - p_1}{q_2 - q_1} =$ _____

a) 2, from $\dfrac{8}{4}$

b) $\dfrac{1}{2}$, from $\dfrac{4}{8}$

Linear Equations 93

94 Linear Equations

118. Each line at the right passes through the origin. The variables are S and V.

Using the two-point formula, we computed the slope of line #1 two ways below.

$$m = \frac{V_2 - V_1}{S_2 - S_1} = \frac{30 - 0}{10 - 0} = \frac{30}{10} = 3$$

$$m = \frac{V_2 - V_1}{S_2 - S_1} = \frac{0 - 30}{0 - 10} = \frac{-30}{-10} = 3$$

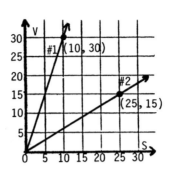

$m = -2$, from:

$$\frac{(-6) - 6}{2 - (-4)} \text{ or}$$

$$\frac{6 - (-6)}{(-4) - 2}$$

Use the two-point formula to compute the slope of line #2.

$$m = \frac{V_2 - V_1}{S_2 - S_1} = \underline{\hspace{4cm}}$$

$m = \frac{3}{5}$, from: $\frac{15 - 0}{25 - 0}$ or $\frac{0 - 15}{0 - 25}$

2-16 SLOPE-INTERCEPT FORM OF LINEAR EQUATIONS WITH OTHER VARIABLES

In this section, we will discuss the slope-intercept form of linear equations with variables other than "x" and "y".

119. The slope-intercept form of linear equations in x and y is: $\boxed{y = mx + b}$
The slope-intercept form of linear equations in other variables has the same pattern. Some examples are given below.

$$h = 2b - 3 \qquad F = \frac{9}{5}C + 32 \qquad p = -t + \frac{5}{4}$$

Which of the following are in slope-intercept form? _____

a) $2c + 3d = 6$ b) $T = -2S + 4$ c) $k - 4q = 12$

120. When putting linear equations in x and y in slope-intercept form, we always solve for "y". However, when putting linear equations with other variables in slope-intercept form, we can solve for either variable.

To put $\boxed{v - 2t = 8}$ in slope-intercept form, we solved for "v" at the left. Put the same equation in slope-intercept form by solving for "t" at the right.

$$v - 2t = 8 \qquad\qquad v - 2t = 8$$
$$v - 2t + 2t = 2t + 8$$
$$v = 2t + 8$$

Only (b)

121. In the last frame, we got the two slope-intercept forms for $\boxed{v - 2t = 8}$ by solving for "v" and "t".

$\boxed{v = 2t + 8}$ $\boxed{t = \frac{1}{2}v - 4}$

We graphed $\boxed{v - 2t = 8}$ below in two different ways. At the left, "v" is scaled on the vertical axis; at the right, "t" is scaled on the vertical axis.

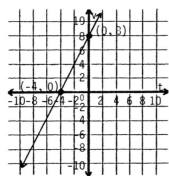 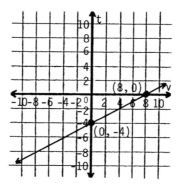

In $\boxed{v = 2t + 8}$: m = 2 and b = 8. From the graphs, you can see that the slope is 2 and the vertical intercept is (0, 8) only if v is scaled on the vertical axis.

In $\boxed{t = \frac{1}{2}v - 4}$: m = $\frac{1}{2}$ and b = -4. From the graphs, you can see that the slope is $\frac{1}{2}$ and the vertical intercept is (0, -4) only if _____ is scaled on the vertical axis.

$t = \frac{1}{2}v - 4$

122. When an equation with variables other than x and y is in slope-intercept form, "m" and "b" tell us the slope and vertical intercept only if the solved-for variable is scaled on the vertical axis. That is:

In $\boxed{F = \frac{9}{5}C + 32}$, the slope is $\frac{9}{5}$ and the vertical intercept is (0, 32) only if F is scaled on the vertical axis.

In $\boxed{P = 3Q - 1}$, the slope is 3 and the vertical intercept is (0, -1) only if _____ is scaled on the vertical axis.

t

123. a) In $\boxed{h = 5d - 3}$, is 5 the slope if h is scaled on the vertical axis? _____

b) In $\boxed{C = -2K + 7}$, is -2 the slope if K is scaled on the vertical axis? _____

P

124. a) In $\boxed{p = -q + 6}$, is (0, 6) the vertical intercept if q is scaled on the vertical axis? _____

b) In $\boxed{R = 3S - 10}$, is (0, -10) the vertical intercept if R is scaled on the vertical axis? _____

a) Yes

b) No. Only if C is scaled on the vertical axis.

96   Linear Equations

125. The equation at the right is linear. $\boxed{3c - 4d = 36}$
To find its slope and vertical intercept when c is scaled on the vertical axis, we must solve for c.

   a) Solve for c and put the solution in slope-intercept form.

   b) The slope is _____.   c) The vertical intercept is _____.

	a) No. Only if p is scaled on the vertical axis.
	b) Yes

126. The equation at the right is linear. $\boxed{4H + 5P = 80}$
To find its slope and vertical intercept when P is scaled on the vertical axis, we must solve for P.

   a) Solve for P and put the solution in slope-intercept form.

   b) The slope is _____.   c) The vertical intercept is _____.

a) $c = \frac{4}{3}d + 12$

b) $m = \frac{4}{3}$

c) (0, 12)

127. If E is scaled on the vertical axis when graphing: $\boxed{E = -8I + 40}$

   The slope is -8.

   The vertical intercept is (0, 40).

   To find the slope and vertical intercept when "I" is scaled on the vertical axis:

   a) Solve for "I" and put the solution in slope-intercept form.

   b) The slope is _____.   c) The vertical intercept is _____.

a) $P = -\frac{4}{5}H + 16$

b) $m = -\frac{4}{5}$

c) (0, 16)

128. Here is a formula showing the relationship between degrees-Fahrenheit (F) and degrees-Celsius (C):

$$\boxed{F = \frac{9}{5}C + 32}$$

If F is scaled on the vertical axis, the slope is $\frac{9}{5}$. That slope stands for many ratios of ΔF to ΔC. That is:

   For any 5-degree increase on the Celsius scale, we get a 9-degree increase on the Fahrenheit scale.

   For any 50-degree increase on the Celsius scale, we get a _____-degree increase on the Fahrenheit scale.

a) $I = -\frac{1}{8}E + 5$

b) $m = -\frac{1}{8}$

c) (0, 5)

90-degree

129. Here is one form of the load-line formula showing the relationship between voltage (E) and current (I) in an electronic circuit.

$$E = -4I + 20$$

If E is scaled on the vertical axis, the slope is -4. That slope stands for many ratios of ΔE to ΔI. That is:

For any 1-milliampere increase in current, we get a 4-volt decrease in voltage.

For any 10-milliampere increase in current, we get a _____-volt decrease in voltage.

40-volt

2-17 LINES THROUGH THE ORIGIN WITH OTHER VARIABLES

In this section, we will discuss the slope-intercept form of the equations of lines through the origin for variables other than "x" and "y".

130. For lines through the origin, the slope-intercept form of equations in x and y is: $y = mx$. The same pattern applies for lines through the origin with other variables. Some examples are:

$$d = 100t \qquad E = 75R \qquad P = 62.4h$$

Since each equation above graphs as a line through the origin, the vertical intercept for each is _____.

131. For $d = 100t$, the slope is 100 only if d is scaled on the vertical axis.

For $E = 75R$, the slope is 75 only if ____ is scaled on the vertical axis.

(0, 0)

132. If F is scaled on the vertical axis when graphing $F = 10s$:

The slope is 10.

The vertical intercept is (0, 0).

To find the slope and vertical intercept when s is scaled on the vertical axis:

a) Solve for "s": _____

b) The slope is _____.   c) The vertical intercept is _____.

E

Linear Equations 97

98   Linear Equations

133. The graph of the relationship between R and V is a straight line through the origin.

   a) If R is plotted as "y", the slope is 20. Write the equation of the line in slope-intercept form. _____

   b) If V is plotted as "y", the equation in slope-intercept form is:
   _____

a) $s = \frac{1}{10}F$

b) $m = \frac{1}{10}$

c) (0, 0)

---

134. Each line at the right passes through the origin. The variables are C and D. Since D is scaled on the vertical axis, the form of each equation is: D = mC.

   For line #1: $m = \frac{\Delta D}{\Delta C} = \frac{60}{30} = 2$;
   the equation is: D = 2C

   For line #2:   a) $m = \frac{\Delta D}{\Delta C} = $ _____

   b) the equation is: _____

a) R = 20V

b) $V = \frac{1}{20}R$

---

135. The graph of the relationship between F and T is a straight line through the origin.

   When F = 50, T = 40.

   Sketch that point at the right if F is plotted as the ordinate. Then write the equation of the line. _____

a) $m = \frac{2}{3}$

b) $D = \frac{2}{3}C$

---

136. Here is the formula showing the relationship between distance traveled (d) and time (t) at a constant velocity of 100 miles per hour.

   $$\boxed{d = 100t}$$

   If d is scaled on the vertical axis, the slope is 100. That slope stands for many ratios of Δd to Δt. That is:

   For any 1-hour increase in time, the distance traveled increases 100 miles.

   For any 2.5-hour increase in time, the distance traveled increases _____ miles.

$F = \frac{5}{4}T$

---

250 miles

Linear Equations 99

137. Here is a formula showing the relationship between the increase in length (L) in inches of a 50-inch metal rod as the temperature (t) increases.

$$L = 0.00065t$$

If L is scaled on the vertical axis, the slope is 0.00065. That slope stands for many ratios of ΔL to Δt. That is:

For any 1-degree increase in temperature, the length increases 0.00065 inch.

For any 10-degree increase in temperature, the length increases _____ inch.

0.0065 inch

2-18 ESTIMATING THE EQUATIONS OF GRAPHED LINES

When we are given the exact coordinates of two points on a line, we can compute "m" and "b" exactly and write the exact equation of the line. When we are not given the exact coordinates of two points, we can only estimate "m" and "b" and write an approximate equation of the line. We will discuss the estimation method in this section.

138. To estimate the equation of a graphed line, we must estimate "m" and "b". To estimate "b", we estimate where the line crosses the vertical axis. To estimate "m", <u>we try to find two points where the graphed line crosses an intersection of the grid lines because the coordinates of such points are approximately whole numbers.</u> An example is discussed below.

Let's estimate the equation of the line at the right.

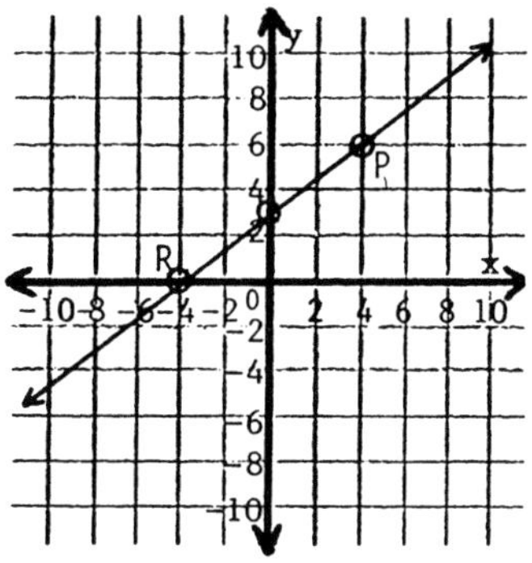

[partially obscured text:]
...ate
...osses

...se
of

...4, 6).
...-4, 0).

$\frac{\ldots 0}{\ldots(-4)} = \frac{6}{8} = \frac{3}{4}$

...e is approximately _____

$y = \frac{3}{4}x + 3$

100   Linear Equations

139. Let's estimate the equation of the line at the right.

1) To estimate "b", we estimate the point where the line crosses the vertical axis.

"b" is approximately -1.5 or $-\frac{3}{2}$.

2) To estimate "m", we can use the estimated coordinates of points A and C.

Point A is approximately (-1, 1).
Point C is approximately (1, -4).

"m" is approximately $\frac{1-(-4)}{(-1)-1} = \frac{5}{-2}$ or $-\frac{5}{2}$

Therefore, the equation of the line is approximately _____

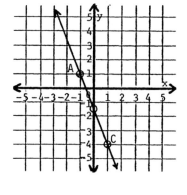

$y = -\frac{5}{2}x - \frac{3}{2}$

or

$y = -2.5x - 1.5$

---

140. On each line below, we circled two points whose coordinates are approximately whole numbers. Let's estimate the equation of each line.

For line #1:

a) "b" is approximately _____

b) "m" is approximately _____

c) The equation is approximately

_____

For line #2:

d) "b" is approximately _____

e) "m" is approximately _____

f) The equation is approximately _____

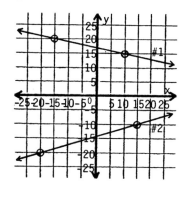

---

141. The variables below are "p" and "t". On each line, we circled two points whose coordinates are approximately whole numbers. Let's estimate the equations of each line.

For line #1:

a) "b" is approximately _____

b) "m" is approximately _____

c) The equation is approximately

_____

For line #2:

d) "b" is approximately _____

e) "m" is approximately _____

f) The equation is approximately _____

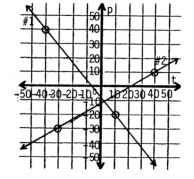

a) 17

b) $-\frac{1}{5}$

c) $y = -\frac{1}{5}x + 17$

d) -14

e) $\frac{2}{7}$

f) $y = \frac{2}{7}x - 14$

142. It looks as if both lines at the right pass through the origin. We circled one point on each line whose coordinates are approximately whole numbers. We can estimate the equation of each line by estimating its slope.

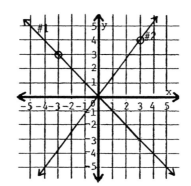

For line #1:

a) "m" is approximately _____

b) The equation is approximately _____

For line #2:

c) "m" is approximately _____

d) The equation is approximately _____

a) -8

b) $-\frac{6}{5}$

c) $p = -\frac{6}{5}t - 8$

d) -13

e) $\frac{4}{7}$

f) $p = \frac{4}{7}t - 13$

---

143. The variables at the right are F and V. Both lines apparently start at the origin. Using the circled points whose coordinates seem to be whole numbers, estimate the equation of each line.

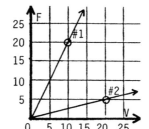

a) The equation of line #1 is approximately _____

b) The equation of line #2 is approximately _____

a) -1

b) $y = -1x$ or $y = -x$

c) $\frac{4}{3}$

d) $y = \frac{4}{3}x$

---

144. When a graphed line does not cross an intersection of the grid lines, we can estimate "b" in the usual way. However, to estimate "m", we pick two points whose abscissas are approximately whole numbers. The two points should be fairly far apart so that any errors in estimating their ordinates only slightly affect the estimated value of the slope. An example is discussed below.

Let's estimate the equation of the line at the right.

$b = -\frac{1}{2}$ or -0.5

Point A is approximately $(4, 3.5)$

Point B is approximately $(-4, -4.5)$

$m \doteq \frac{3.5 - (-4.5)}{4 - (-4)}$ or $\frac{8}{8}$ or 1

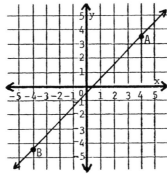

Therefore, the approximate equation of the line is _____

a) $F = 2V$

b) $F = \frac{1}{4}V$

102   Linear Equations

145.   The variables at the right are P and V.
Let's estimate the equation of the line.

b ≐ 2.5

Point C is approximately (-2, 4.6)

Point D is approximately (4, -1.8)

m ≐ $\frac{(-1.8) - 4.6}{4 - (-2)}$ or $\frac{-6.4}{6}$ or   -1.07

Therefore, the equation of the line is approximately _____

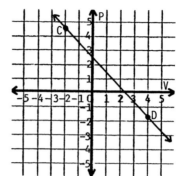

$y = 1x - \frac{1}{2}$

or

$y = x - \frac{1}{2}$

or

$y = x - 0.5$

---

146.   The variables at the right are d and t.
The line seems to begin at the origin.
Let's estimate the equation of the line.

a) Point B is approximately (,   ).

b) m ≐ _____

c) The equation is approximately:
_____

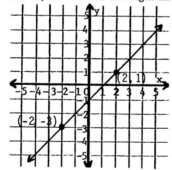

P = -1.07V + 2.5

a) (5, 3.5)   b) 0.7   c) d = 0.7t

2-19   SLOPE AND AXES WITH DIFFERENT SCALES

In this section, we will discuss the slope of lines when the scales on the axes are different.

147.   Two graphed lines are shown below. The axes at the left have the same scales; the axes at the right have different scales.

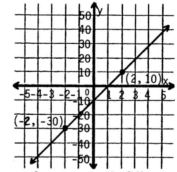

Though the slopes of the lines look the same, they are actually different.
Let's use the two given points to compute the slope of each.

a) For the line at the left, the slope is _____.

b) For the line at the right, the slope is _____.

148. Two graphed lines are shown below. The axes at the left have the same scales; the axes at the right have different scales.

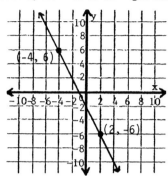 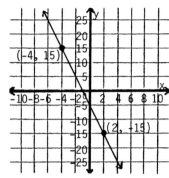

Though the slopes of the lines look the same, they are actually different. Let's use the two given points to compute the slope of each.

a) For the line at the left, the slope is _____.

b) For the line at the right, the slope is _____.

a) 1, from: $\dfrac{1 - (-3)}{2 - (-2)}$

b) 10, from: $\dfrac{10 - (-30)}{2 - (-2)}$

---

149. When the scales on the axes are different, the apparent slope of a line can be deceiving. As an example, two graphed lines are shown below.

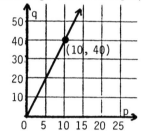 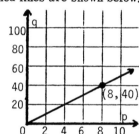

From the graphs, it looks as if the slope at the left is greater than the slope at the right. Let's compute each slope and compare them.

a) For the line at the left, the slope is _____.

b) For the line at the right, the slope is _____.

c) Which line has the greater slope? _____

a) -2, from: $\dfrac{(-6) - 6}{2 - (-4)}$

b) -5, from: $\dfrac{(-15) - 15}{2 - (-4)}$

---

150. The variables at the right are d and t. The scales on the axes are different. Estimate the equation of each line.

a) line #1: _____

b) line #2: _____

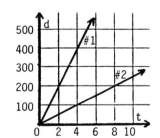

a) 4

b) 5

c) The one at the right

a) d = 100t   b) d = 25t

SELF-TEST 7 (pp. 89-104)

For $\boxed{3d - 4t = 24}$, $\underline{d}$ is plotted as the abscissa. Find the coordinates of the:

1. Horizontal intercept _____
2. Vertical intercept. _____

3. Graph $\boxed{P + 3V = 12}$ by finding and plotting its intercepts.

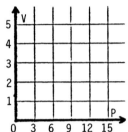

For $\boxed{2F - 3S = 8}$, F is plotted on the vertical axis.

4. Put the equation in slope-intercept form. _____
5. Find the slope. _____
6. Find the coordinates of the vertical intercept. _____

For $\boxed{w = \frac{3}{2}h + 5}$, $\underline{w}$ is scaled on the vertical axis.

7. For a 2-unit increase in $\underline{h}$, what is the increase in $\underline{w}$? _____
8. For a 12-unit increase in $\underline{h}$, what is the increase in $\underline{w}$? _____

On the graph below, lines #1 and #2 pass through the origin. The variables are G and R.

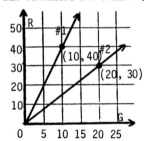

For line #1: 9. The slope is _____.
10. The equation is _____.
For line #2: 11. The slope is _____.
12. The equation is _____.

A straight line is shown on the graph below. The variables are $\underline{r}$ and $\underline{t}$.

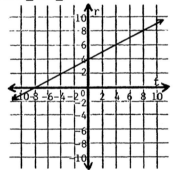

13. The coordinates of the vertical intercept are _____.
14. The slope is _____.
15. The equation of the line is _____.

ANSWERS:
1. (8, 0)
2. (0, -6)
3. V, (0, 4), (12, 0) P
4. $F = \frac{3}{2}S + 4$
5. $\frac{3}{2}$
6. (0, 4)
7. 3-unit increase
8. 18-unit increase
9. 4
10. $R = 4G$
11. $\frac{3}{2}$
12. $R = \frac{3}{2}G$
13. (0, 4)
14. $\frac{1}{2}$
15. $r = \frac{1}{2}t + 4$

SUPPLEMENTARY PROBLEMS - CHAPTER 2

Assignment 5

State whether each equation is <u>linear</u> or <u>non-linear</u>.

1. $y = \dfrac{9}{x}$
2. $\dfrac{y}{3x} = 5$
3. $4x - 3y = 6$
4. $\dfrac{x-2}{y} = 2y$
5. $\dfrac{x}{y-1} = 7$

Find the coordinates of the <u>x-intercept</u> of each equation.

6. $y = 2x + 6$
7. $3x + 5y = 15$
8. $4x - 3y = 2$
9. $2y = 5x + 8$

Find the coordinates of the <u>y-intercept</u> of each equation.

10. $y = 3x - 9$
11. $2x + 5y = 10$
12. $3y = 4x + 1$
13. $6x - 2y = 5$

Find the intercepts of each equation and use them to graph the equation.

14. $y = x + 2$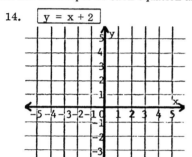
15. $x + 3y = 3$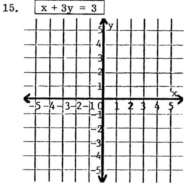
16. $3x - 2y = 6$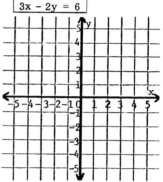

Find the slope of each line graphed below.

17. Line #1
18. Line #2

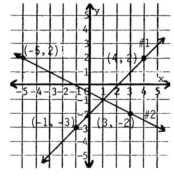

19. Line #3
20. Line #4

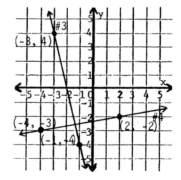

Five lines have these slopes: Line A: 4   Line B: -1   Line C: $-\dfrac{9}{2}$ Line D: $\dfrac{3}{5}$ Line E: -3

21. Which lines <u>rise</u> from left to right?
22. Which line has the greatest rise?
23. Which lines <u>fall</u> from left to right?
24. Which line has the greatest fall?
25. Which line is steepest?

26. The slope of a line is 4. If <u>x</u> increases 8 units, <u>y</u> increases how many units?
27. The slope of a line is $\dfrac{1}{2}$. If <u>x</u> increases 6 units, <u>y</u> increases how many units?
28. The slope of a line is -3.95. If <u>x</u> increases 1 unit, <u>y</u> decreases how many units?

Assignment 5 (Continued)

29. The slope of a line is -2. If $\Delta x = 4$, find Δy.
30. The slope of a line is $\frac{5}{2}$. If $\Delta x = 6$, find Δy.
31. The slope of a line is -1.5. If $\Delta x = 1$, find Δy.

For each equation below, find the slope and the coordinates of the y-intercept.

32. $y = 5x + 2$
33. $y = x - 3$
34. $y = -\frac{2}{5}x + \frac{3}{2}$
35. $y = -4.2x - 7.5$

Write the slope-intercept form of the linear equation whose:

36. Slope is -1 and y-intercept is $(0, 2)$.
37. Slope is 3 and y-intercept is $(0, -5)$.
38. Slope is $\frac{1}{2}$ and y-intercept is $(0, -1)$
39. Slope is $\frac{8}{3}$ and y-intercept is $(0, \frac{7}{4})$.

Assignment 6

Put each equation in slope-intercept form; then list its slope and the coordinates of its y-intercept.

1. $y = 6 - 4x$
2. $y - 3x = 2$
3. $x - y = 5$
4. $4y = x - 8$
5. $6x + 2y = 3$
6. $3x - 6y = 10$

Find the <u>slope</u> of the line through each pair of points.

7. $(2, 3)$ and $(-1, -3)$
8. $(-4, 2)$ and $(2, -4)$
9. $(6, 8)$ and $(10, 2)$
10. $(-3, -4)$ and $(5, -2)$
11. $(-4, 4)$ and $(0, 8)$
12. $(0, 6)$ and $(10, 0)$

Find the equation of the line whose:

13. Slope is 3 and which passes through $(2, 0)$.
14. Slope is -1 and which passes through $(3, -1)$.
15. Slope is $\frac{1}{3}$ and which passes through $(-6, 3)$.
16. Slope is $\frac{5}{2}$ and which passes through $(-4, -2)$.
17. Slope is -4 and which passes through $(1, 2)$.
18. Slope is $-\frac{1}{2}$ and which passes through $(-3, 5)$.

Find the equation of the line through each pair of points.

19. $(1, 1)$ and $(2, 3)$
20. $(0, 6)$ and $(2, 0)$
21. $(-2, 3)$ and $(3, -2)$
22. $(6, 1)$ and $(-2, -3)$
23. $(-5, 4)$ and $(4, -2)$
24. $(8, 0)$ and $(0, -20)$

25. Which equations below have graphs which pass through the origin?

   a) $y = 7x$
   b) $y = 3x + 5$
   c) $x - y = 2$
   d) $x + 2y = 0$
   e) $y = -\frac{3}{4}x$

26. Find the equation of the line whose slope is $\frac{3}{4}$ and which passes through the origin.

Find the equation of the line through each pair of points.

27. $(0, 0)$ and $(4, 8)$
28. $(0, 0)$ and $(-2, 6)$
29. $(0, 0)$ and $(-4, -2)$

30. Which of the following lines are parallel?

   a) $2y = x$
   b) $2x + y = 0$
   c) $y = 2x$
   d) $2x - y = 0$
   e) $3y = 6x - 2$

Find the equation of each of the following lines.

31. The line parallel to the x-axis and 6 units above it.
32. The line parallel to the y-axis and 3 units to the left of it.
33. The vertical line through the point $(5, -2)$.
34. The horizontal line through the point $(-4, -8)$.

Find the <u>slope</u> of: 35. Any horizontal line.   36. Any vertical line.

Assignment 7

Find the horizontal and vertical intercepts for each equation.

1. $3s + 2t = 18$
   (s is plotted as abscissa)

2. $P - 5R = 30$
   (P is plotted as abscissa)

3. $2d = 12 + 3h$
   (d is plotted as abscissa)

4. $2A + 14 = 7B$
   (A is plotted as abscissa)

5. a) Graph $\boxed{2s + 6 = 3f}$ below by finding and plotting its intercepts.
   b) Then find its slope from the graph.

6. a) Graph $\boxed{2d + w = 8}$ below by finding and plotting its intercepts.
   b) Then find its slope from the graph.

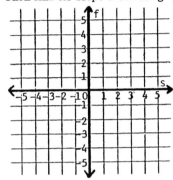
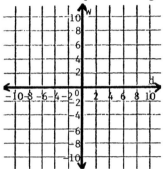

For each equation below:
a) Put the equation in slope-intercept form.
b) Find the slope.
c) Find the coordinates of the vertical intercept.

7. $v - 3t = 5$
   (v is scaled on the vertical axis)

8. $3h - 4p = 20$
   (p is scaled on the vertical axis)

9. $r + 2s = 2$
   (s is scaled on the vertical axis)

10. $4V = 50 - 5T$
    (T is scaled on the vertical axis)

11. For $\boxed{p = 5w - 4}$, p is scaled on the vertical axis.
    a) For a 1-unit increase in w, find the increase in p.
    b) For a 4-unit increase in w, find the increase in p.

12. For $\boxed{F = \frac{4}{3}R + \frac{7}{2}}$, F is scaled on the vertical axis.
    a) For a 1-unit increase in R, find the increase in F.
    b) For a 12-unit increase in R, find the increase in F.

13. The graph of the relationship between d and w is a straight line through the origin. If d is plotted as "y", the slope is 8. Write the equation of the line.

14. The graph of the relationship between P and R is a straight line through the origin. P is plotted as the ordinate. When R = 40, P = 16. Write the equation of the line.

15. The equation $\boxed{d = 20t}$ shows the relationship between distance traveled (d) and time (t) at a constant velocity of 20 meters per second. For any 4.5-second increase in time, the distance traveled increases how many meters?

Each line below passes through the origin. Write the equation of each line.

16.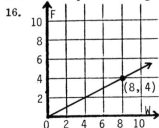

17.

18.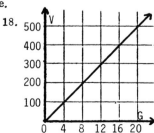

108   Linear Equations

Assignment 7 (Continued)

Using estimation, write the equation of each line below in slope-intercept form.

19.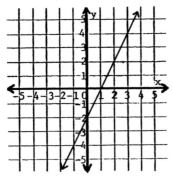

20.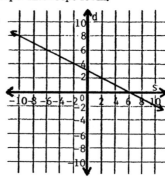

Chapter 3 SQUARE ROOT RADICALS

In this chapter, we will discuss the basic operations with square root radicals. Methods for simplifying radicals and methods for rationalizing denominators and numerators are included. Both imaginary and complex numbers are defined, and the basic operations with numbers of that type are discussed.

3-1 SQUARE ROOTS OF POSITIVE NUMBERS

In this section, we will discuss what is meant by the "<u>square root</u>" of a positive number. We will show that any positive number has two square roots.

1. A "<u>square root</u>" of a positive number N is "<u>a number whose square is N</u>".
   Therefore, any positive number has two square roots, one positive and one negative. For example:

   $\quad$ 4 is a square root of 16, since $(4)^2 = 16$

   $\quad$ -4 is a square root of 16, since $(-4)^2 = 16$

   Complete: a) The <u>positive</u> square root of 36 is _____.

   $\qquad\quad\;\;$ b) The <u>negative</u> square root of 100 is _____.

   $\qquad\quad\;\;$ c) The two square roots of 49 are _____ and _____.

2. Though negative square roots are called "<u>negative</u>" square roots, positive square roots are called "<u>principal</u>" square roots.

   The symbols $+\sqrt{}$ or $\sqrt{}$ are used for <u>principal</u> square roots. That is:

   $\quad +\sqrt{25} = 5 \qquad \sqrt{81} = 9 \qquad \sqrt{400} = 20$

   The symbol $-\sqrt{}$ is used for <u>negative</u> square roots. That is:

   $\quad -\sqrt{9} = -3 \qquad -\sqrt{1} = -1 \qquad -\sqrt{900} = -30$

   Complete these:

   a) $+\sqrt{1} =$ _____   b) $-\sqrt{64} =$ _____   c) $\sqrt{9} =$ _____

a) 6

b) -10

c) 7 and -7

a) 1

b) -8

c) 3

Square Root Radicals

3. For expressions like $\sqrt{25}$, the following terminology is used.
   1) The entire expression $\sqrt{25}$ is called a <u>square root radical</u>.
   2) The symbol $\sqrt{}$ is called a <u>radical sign</u>.
   3) The number "25" under the radical is called the <u>radicand</u>.

   Write the square root radical whose radicand is 144. _____

	$\sqrt{144}$

4. The symbol "±" means "<u>both + and −</u>". It can be written either in front of a radical or a number. That is:

   $\pm\sqrt{25}$ means both $+\sqrt{25}$ and $-\sqrt{25}$.

   ± 5 means both $+5$ and -5.

   Therefore: $\pm\sqrt{25} = \pm 5$ means $+\sqrt{25} = +5$ and $-\sqrt{25} = -5$.

   Use the "±" symbol to complete these.

   a) $\pm\sqrt{4} =$ _____   b) $\pm\sqrt{1} =$ _____   c) $\pm\sqrt{81} =$ _____

5. Any whole number that is a perfect square has two whole numbers, one positive and one negative, as its square roots. For example:

   Since both $(15)^2$ and $(-15)^2 = 225$, $\pm\sqrt{225} = \pm 15$.

   However, the number "0" has only one square root. That is:

   Since $(0)^2 = 0$, $\sqrt{0} =$ _____

   a) ±2
   b) ±1
   c) ±9

6. Any decimal number that is a perfect square has two decimal numbers as its square roots. That is:

   Since $(2.5)^2$ and $(-2.5)^2 = 6.25$, $\pm\sqrt{6.25} = \pm 2.5$
   Since $(.08)^2$ and $(-.08)^2 = .0064$, $\pm\sqrt{.0064} =$ _____

   0

7. To find the square root of a number on a calculator, we use the $\boxed{\sqrt{x}}$ key. Use a calculator for these.

   a) $\sqrt{169} =$ _____   c) $\sqrt{129.96} =$ _____
   b) $\sqrt{348,100} =$ _____   d) $\sqrt{.000729} =$ _____

   ±.08

8. Though only principal square roots are given by a calculator, it can be used to get both square roots. For example:

   $\pm\sqrt{196} = \pm 14$ $\pm\sqrt{5.76} = \pm 2.4$ $\pm\sqrt{.0289} =$ _____

   a) 13   c) 11.4
   b) 590   d) .027

   ±.17

9. The square roots of numbers that are not perfect squares are unending decimal numbers. When finding square roots of that type, a calculator gives many decimal places. For example:

$$\sqrt{65} = 8.0622577$$
$$\sqrt{41.8} = 6.465292$$
$$\sqrt{.069} = .26267851$$

Usually square roots like those above are rounded. For example, rounding to two decimal places, we get:

$\sqrt{65} = 8.06$ $\qquad$ $\sqrt{41.8} = 6.47$ $\qquad$ $\sqrt{.069} =$ _____

---

10. Since the square of any number (positive or negative) is <u>positive</u>, there is no ordinary number that is the square root of a <u>negative</u> number. For example:

$$\sqrt{-16} \neq +4, \text{ since } (+4)^2 = +16$$
$$\sqrt{-16} \neq -4, \text{ since } (-4)^2 = +16$$

Mathematicians have defined "<u>imaginary</u>" numbers to handle the square roots of negative numbers. We will discuss "imaginary" numbers at the end of this chapter. <u>Until that time, however, we will assume that all radicands are positive quantities.</u>

.26

3-2 FORMULA EVALUATIONS

In this section, we will discuss evaluations with formulas containing square root radicals.

11. Just like numbers, expressions containing one or more variables have both a principal and a negative square root. For example, we could write:

$\pm\sqrt{x}$ $\qquad$ $\pm\sqrt{\dfrac{2s}{a}}$ $\qquad$ $\pm\sqrt{p^2 + q^2}$

However, negative square roots have little use in applied mathematics. Therefore, <u>the discussion in the rest of this chapter and book will be limited to principal square roots.</u>

12. In $\boxed{s = \sqrt{A}}$, we can find "s" by finding the square root of A. That is:

If $A = 64$, $s = \sqrt{A} = \sqrt{64} = 8$

If $A = 100$, $s = \sqrt{A} = \sqrt{100} =$ _____

10

112   Square Root Radicals

13. When the radicand is a more complex expression, we simplify before finding the square root. For example:

   In $\boxed{v = \sqrt{2as}}$, find "v" when a = 3 and s = 6.

   $v = \sqrt{2as} = \sqrt{2(3)(6)} = \sqrt{36} = 6$

   In $\boxed{E = \sqrt{PR}}$, find E when P = 3 and R = 27.

   $E = \sqrt{PR} = $ _____

---

14. Following the example, complete the other evaluation.

   In $\boxed{I = \sqrt{\dfrac{P}{R}}}$, find I when P = 400 and R = 4.

   $I = \sqrt{\dfrac{P}{R}} = \sqrt{\dfrac{400}{4}} = \sqrt{100} = 10$

   In $\boxed{t = \sqrt{\dfrac{2s}{g}}}$, find "t" when s = 64 and g = 32.

   $t = \sqrt{\dfrac{2s}{g}} = $ _____

E = 9, from: $\sqrt{81}$

---

15. Following the example, complete the other evaluation.

   In $\boxed{c = \sqrt{a^2 + b^2}}$, find "c" when a = 3 and b = 4.

   $c = \sqrt{a^2 + b^2} = \sqrt{3^2 + 4^2} = \sqrt{9 + 16} = \sqrt{25} = 5$

   In $\boxed{R = \sqrt{Z^2 - X^2}}$, find R when Z = 10 and X = 8.

   $R = \sqrt{Z^2 - X^2} = $ _____

t = 2, from: $\sqrt{4}$

---

16. In the formula below, the radical is a factor in a multiplication. Notice that we found the square root <u>before</u> <u>multiplying</u>. Complete the other evaluation.

   In $\boxed{r = a\sqrt{w}}$, find "r" when a = 3 and w = 16.

   $r = a\sqrt{w} = 3\sqrt{16} = 3(4) = 12$

   In $\boxed{N = K\sqrt{\dfrac{V}{H}}}$, find N when K = 7, V = 18, and H = 2,

   $N = K\sqrt{\dfrac{V}{H}} = $ _____

R = 6, from: $\sqrt{36}$

N = 21, from: $7\sqrt{9}$

17. Complete each evaluation.

   a) In $\boxed{p = m + r\sqrt{t}}$, find "p" when m = 50, r = 6, and t = 25.

   $p = m + r\sqrt{t} = $ _____

   b) In $\boxed{K = \dfrac{L}{\sqrt{MT}}}$, find K when L = 100, M = 2, and T = 8.

   $K = \dfrac{L}{\sqrt{MT}} = $ _____

   a) p = 80, from:
   50 + 30

   b) K = 25, from:
   $\dfrac{100}{\sqrt{16}}$

18. Complete each evaluation.

   a) In $\boxed{a = \sqrt{\dfrac{b-c}{c}}}$, find "a" when b = 100 and c = 10.

   $a = \sqrt{\dfrac{b-c}{c}} = $ _____

   b) In $\boxed{s = \dfrac{\sqrt{h+2r}}{d}}$, find "s" when h = 50, r = 7, and d = 4.

   $s = \dfrac{\sqrt{h+2r}}{d} = $ _____

   a) a = 3, from: $\sqrt{9}$ b) s = 2, from: $\dfrac{\sqrt{64}}{4}$

3-3 MULTIPLYING RADICALS

In this section, we will discuss the procedure for multiplying square root radicals.

19. To indicate a multiplication of radicals, we ordinarily use a dot. For example:

   $\sqrt{2} \cdot \sqrt{10}$ means "multiply $\sqrt{2}$ and $\sqrt{10}$".

   To multiply radicals, we multiply their radicands. That is:

   $\sqrt{3} \cdot \sqrt{7} = \sqrt{(3)(7)}$ $\sqrt{p} \cdot \sqrt{q} = \sqrt{(p)(q)}$

   To show that the multiplication procedure above makes sense, two examples involving perfect-square radicands are shown below.

   $\sqrt{9} \cdot \sqrt{4} = \sqrt{(9)(4)}$ $\sqrt{16} \cdot \sqrt{25} = \sqrt{(16)(25)}$
   $\quad 3 \cdot 2 = \sqrt{36}$ $\quad 4 \cdot 5 = \sqrt{400}$
   $\qquad 6 = 6$ $\qquad 20 = 20$

114 Square Root Radicals

20. When multiplying radicals, we simplify the radicand of the product as much as possible. For example:

$$\sqrt{2} \cdot \sqrt{3} = \sqrt{(2)(3)} = \sqrt{6}$$
$$\sqrt{7} \cdot \sqrt{y^5} = \sqrt{(7)(y^5)} = \sqrt{7y^5}$$
$$\sqrt{5x} \cdot \sqrt{6y} = \sqrt{(5x)(6y)} = \sqrt{}$$

---

21. Sometimes the law of exponents for multiplication is used to simplify the radicand of the product. For example:

$$\sqrt{x^4} \cdot \sqrt{x} = \sqrt{(x^4)(x)} = \sqrt{x^5}$$
$$\sqrt{3y} \cdot \sqrt{5y} = \sqrt{(3y)(5y)} = \sqrt{15y^2}$$
$$\sqrt{at^3} \cdot \sqrt{a^2t^2} = \sqrt{(at^3)(a^2t^2)} = \sqrt{}$$

$\sqrt{30xy}$

---

22. Complete these:

a) $\sqrt{7} \cdot \sqrt{11} = \sqrt{}$ c) $\sqrt{m^4} \cdot \sqrt{m^4} = \sqrt{}$

b) $\sqrt{3a} \cdot \sqrt{8b^2} = \sqrt{}$ d) $\sqrt{2x} \cdot \sqrt{3x^2y} = \sqrt{}$

$\sqrt{a^3t^5}$

---

23. To multiply three or more radicals, we also multiply their radicands. For example:

$$\sqrt{2} \cdot \sqrt{3} \cdot \sqrt{5} = \sqrt{(2)(3)(5)} = \sqrt{30}$$
$$\sqrt{a} \cdot \sqrt{b^2} \cdot \sqrt{7} \cdot \sqrt{b^2} = \sqrt{(a)(b^2)(7)(b^2)} = \sqrt{7ab^4}$$

Following the examples, complete these:

a) $\sqrt{y} \cdot \sqrt{2y} \cdot \sqrt{3} = \sqrt{}$

b) $\sqrt{5pq} \cdot \sqrt{10} \cdot \sqrt{p^2} \cdot \sqrt{q^5} = \sqrt{}$

a) $\sqrt{77}$ c) $\sqrt{m^8}$
b) $\sqrt{24ab^2}$ d) $\sqrt{6x^3y}$

---

24. To indicate a multiplication of a non-radical and a radical, we write the non-radical in front of the radical. That is:

$2\sqrt{5}$ means "multiply 2 and $\sqrt{5}$".

To simplify the expressions below, we multiplied the non-radical factors.

$$2 \cdot 4\sqrt{7} = 8\sqrt{7} \qquad\qquad y \cdot 6\sqrt{y^3} = 6y\sqrt{y^3}$$

Following the examples, simplify these:

a) $3x \cdot 5\sqrt{x} = $ _____   b) $t^2 \cdot t^3\sqrt{7} = $ _____

a) $\sqrt{6y^2}$
b) $\sqrt{50p^3q^6}$

---

a) $15x\sqrt{x}$
b) $t^5\sqrt{7}$

25. To simplify the expressions below, we multiplied the radical factors.

$$7\sqrt{3} \cdot \sqrt{2} = 7\sqrt{6} \qquad a\sqrt{x} \cdot \sqrt{2x} = a\sqrt{2x^2}$$

Following the examples, simplify these.

a) $9\sqrt{y^3} \cdot \sqrt{y^3} = $ _____ b) $at\sqrt{t} \cdot \sqrt{a^2t^2} = $ _____

---

26. To simplify the expressions below, we multiplied both the non-radical and the radical factors.

$$2\sqrt{3} \cdot 5\sqrt{7} = 2 \cdot 5 \cdot \sqrt{3} \cdot \sqrt{7} = 10\sqrt{21}$$
$$a\sqrt{b} \cdot a\sqrt{b^4} = a \cdot a \cdot \sqrt{b} \cdot \sqrt{b^4} = a^2\sqrt{b^5}$$

Following the examples, simplify these:

a) $3\sqrt{5x} \cdot x\sqrt{2} = $ _____

b) $2d\sqrt{p} \cdot d^3\sqrt{dp} = $ _____

a)	$9\sqrt{y^6}$
b)	$at\sqrt{a^2t^3}$

---

27. Any positive number is equal to the square root radical whose radicand is the square of the number. For example:

$$2 = \sqrt{4} \qquad 5 = \sqrt{25} \qquad 8 = \sqrt{64}$$

Using the fact above, we can convert $3\sqrt{7}$ to a single radical. To do so, we substitute $\sqrt{9}$ for 3 and multiply. That is:

$$3\sqrt{7} = \sqrt{9} \cdot \sqrt{7} = \sqrt{63}$$

Following the example, convert each multiplication below to a single radical.

a) $2\sqrt{5} = \sqrt{} \cdot \sqrt{5} = \sqrt{}$

b) $6\sqrt{x} = \sqrt{} \cdot \sqrt{x} = \sqrt{}$

a)	$3x\sqrt{10x}$
b)	$2d^4\sqrt{dp^2}$

---

28. Any letter is equal to the square root radical whose radicand is the square of the letter. For example:

$$x = \sqrt{x^2} \qquad P = \sqrt{P^2}$$

Using the fact above, we can convert $a\sqrt{b}$ to a single radical. To do so, we substitute $\sqrt{a^2}$ for "a" and multiply. That is:

$$a\sqrt{b} = \sqrt{a^2} \cdot \sqrt{b} = \sqrt{a^2b}$$

Following the example, convert each multiplication below to a single radical.

a) $x\sqrt{7} = \sqrt{} \cdot \sqrt{7} = \sqrt{}$

b) $D\sqrt{F} = \sqrt{} \cdot \sqrt{F} = \sqrt{}$

a)	$\sqrt{4} \cdot \sqrt{5} = \sqrt{20}$
b)	$\sqrt{36} \cdot \sqrt{x} = \sqrt{36x}$

116   Square Root Radicals

29. Write each of the following as a single radical.

a) $4\sqrt{2}$ = _____   c) $c\sqrt{t}$ = _____

b) $9\sqrt{x}$ = _____   d) $m\sqrt{3}$ = _____

a) $\sqrt{x^2} \cdot \sqrt{7} = \sqrt{7x^2}$
b) $\sqrt{D^2} \cdot \sqrt{F} = \sqrt{D^2F}$

---

30. Convert each of these to a single radical.

a) $3\sqrt{10}$ = _____   c) $p\sqrt{q^5}$ = _____

b) $10\sqrt{T}$ = _____   d) $R\sqrt{11}$ = _____

a) $\sqrt{32}$ c) $\sqrt{c^2 t}$
b) $\sqrt{81x}$ d) $\sqrt{3m^2}$

a) $\sqrt{90}$ b) $\sqrt{100T}$ c) $\sqrt{p^2 q^5}$ d) $\sqrt{11R^2}$

3-4   SIMPLIFYING RADICALS BY FACTORING OUT PERFECT-SQUARE NUMBERS

In this section, we will show how radicals can sometimes be simplified by factoring out perfect-square numbers.

31. We can factor a radical by reversing the procedure for multiplying. That is:

$\sqrt{6} = \sqrt{2} \cdot \sqrt{3}$ $\sqrt{35} = \sqrt{5} \cdot \sqrt{7}$

Following the examples, complete these.

a) $\sqrt{14} = \sqrt{} \cdot \sqrt{}$ b) $\sqrt{15} = \sqrt{} \cdot \sqrt{}$

---

32. A radical is not factored when the radicand can only be factored into itself and "1". For example:

$\sqrt{2}$ is not factored, since 2 is only equal to (2)(1).

$\sqrt{11}$ is not factored, since 11 is only equal to (11)(1).

Which of the following would not be factored? _____

a) $\sqrt{7}$ b) $\sqrt{21}$ c) $\sqrt{13}$ d) $\sqrt{34}$

a) $\sqrt{2} \cdot \sqrt{7}$
b) $\sqrt{3} \cdot \sqrt{5}$

---

33. When a radical factor contains a perfect square, we can simplify. For example:

$\sqrt{12} = \sqrt{4} \cdot \sqrt{3} = 2\sqrt{3}$ $\sqrt{50} = \sqrt{25} \cdot \sqrt{2} = 5\sqrt{2}$

The process above is called "simplifying by factoring out a perfect square." Using that process, complete these.

a) $\sqrt{700} = \sqrt{100} \cdot \sqrt{7} =$ _____   b) $\sqrt{27} = \sqrt{9} \cdot \sqrt{3} =$ _____

(a) and (c)

34. When the radicand can be factored in various ways, we always look for a perfect-square factor in order to simplify. That is:

   Instead of $\sqrt{18} = \sqrt{6} \cdot \sqrt{3}$, we use $\sqrt{18} = \sqrt{9} \cdot \sqrt{2} = 3\sqrt{2}$
   Instead of $\sqrt{28} = \sqrt{2} \cdot \sqrt{14}$, we use $\sqrt{28} = \sqrt{4} \cdot \sqrt{7} = 2\sqrt{7}$

   Simplify each of these by factoring out a perfect square.

   a) $\sqrt{8}$ = _____   c) $\sqrt{45}$ = _____
   b) $\sqrt{20}$ = _____   d) $\sqrt{75}$ = _____

   a) $10\sqrt{7}$
   b) $3\sqrt{3}$

35. Some numbers have more than one perfect-square factor. For example, 32 has 4 and 16; 72 has 4, 9, and 36. When a number of that type is the radicand, we can simplify the radical completely in one step only by factoring out the largest perfect square. We get:

   $\sqrt{32} = \sqrt{16} \cdot \sqrt{2} = 4\sqrt{2}$ $\sqrt{72} = \sqrt{36} \cdot \sqrt{2} = 6\sqrt{2}$

   If we do not factor out the largest perfect square, the remaining radical (like $\sqrt{8}$ and $\sqrt{18}$ below) will still contain a perfect-square factor.

   $\sqrt{32} = \sqrt{4} \cdot \sqrt{8} = 2\sqrt{8}$ $\sqrt{72} = \sqrt{4} \cdot \sqrt{18} = 2\sqrt{18}$

   Therefore, to simplify completely, we have to factor again. We get:

   $\sqrt{32} = 2\sqrt{8} = 2\sqrt{4} \cdot \sqrt{2} = 2 \cdot 2\sqrt{2} = 4\sqrt{2}$
   $\sqrt{72} = 2\sqrt{18} = 2\sqrt{9} \cdot \sqrt{2} =$ _____

   a) $2\sqrt{2}$ c) $3\sqrt{5}$
   b) $2\sqrt{5}$ d) $5\sqrt{3}$

36. Simplify completely. Try to do so in one step by factoring out the largest perfect square.

   a) $\sqrt{48}$ = _____   c) $\sqrt{80}$ = _____
   b) $\sqrt{200}$ = _____   d) $\sqrt{300}$ = _____

   $2 \cdot 3\sqrt{2} = 6\sqrt{2}$

37. We can use the same procedure to factor a numerical coefficient out of a term. For example:

   $\sqrt{5x} = \sqrt{5} \cdot \sqrt{x}$ $\sqrt{7ab} = \sqrt{7} \cdot \sqrt{ab}$

   If the coefficient is a perfect square, we can simplify after factoring. For example:

   $\sqrt{9t} = \sqrt{9} \cdot \sqrt{t} = 3\sqrt{t}$ $\sqrt{49pq} = \sqrt{49} \cdot \sqrt{pq} =$ _____

   a) $4\sqrt{3}$ c) $4\sqrt{5}$
   b) $10\sqrt{2}$ d) $10\sqrt{3}$

   $7\sqrt{pq}$

118   Square Root Radicals

38. If the coefficient is not a perfect square, we factor out a perfect square if possible. The remaining factors are left under a single radical sign.
For example:

$$\sqrt{12t} = \sqrt{4} \cdot \sqrt{3t} = 2\sqrt{3t}$$
$$\sqrt{50xy} = \sqrt{25} \cdot \sqrt{2xy} = 5\sqrt{2xy}$$

Simplify by factoring out the largest possible perfect square.

a) $\sqrt{8y}$ = _____   c) $\sqrt{27pq}$ = _____

b) $\sqrt{32b}$ = _____   d) $\sqrt{24SV}$ = _____

---

39. Since the radicands below do not contain a perfect-square factor, the radicals cannot be simplified.

$$\sqrt{14} \qquad \sqrt{33x} \qquad \sqrt{58cd}$$

Simplify each radical if possible.

a) $\sqrt{40}$ = _____   c) $\sqrt{21y}$ = _____

b) $\sqrt{55}$ = _____   d) $\sqrt{45pq}$ = _____

a) $2\sqrt{2y}$ c) $3\sqrt{3pq}$
b) $4\sqrt{2b}$ d) $2\sqrt{6SV}$

---

40. When multiplying radicals, we simplify the product (if possible) by factoring out the largest perfect square. For example:

$$\sqrt{2} \cdot \sqrt{6} = \sqrt{12} = 2\sqrt{3}$$
$$\sqrt{5x} \cdot \sqrt{40y} = \sqrt{200xy} = 10\sqrt{2xy}$$

Following the examples, multiply and simplify each product.

a) $\sqrt{5} \cdot \sqrt{10}$ = _____   b) $\sqrt{2} \cdot \sqrt{8t}$ = _____

a) $2\sqrt{10}$
b) Not possible
c) Not possible
d) $3\sqrt{5pq}$

---

41. When multiplying radicals, it is not always possible to simplify the product. Some examples are shown.

$$\sqrt{2} \cdot \sqrt{3} = \sqrt{6} \qquad \sqrt{3a} \cdot \sqrt{7b} = \sqrt{21ab}$$

Multiply and simplify each product if possible.

a) $\sqrt{2} \cdot \sqrt{32}$ = _____   c) $\sqrt{2x} \cdot \sqrt{14y}$ = _____

b) $\sqrt{5x} \cdot \sqrt{2}$ = _____   d) $\sqrt{8} \cdot \sqrt{6t}$ = _____

a) $5\sqrt{2}$
b) $4\sqrt{t}$

---

a) 8
b) $\sqrt{10x}$
c) $2\sqrt{7xy}$
d) $4\sqrt{3t}$

3-5 SIMPLIFYING RADICALS BY FACTORING OUT PERFECT-SQUARE POWERS

In this section, we will show how radicals can sometimes be simplified by factoring out perfect-square powers.

42. Since we square a power by multiplying its exponent by 2, the square of any power has an <u>even</u> exponent. For example:

    $(x^1)^2 = x^2$ $(x^2)^2 = x^4$ $(x^5)^2 = x^{10}$

    Therefore, powers with <u>even</u> exponents are called "<u>perfect-square</u>" powers. We can find the square root of any perfect-square power. That is:

    Since $y^3 \cdot y^3 = y^6$, $\sqrt{y^6} = y^3$
    Since $t^6 \cdot t^6 = t^{12}$, $\sqrt{t^{12}} = $ _____

43. The two square roots from the last frame are shown below.

    $\sqrt{y^6} = y^3$ $\sqrt{t^{12}} = t^6$

    To find the square root of a perfect-square power, we divide the exponent by 2. That is:

    $\sqrt{y^6} = y^{\frac{6}{2}} = y^3$ $\sqrt{t^{12}} = t^{\frac{12}{2}} = t^6$

    By dividing the exponent by 2, find each square root.

    a) $\sqrt{m^2} = $ _____   b) $\sqrt{p^8} = $ _____   c) $\sqrt{b^{20}} = $ _____

    | t^6 |

44. Radicals containing perfect-square powers can also be simplifed by factoring out perfect squares. For example:

    $\sqrt{4x^2} = \sqrt{4} \cdot \sqrt{x^2} = 2x$
    $\sqrt{p^4 q^8} = \sqrt{p^4} \cdot \sqrt{q^8} = p^2 q^4$

    Following the examples, simplify these.

    a) $\sqrt{25d^6} = \sqrt{25} \cdot \sqrt{d^6} = $ _____   b) $\sqrt{a^2 b^{10}} = \sqrt{a^2} \cdot \sqrt{b^{10}} = $ _____

    a) m^1 or m
    b) p^4
    c) b^{10}

45. The radicals below have been simplified by factoring out perfect squares.

    $\sqrt{5x^4} = \sqrt{5} \cdot \sqrt{x^4} = x^2\sqrt{5}$
    $\sqrt{7t^{10}v^2} = \sqrt{7} \cdot \sqrt{t^{10}} \cdot \sqrt{v^2} = t^5 v\sqrt{7}$

    Following the examples, simplify these.

    a) $\sqrt{3y^8} = \sqrt{3} \cdot \sqrt{y^8} = $ _____   b) $\sqrt{cd^{12}} = \sqrt{c} \cdot \sqrt{d^{12}} = $ _____

    a) $5d^3$
    b) ab^5

a) $y^4\sqrt{3}$ b) $d^6\sqrt{c}$

46. The radicals below have been simplified by factoring out perfect squares.

$$\sqrt{12y^6} = \sqrt{4} \cdot \sqrt{3} \cdot \sqrt{y^6} = 2y^3\sqrt{3}$$
$$\sqrt{50ax^2} = \sqrt{25} \cdot \sqrt{2a} \cdot \sqrt{x^2} = 5x\sqrt{2a}$$

Following the examples, simplify these.

a) $\sqrt{18t^4} = \sqrt{9} \cdot \sqrt{2} \cdot \sqrt{t^4} = $ _____

b) $\sqrt{24m^{10}v^8} = \sqrt{4} \cdot \sqrt{6} \cdot \sqrt{m^{10}} \cdot \sqrt{v^8} = $ _____

---

47. Simplify by factoring out as many perfect squares as possible.

a) $\sqrt{49x^2y^4} = $ _____   c) $\sqrt{28a^2} = $ _____

b) $\sqrt{11y^8} = $ _____   d) $\sqrt{32h^6t} = $ _____

a) $3t^2\sqrt{2}$
b) $2m^5v^4\sqrt{6}$

---

48. We can factor any non-perfect-square power whose exponent is greater than 2 into a perfect-square power and a first power. That is:

$$x^3 = x^2 \cdot x^1 \qquad t^7 = t^6 \cdot t^1$$

Using the factoring process above, we simplified each radical below by factoring out the largest perfect square.

$$\sqrt{y^5} = \sqrt{y^4 \cdot y^1} = \sqrt{y^4} \cdot \sqrt{y^1} = y^2\sqrt{y}$$
$$\sqrt{p^{11}} = \sqrt{p^{10} \cdot p^1} = \sqrt{p^{10}} \cdot \sqrt{p^1} = p^5\sqrt{p}$$

Simplify these by factoring out the largest perfect square.

a) $\sqrt{x^3} = $ _____   b) $\sqrt{m^9} = $ _____   c) $\sqrt{b^{13}} = $ _____

a) $7xy^2$
b) $y^4\sqrt{11}$
c) $2a\sqrt{7}$
d) $4h^3\sqrt{2t}$

---

49. We simplified each radical below by factoring out perfect squares.

$$\sqrt{16x^3} = \sqrt{16} \cdot \sqrt{x^2} \cdot \sqrt{x^1} = 4x\sqrt{x}$$
$$\sqrt{a^4b^5} = \sqrt{a^4} \cdot \sqrt{b^4} \cdot \sqrt{b^1} = a^2b^2\sqrt{b}$$

Following the examples, complete these.

a) $\sqrt{9t^7} = \sqrt{9} \cdot \sqrt{t^6} \cdot \sqrt{t^1} = $ _____

b) $\sqrt{81c^8d^9} = \sqrt{81} \cdot \sqrt{c^8} \cdot \sqrt{d^8} \cdot \sqrt{d^1} = $ _____

a) $x\sqrt{x}$
b) $m^4\sqrt{m}$
c) $b^6\sqrt{b}$

---

50. We simplified below by factoring out perfect squares.

$$\sqrt{3x^5} = \sqrt{x^4} \cdot \sqrt{3x^1} = x^2\sqrt{3x}$$
$$\sqrt{a^7b} = \sqrt{a^6} \cdot \sqrt{a^1b} = a^3\sqrt{ab}$$

a) $3t^3\sqrt{t}$
b) $9c^4d^4\sqrt{d}$

Continued on following page.

Square Root Radicals   121

50. Continued

Following the examples, complete these.

a) $\sqrt{7t^3} = \sqrt{t^2} \cdot \sqrt{7t^1} =$ _____

b) $\sqrt{2xy^9} = \sqrt{y^8} \cdot \sqrt{2xy^1} =$ _____

---

51. We simplified below by factoring out perfect squares.

$\sqrt{8x^3} = \sqrt{4} \cdot \sqrt{x^2} \cdot \sqrt{2x^1} = 2x\sqrt{2x}$

$\sqrt{45a^5y^7} = \sqrt{9} \cdot \sqrt{a^4} \cdot \sqrt{y^6} \cdot \sqrt{5a^1y^1} = 3a^2y^3\sqrt{5ay}$

Following the examples, complete these.

a) $\sqrt{27p^{11}} = \sqrt{9} \cdot \sqrt{p^{10}} \cdot \sqrt{3p^1} =$ _____

b) $\sqrt{50b^3d^9} = \sqrt{25} \cdot \sqrt{b^2} \cdot \sqrt{d^8} \cdot \sqrt{2b^1d^1} =$ _____

a) $t\sqrt{7t}$

b) $y^4\sqrt{2xy}$

---

52. Simplify by factoring out as many perfect squares as possible.

a) $\sqrt{64t^3} =$ _____   c) $\sqrt{40x^7} =$ _____

b) $\sqrt{14bd^5} =$ _____   d) $\sqrt{75a^3y^{11}} =$ _____

a) $3p^5\sqrt{3p}$

b) $5bd^4\sqrt{2bd}$

---

53. After multiplying below, we simplified each product by factoring out perfect squares.

$\sqrt{2x} \cdot \sqrt{10x} = \sqrt{20x^2} = 2x\sqrt{5}$

$\sqrt{3ab^3} \cdot \sqrt{12ab^2} = \sqrt{36a^2b^5} = 6ab^2\sqrt{b}$

Following the examples, multiply and simplify each product.

a) $\sqrt{5y^2} \cdot \sqrt{6y^4} =$ _____   b) $\sqrt{3x} \cdot \sqrt{6x^2} =$ _____

a) $8t\sqrt{t}$

b) $d^2\sqrt{14bd}$

c) $2x^3\sqrt{10x}$

d) $5ay^5\sqrt{3ay}$

---

54. Multiply and simplify each product.

a) $\sqrt{20p^3q^2} \cdot \sqrt{5p^5q^2} =$ _____

b) $\sqrt{ab^4} \cdot \sqrt{8a^2b} =$ _____

a) $y^3\sqrt{30}$

b) $3x\sqrt{2x}$

---

a) $10p^4q^2$

b) $2ab^2\sqrt{2ab}$

122 Square Root Radicals

3-6 DIVIDING RADICALS

In this section, we will discuss the procedure for dividing square root radicals.

55. To indicate a division of radicals, we use a fraction. For example:

$\dfrac{\sqrt{36}}{\sqrt{9}}$ means "divide $\sqrt{36}$ by $\sqrt{9}$".

Any division of two radicals is equal to the square root of a fraction. That is:

$\dfrac{\sqrt{36}}{\sqrt{9}} = \sqrt{\dfrac{36}{9}}$ $\qquad$ $\dfrac{\sqrt{64}}{\sqrt{4}} = \sqrt{\dfrac{64}{4}}$

To show that the statements above are true, we evaluated both sides below.

$\dfrac{\sqrt{36}}{\sqrt{9}} = \sqrt{\dfrac{36}{9}}$ $\qquad$ $\dfrac{\sqrt{64}}{\sqrt{4}} = \sqrt{\dfrac{64}{4}}$

$\dfrac{6}{3} = \sqrt{4}$ $\qquad$ $\dfrac{8}{2} = \sqrt{16}$

$2 = 2$ $\qquad$ $4 = 4$

56. To divide radicals, we convert to the square root of a fraction and simplify. For example:

$\dfrac{\sqrt{50}}{\sqrt{2}} = \sqrt{\dfrac{50}{2}} = \sqrt{25} = 5$ $\qquad$ $\dfrac{\sqrt{27}}{\sqrt{3}} = \sqrt{\dfrac{27}{3}} = \sqrt{} = \underline{}$

57. When the radicand of the quotient is not a perfect square, we simplify it as much as possible by factoring out perfect squares. For example:

$\dfrac{\sqrt{25}}{\sqrt{5}} = \sqrt{\dfrac{25}{5}} = \sqrt{5}$ $\qquad$ $\dfrac{\sqrt{36}}{\sqrt{3}} = \sqrt{\dfrac{36}{3}} = \sqrt{12} = 2\sqrt{3}$

Simplify each quotient as much as possible.

a) $\dfrac{\sqrt{20}}{\sqrt{2}} = \sqrt{\dfrac{20}{2}} = \underline{}$

b) $\dfrac{\sqrt{150}}{\sqrt{3}} = \sqrt{\dfrac{150}{3}} = \underline{}$

58. The same procedure is used for divisions of radicals containing powers. For example:

$\dfrac{\sqrt{18x^5}}{\sqrt{2x^3}} = \sqrt{\dfrac{18x^5}{2x^3}} = \sqrt{9x^2} = 3x$

Continued on following page.

$\sqrt{9} = 3$

a) $\sqrt{10}$

b) $5\sqrt{2}$, from: $\sqrt{50}$

Square Root Radicals   123

58. Continued

Following the example, complete these.

a) $\dfrac{\sqrt{c^7 d^9}}{\sqrt{c^3 d^3}} = \sqrt{\dfrac{c^7 d^9}{c^3 d^3}} =$ _____

b) $\dfrac{\sqrt{75 y^{10}}}{\sqrt{3 y^6}} = \sqrt{\dfrac{75 y^{10}}{3 y^6}} =$ _____

---

59. In the division below, we simplified the quotient as much as possible.

$$\dfrac{\sqrt{40 y^4}}{\sqrt{2y}} = \sqrt{\dfrac{40 y^4}{2y}} = \sqrt{20 y^3} = 2y\sqrt{5y}$$

Simplify each quotient as much as possible.

a) $\dfrac{\sqrt{6 x^6}}{\sqrt{2 x^4}} =$ _____

b) $\dfrac{\sqrt{p^6 q^8}}{\sqrt{pq}} =$ _____

a) $c^2 d^3$, from: $\sqrt{c^4 d^6}$

b) $5y^2$, from: $\sqrt{25 y^4}$

---

60. After converting a division of two radicals to the square root of a fraction, we cannot always simplify. For example:

$$\dfrac{\sqrt{37}}{\sqrt{11}} = \sqrt{\dfrac{37}{11}} \qquad \dfrac{\sqrt{cd}}{\sqrt{V}} = \underline{\qquad}$$

a) $x\sqrt{3}$, from: $\sqrt{3 x^2}$

b) $p^2 q^3 \sqrt{pq}$, from: $\sqrt{p^5 q^7}$

$\sqrt{\dfrac{cd}{V}}$

SELF-TEST 8 (pages 109-124)

2. Find the principal square root of 0.09.

4. In $\boxed{b = \sqrt{c^2 - a^2}}$, find "b" when c = 15 and a = 12.

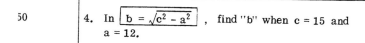

7. $3p\sqrt{2m} \cdot 2w\sqrt{s}$

124   Square Root Radicals

SELF-TEST 8 (pages 109-124) - Continued		
8. Convert $b\sqrt{d}$ to a single radical.		
Simplify each radical by factoring out all perfect squares.		
9. $\sqrt{80}$	10. $\sqrt{54x^3}$	11. $\sqrt{75a^2b^5}$
Multiply and simplify each product by factoring out all perfect squares.		Divide and simplify each quotient by factoring out all perfect squares.
12. $\sqrt{6y} \cdot \sqrt{12}$		14. $\dfrac{\sqrt{60}}{\sqrt{3}}$
13. $\sqrt{2r^3} \cdot \sqrt{32s^4}$		15. $\dfrac{\sqrt{8x^3y^2}}{\sqrt{2xy}}$

ANSWERS:
1. +12 and -12
2. +0.3
3. H = 10
4. b = 9
5. $\sqrt{6rt}$
6. $\sqrt{10xy}$
7. $6pw\sqrt{2ms}$
8. $\sqrt{b^2d}$
9. $4\sqrt{5}$
10. $3x\sqrt{6x}$
11. $5ab^2\sqrt{3b}$
12. $6\sqrt{2y}$
13. $8rs^2\sqrt{r}$
14. $2\sqrt{5}$
15. $2x\sqrt{y}$

3-7   SIMPLIFYING RADICALS CONTAINING A FRACTION

In this section, we will discuss the procedure for simplifying square root radicals containing a fraction.

61. To simplify a radical containing a fraction, we can reverse the procedure for dividing radicals. That is, we can convert the radical to a division of two radicals and then simplify both terms. For example: $$\sqrt{\dfrac{9}{25}} = \dfrac{\sqrt{9}}{\sqrt{25}} = \dfrac{3}{5} \qquad \sqrt{\dfrac{a^4}{b^6}} = \dfrac{\sqrt{a^4}}{\sqrt{b^6}} = \underline{}$$	
62. After converting to a division of two radicals, we factor out the perfect squares in each term. For example: $$\sqrt{\dfrac{5y^2}{4}} = \dfrac{\sqrt{5y^2}}{\sqrt{4}} = \dfrac{y\sqrt{5}}{2} \qquad \sqrt{\dfrac{x^8}{ay}} = \dfrac{\sqrt{x^8}}{\sqrt{ay}} = \dfrac{x^4}{\sqrt{ay}}$$ Factor out the perfect squares in these. a) $\sqrt{\dfrac{3p^2}{16q^4}} = \dfrac{\sqrt{3p^2}}{\sqrt{16q^4}} = \underline{}$ b) $\sqrt{\dfrac{c^2q^2}{4d}} = \dfrac{\sqrt{c^2q^2}}{\sqrt{4d}} = \underline{}$	$\dfrac{a^2}{b^3}$
	a) $\dfrac{p\sqrt{3}}{4q^2}$ b) $\dfrac{cq}{2\sqrt{d}}$

Square Root Radicals   125

63. Following the example, simplify the other radical.

$$\sqrt{\frac{18x}{y^2}} = \frac{\sqrt{18x}}{\sqrt{y^2}} = \frac{3\sqrt{2x}}{y} \qquad \sqrt{\frac{7a}{50b^4}} = \frac{\sqrt{7a}}{\sqrt{50b^4}} = \underline{}$$

---

64. When simplifying a radical, remember that "1" is a perfect square. For example:

$$\sqrt{\frac{1}{4}} = \frac{\sqrt{1}}{\sqrt{4}} = \frac{1}{2} \qquad \sqrt{\frac{1}{5x}} = \frac{\sqrt{1}}{\sqrt{5x}} = \underline{}$$

$\dfrac{\sqrt{7a}}{5b^2\sqrt{2}}$

---

65. Simplify each radical.

a) $\sqrt{\dfrac{25x^2}{16y^4}} = $ _____   c) $\sqrt{\dfrac{cd^3}{7f^5}} = $ _____

b) $\sqrt{\dfrac{1}{m^2b^6}} = $ _____   d) $\sqrt{\dfrac{1}{a^3b^7}} = $ _____

$\dfrac{1}{\sqrt{5x}}$

---

66. After factoring out perfect squares, we sometimes obtain a fraction with a radical in each term. Usually, this division of two radicals is converted back to a single radical. For example:

$$\frac{a\sqrt{b}}{\sqrt{c}} \text{ is converted to } a\sqrt{\frac{b}{c}}$$

$$\frac{5\sqrt{m}}{6\sqrt{t}} \text{ is converted to } \frac{5}{6}\sqrt{\frac{m}{t}}$$

Following the examples, convert each division of two radicals to the square root of a fraction.

a) $\dfrac{4\sqrt{x}}{3\sqrt{y}} = $ _____   b) $\dfrac{cd\sqrt{S}}{\sqrt{R}} = $ _____

a) $\dfrac{5x}{4y^2}$

b) $\dfrac{1}{mb^3}$

c) $\dfrac{d\sqrt{cd}}{f^2\sqrt{7f}}$

d) $\dfrac{1}{ab^3\sqrt{ab}}$

---

67. To convert the division of two radicals below back to a single radical, we inserted a "1" in the numerator.

$$\frac{\sqrt{x}}{a\sqrt{y}} = \frac{1\sqrt{x}}{a\sqrt{y}} = \frac{1}{a}\sqrt{\frac{x}{y}}$$

Following the example, convert each division of two radicals to a single radical.

a) $\dfrac{\sqrt{cd}}{m\sqrt{t}} = $ _____   b) $\dfrac{\sqrt{b}}{3\sqrt{c}} = $ _____

a) $\dfrac{4}{3}\sqrt{\dfrac{x}{y}}$

b) $cd\sqrt{\dfrac{S}{R}}$

---

68. Simplify and convert any division of two radicals to a single radical.

a) $\sqrt{\dfrac{x^2y}{d}} = $ _____

b) $\sqrt{\dfrac{a}{5b^3}} = $ _____

a) $\dfrac{1}{m}\sqrt{\dfrac{cd}{t}}$

b) $\dfrac{1}{3}\sqrt{\dfrac{b}{c}}$

126 Square Root Radicals

69. Before simplifying a radical containing a fraction, <u>the fraction should always be reduced to lowest terms</u>. For example:

$$\sqrt{\frac{5c^2}{10d}} = \sqrt{\frac{c^2}{2d}} = \frac{\sqrt{c^2}}{\sqrt{2d}} = \frac{c}{\sqrt{2d}}$$

$$\sqrt{\frac{ax^2y^2}{bt^4y^2}} = \sqrt{\frac{ax^2}{bt^4}} = \frac{\sqrt{ax^2}}{\sqrt{bt^4}} = \frac{x\sqrt{a}}{t^2\sqrt{b}} = \frac{x}{t^2}\sqrt{\frac{a}{b}}$$

Following the examples, simplify these:

a) $\sqrt{\dfrac{8mb}{2m}} = $ _____

b) $\sqrt{\dfrac{b^3 V}{ab^3 T^2}} = $ _____

a) $x\sqrt{\dfrac{y}{d}}$

b) $\dfrac{1}{b}\sqrt{\dfrac{a}{5b}}$

70. Notice how we reduced each fraction to lowest terms before simplifying.

$$\sqrt{\frac{c^2 t^2}{dt}} = \sqrt{\frac{c^2 t}{d}} = \sqrt{\frac{c^2 t}{d}} = \frac{c\sqrt{t}}{\sqrt{d}} = c\sqrt{\frac{t}{d}}$$

$$\sqrt{\frac{b}{ab^2}} = \sqrt{\frac{1}{ab}} = \sqrt{\frac{1}{ab}} = \frac{1}{\sqrt{ab}}$$

Following the examples, simplify these:

a) $\sqrt{\dfrac{px^2}{q^4 x}} = $ _____

b) $\sqrt{\dfrac{c^2 y}{y^2}} = $ _____

a) $2\sqrt{b}$

b) $\dfrac{1}{T}\sqrt{\dfrac{V}{a}}$

71. Reduce each fraction to lowest terms and then simplify if possible.

a) $\sqrt{\dfrac{3xy^2}{9xy}} = $ _____

b) $\sqrt{\dfrac{8a^4 c}{4b^2 c^2}} = $ _____

a) $\dfrac{\sqrt{px}}{q^2}$

b) $\dfrac{c}{\sqrt{y}}$

72. When simplifying a radical, we sometimes get a non-radical expression. For example:

$$\sqrt{\frac{px^4}{py^6}} = \sqrt{\frac{x^4}{y^6}} = \frac{\sqrt{x^4}}{\sqrt{y^6}} = \frac{x^2}{y^3}$$

Following the example, simplify these.

a) $\sqrt{\dfrac{5m^2}{5}} = $ _____

b) $\sqrt{\dfrac{2y}{8x^2 y}} = $ _____

a) $\sqrt{\dfrac{y}{3}}$

b) $\dfrac{a^2}{b}\sqrt{\dfrac{2}{c}}$

a) m   b) $\dfrac{1}{2x}$

Square Root Radicals   127

73. After factoring perfect squares out of a radical in one term of a fraction, always check to see that the fraction is reduced to lowest terms. For example:

$$\frac{\sqrt{x^2 y}}{3xy} = \frac{\cancel{x}\sqrt{y}}{3\cancel{x}y} = \frac{\sqrt{y}}{3y}$$

Factor out perfect squares and then reduce to lowest terms.

a) $\dfrac{ab}{\sqrt{a^2 b}} =$ _____

b) $\dfrac{S}{\sqrt{S^2 T}} =$ _____

74. Following the example, reduce the other fractions to lowest terms.

$$\frac{c\sqrt{y}}{d\sqrt{y}} = \frac{c}{d}$$ a) $\dfrac{a\sqrt{x}}{2b\sqrt{x}} =$ _____   b) $\dfrac{\sqrt{TV}}{3\sqrt{TV}} =$ _____

a) $\dfrac{b}{\sqrt{b}}$

b) $\dfrac{1}{\sqrt{T}}$

a) $\dfrac{a}{2b}$ b) $\dfrac{1}{3}$

3-8 SIMPLIFYING RADICALS CONTAINING INDICATED SQUARES

In this section, we will discuss the procedure for simplifying radicals containing an indicated square.

75. To indicate the square of a monomial containing more than one factor, we put parentheses around the monomial. For example:

$(3x)^2$ means "square $3x$".

To square any monomial, we multiply the monomial by itself. That is:

$(3x)^2 = (3x)(3x) = 9x^2$ $(ab)^2 = (ab)(ab) =$ _____

76. The square root of the indicated square of a monomial is the monomial itself. For example:

$\sqrt{(5y)^2} = 5y$, since $\sqrt{(5y)^2} = \sqrt{25y^2} = 5y$

$\sqrt{(pq)^2} = pq$, since $\sqrt{(pq)^2} = \sqrt{p^2 q^2} = pq$

Find each square root.

a) $\sqrt{(2x)^2} =$ _____   b) $\sqrt{(MV)^2} =$ _____   c) $\sqrt{(7ab)^2} =$ _____

$a^2 b^2$

a) $2x$

b) MV

c) $7ab$

128   Square Root Radicals

77. Simplify. Be careful because some radicals are not indicated squares.

   a) $\sqrt{(4x)^2}$ = _____   c) $\sqrt{(36t)^2}$ = _____

   b) $\sqrt{4x^2}$ = _____   d) $\sqrt{36t^2}$ = _____

---

78. To indicate the square of a fraction, we put parentheses around the fraction. For example:

   $\left(\dfrac{x}{y}\right)^2$ means "square $\dfrac{x}{y}$."

   To square a fraction, we multiply the fraction by itself. That is:

   $\left(\dfrac{1}{3}\right)^2 = \left(\dfrac{1}{3}\right)\left(\dfrac{1}{3}\right) = \dfrac{1}{9}$ $\left(\dfrac{x}{y}\right)^2 = \left(\dfrac{x}{y}\right)\left(\dfrac{x}{y}\right) =$ _____

   a) 4x   c) 36t
   b) 2x   d) 6t

---

79. Squaring a fraction is the same as squaring both its numerator and its denominator. For example:

   $\left(\dfrac{3}{4}\right)^2 = \dfrac{3^2}{4^2} = \dfrac{9}{16}$ $\left(\dfrac{6a}{5b}\right)^2 = \dfrac{(6a)^2}{(5b)^2} = \dfrac{36a^2}{25b^2}$

   Find the following squares.

   a) $\left(\dfrac{1}{8}\right)^2 =$ _____   b) $\left(\dfrac{5x}{7}\right)^2 =$ _____   c) $\left(\dfrac{x^2y}{a^3b}\right)^2 =$ _____

   $\dfrac{x^2}{y^2}$

---

80. The square root of the indicated square of a fraction is the fraction itself. For example:

   $\sqrt{\left(\dfrac{x}{y}\right)^2} = \dfrac{x}{y}$, since $\sqrt{\left(\dfrac{x}{y}\right)^2} = \sqrt{\dfrac{x^2}{y^2}} = \dfrac{x}{y}$

   $\sqrt{\left(\dfrac{3a}{2b}\right)^2} = \dfrac{3a}{2b}$, since $\sqrt{\left(\dfrac{3a}{2b}\right)^2} = \sqrt{\dfrac{9a^2}{4b^2}} = \dfrac{3a}{2b}$

   Find each square root.

   a) $\sqrt{\left(\dfrac{1}{5}\right)^2} =$ _____   b) $\sqrt{\left(\dfrac{7m}{9}\right)^2} =$ _____

   a) $\dfrac{1}{64}$

   b) $\dfrac{25x^2}{49}$

   c) $\dfrac{x^4y^2}{a^6b^2}$

---

81. Simplify. Be careful because some radicals are not indicated squares.

   a) $\sqrt{\left(\dfrac{p}{q}\right)^2} =$ _____   c) $\sqrt{\left(\dfrac{7a}{4b}\right)^2} =$ _____

   b) $\sqrt{\dfrac{p^2}{q}} =$ _____   d) $\sqrt{\dfrac{7a^2}{4b}} =$ _____

   a) $\dfrac{1}{5}$

   b) $\dfrac{7m}{9}$

---

a) $\dfrac{p}{q}$ c) $\dfrac{7a}{4b}$

b) $\dfrac{p}{\sqrt{q}}$ d) $\dfrac{a}{2}\sqrt{\dfrac{7}{b}}$

3-9 SQUARING RADICALS

In this section, we will discuss the procedure for squaring square root radicals.

82. To indicate the square of a radical, we put parentheses around the radical. For example:

$(\sqrt{7})^2$ means "square $\sqrt{7}$".

To square a radical, we multiply the radical by itself. That is:

$(\sqrt{7})^2 = \sqrt{7} \cdot \sqrt{7}$ $(\sqrt{xy})^2 = \sqrt{} \cdot \sqrt{}$

83. The square of a radical is the radicand. For example:

$(\sqrt{5})^2 = 5$, since $(\sqrt{5})^2 = \sqrt{5} \cdot \sqrt{5} = \sqrt{25} = 5$

$(\sqrt{3y})^2 = 3y$, since $(\sqrt{3y})^2 = \sqrt{3y} \cdot \sqrt{3y} = \sqrt{9y^2} = 3y$

Write each square.

a) $(\sqrt{10})^2 = $ _____   b) $(\sqrt{9x})^2 = $ _____   c) $(\sqrt{PV})^2 = $ _____

Answer: $\sqrt{xy} \cdot \sqrt{xy}$

84. The square of a radical is also the radicand when the radicand is a fraction. For example:

$\left(\sqrt{\dfrac{x}{3}}\right)^2 = \dfrac{x}{3}$, since $\left(\sqrt{\dfrac{x}{3}}\right)^2 = \sqrt{\dfrac{x}{3}} \cdot \sqrt{\dfrac{x}{3}} = \sqrt{\dfrac{x^2}{9}} = \dfrac{x}{3}$

Write each square.

a) $\left(\sqrt{\dfrac{1}{5}}\right)^2 = $ _____   b) $\left(\sqrt{\dfrac{7}{y}}\right)^2 = $ _____   c) $\left(\sqrt{\dfrac{ab}{cd}}\right)^2 = $ _____

Answers:
a) 10
b) 9x
c) PV

85. Multiplying a radical by itself is the same as squaring the radical. For example:

$\sqrt{x} \cdot \sqrt{x} = (\sqrt{x})^2 = x$ $\sqrt{\dfrac{y}{4}} \cdot \sqrt{\dfrac{y}{4}} = \left(\sqrt{\dfrac{y}{4}}\right)^2 = \dfrac{y}{4}$

Complete these:

a) $\sqrt{7} \cdot \sqrt{7} = $ _____   b) $\sqrt{2t} \cdot \sqrt{2t} = $ _____

Answers:
a) $\dfrac{1}{5}$
b) $\dfrac{7}{y}$
c) $\dfrac{ab}{cd}$

86. All three expressions below are equivalent because each equals 3x.

$(\sqrt{3x})^2$ $\sqrt{(3x)^2}$ $\sqrt{3x} \cdot \sqrt{3x}$

Complete these:

a) $\sqrt{V_1} \cdot \sqrt{V_1} = $ _____   b) $\sqrt{(140)^2} = $ _____   c) $(\sqrt{2xy})^2 = $ _____

Answers:
a) 7
b) 2t

a) V_1 b) 140   c) 2xy

Square Root Radicals 129

130 Square Root Radicals

3-10 RATIONALIZING DENOMINATORS

When a fraction contains a radical in its denominator, it can be converted to an equivalent form <u>without a radical</u> in the denominator. For example:

$$\frac{3}{\sqrt{7}} \text{ can be converted to } \frac{3\sqrt{7}}{7}$$

The conversion process is called "<u>rationalizing the denominator</u>". We will discuss that process in this section.

---

87. The denominator of a fraction is "rationalized" only if it <u>does not contain a radical</u>. That is:

    $\frac{\sqrt{2}}{3}$ and $\frac{a\sqrt{b}}{c}$ <u>are</u> rationalized.

    $\frac{2}{\sqrt{3}}$ and $\frac{1}{x\sqrt{y}}$ <u>are not</u> rationalized.

    Which fractions below have denominators that <u>are</u> rationalized? _____

    a) $\frac{a}{\sqrt{b}}$ b) $\frac{\sqrt{c}}{d}$ c) $\frac{2\sqrt{x}}{5}$ d) $\frac{7}{9\sqrt{y}}$

---

88. The procedure for rationalizing a denominator that contains a radical is based on these facts.

    1) Any radical divided by itself equals "1".

    $\frac{\sqrt{5}}{\sqrt{5}} = 1 \qquad \frac{\sqrt{x}}{\sqrt{x}} = 1 \qquad \frac{\sqrt{3t}}{\sqrt{3t}} = 1$

    2) To obtain equivalent forms of a fraction, we multiply by a fraction that equals "1".

    $\frac{2}{\sqrt{3}} = \frac{2}{\sqrt{3}}\left(\frac{\sqrt{5}}{\sqrt{5}}\right) = \frac{2\sqrt{5}}{\sqrt{15}} \qquad \frac{1}{\sqrt{a}} = \frac{1}{\sqrt{a}}\left(\frac{\sqrt{b}}{\sqrt{b}}\right) = \frac{\sqrt{b}}{\sqrt{ab}}$

    Find an equivalent form of each fraction by completing these:

    a) $\frac{8}{\sqrt{7}} = \frac{8}{\sqrt{7}}\left(\frac{\sqrt{2}}{\sqrt{2}}\right) =$ _____   b) $\frac{x}{\sqrt{y}} = \frac{x}{\sqrt{y}}\left(\frac{\sqrt{c}}{\sqrt{c}}\right) =$ _____

Answer (right column): Only (b) and (c)

---

89. When a denominator is "not rationalized", we can rationalize it <u>by multiplying by a fraction whose terms are identical to the radical in the denominator</u>.

    $\frac{5}{\sqrt{7}} = \frac{5}{\sqrt{7}}\left(\frac{\sqrt{7}}{\sqrt{7}}\right) = \frac{5\sqrt{7}}{(\sqrt{7})^2} = \frac{5\sqrt{7}}{7}$

    $\frac{1}{\sqrt{ab}} = \frac{1}{\sqrt{ab}}\left(\frac{\sqrt{ab}}{\sqrt{ab}}\right) = \frac{\sqrt{ab}}{(\sqrt{ab})^2} = \frac{\sqrt{ab}}{ab}$

Answers (right column):
a) $\frac{8\sqrt{2}}{\sqrt{14}}$
b) $\frac{x\sqrt{c}}{\sqrt{cy}}$

Continued on following page.

89. Continued

    Notice that we got rid of the radical in each denominator by squaring it. Rationalize each denominator below.

    a) $\dfrac{1}{\sqrt{15}} = $ _____

    b) $\dfrac{3}{\sqrt{2y}} = $ _____

90. Following the example, rationalize the other denominators.

    $$\dfrac{1}{5\sqrt{2}} = \dfrac{1}{5\sqrt{2}}\left(\dfrac{\sqrt{2}}{\sqrt{2}}\right) = \dfrac{\sqrt{2}}{5(2)} = \dfrac{\sqrt{2}}{10}$$

    a) $\dfrac{3}{8\sqrt{5}} = $ _____

    b) $\dfrac{x}{a\sqrt{y}} = $ _____

    a) $\dfrac{\sqrt{15}}{15}$

    b) $\dfrac{3\sqrt{2y}}{2y}$

91. Following the example, rationalize the other denominator.

    $$\dfrac{3}{4\sqrt{5x}} = \dfrac{3}{4\sqrt{5x}}\left(\dfrac{\sqrt{5x}}{\sqrt{5x}}\right) = \dfrac{3\sqrt{5x}}{4(5x)} = \dfrac{3\sqrt{5x}}{20x}$$

    $\dfrac{1}{2\sqrt{7x}} = $ _____

    a) $\dfrac{3\sqrt{5}}{40}$

    b) $\dfrac{x\sqrt{y}}{ay}$

92. Following the example, rationalize the other denominator.

    $$\dfrac{2\sqrt{3}}{\sqrt{5}} = \dfrac{2\sqrt{3}}{\sqrt{5}}\left(\dfrac{\sqrt{5}}{\sqrt{5}}\right) = \dfrac{2\sqrt{15}}{5}$$

    $\dfrac{7\sqrt{x}}{\sqrt{y}} = $ _____

    $\dfrac{\sqrt{7x}}{14x}$

93. When a denominator is rationalized, sometimes the new fraction can be reduced to lowest terms. For example:

    $$\dfrac{7}{2\sqrt{7}} = \dfrac{7}{2\sqrt{7}}\left(\dfrac{\sqrt{7}}{\sqrt{7}}\right) = \dfrac{\cancel{7}\sqrt{7}}{2(\cancel{7})} = \dfrac{\sqrt{7}}{2}$$

    $$\dfrac{x}{\sqrt{x}} = \dfrac{x}{\sqrt{x}}\left(\dfrac{\sqrt{x}}{\sqrt{x}}\right) = \dfrac{\cancel{x}\sqrt{x}}{\cancel{x}} = \sqrt{x}$$

    Rationalize each denominator and then reduce to lowest terms.

    a) $\dfrac{3}{\sqrt{3}} = $ _____

    b) $\dfrac{c}{d\sqrt{c}} = $ _____

    $\dfrac{7\sqrt{xy}}{y}$

132 Square Root Radicals

94. After simplifying a radical containing a fraction, we can frequently rationalize the denominator of the new fraction. For example:

$$\sqrt{\frac{c^2x}{d^2y}} = \frac{c\sqrt{x}}{d\sqrt{y}} = \frac{c\sqrt{x}}{d\sqrt{y}}\left(\frac{\sqrt{y}}{\sqrt{y}}\right) = \frac{c\sqrt{xy}}{dy}$$

Simplify and then rationalize the denominator of the new fraction.

a) $\sqrt{\dfrac{8}{27}} =$ _____

b) $\sqrt{\dfrac{m^2}{bt^4}} =$ _____

a) $\sqrt{3}$

b) $\dfrac{\sqrt{c}}{d}$

---

95. Even when we cannot simplify a radical containing a fraction, we can still rationalize its denominator. For example:

$$\sqrt{\frac{5}{7}} = \frac{\sqrt{5}}{\sqrt{7}} = \frac{\sqrt{5}}{\sqrt{7}}\left(\frac{\sqrt{7}}{\sqrt{7}}\right) = \frac{\sqrt{35}}{7}$$

$$\sqrt{\frac{x}{ay}} = \frac{\sqrt{x}}{\sqrt{ay}} = \frac{\sqrt{x}}{\sqrt{ay}}\left(\frac{\sqrt{ay}}{\sqrt{ay}}\right) = \frac{\sqrt{axy}}{ay}$$

Following the examples, rationalize each denominator.

a) $\sqrt{\dfrac{3}{2x}} =$ _____

b) $\sqrt{\dfrac{cp}{q}} =$ _____

a) $\dfrac{2\sqrt{6}}{9}$

b) $\dfrac{m\sqrt{b}}{bt^2}$

---

a) $\dfrac{\sqrt{6x}}{2x}$ b) $\dfrac{\sqrt{cpq}}{q}$

3-11 RATIONALIZING NUMERATORS

When a fraction contains a radical in its numerator, it can be converted to an equivalent form <u>without a radical</u> in the numerator. For example:

$$\frac{\sqrt{3}}{7} \text{ can be converted to } \frac{3}{7\sqrt{3}}$$

The conversion process is called "<u>rationalizing the numerator</u>". We will discuss that process in this section.

96. The numerator of a fraction is "rationalized" only if it does not contain a radical. For example:

$$\frac{2}{\sqrt{3}} \text{ and } \frac{1}{x\sqrt{y}} \text{ are rationalized.}$$

Continued on following page.

Square Root Radicals   133

96. Continued

When a numerator is not rationalized, we can rationalize it by multiplying by a fraction whose terms are identical to the radical in the numerator.

$$\frac{\sqrt{b}}{c} = \frac{\sqrt{b}}{c}\left(\frac{\sqrt{b}}{\sqrt{b}}\right) = \frac{b}{c\sqrt{b}}$$

$$\frac{2\sqrt{3}}{5} = \frac{2\sqrt{3}}{5}\left(\frac{\sqrt{3}}{\sqrt{3}}\right) = \frac{2(3)}{5\sqrt{3}} = \frac{6}{5\sqrt{3}}$$

Following the examples, rationalize each numerator.

a) $\dfrac{\sqrt{5x}}{2} =$ _____

b) $\dfrac{m\sqrt{t}}{k} =$ _____

97. Using the same steps, rationalize each numerator.

a) $\dfrac{5\sqrt{2y}}{3} =$ _____

b) $\dfrac{7\sqrt{x}}{\sqrt{y}} =$ _____

a) $\dfrac{5x}{2\sqrt{5x}}$

b) $\dfrac{mt}{k\sqrt{t}}$

98. When a numerator is rationalized, sometimes the new fraction can be reduced to lowest terms. For example:

$$\frac{3\sqrt{5}}{5} = \frac{3\sqrt{5}}{5}\left(\frac{\sqrt{5}}{\sqrt{5}}\right) = \frac{3(\cancel{5})}{\cancel{5}\sqrt{5}} = \frac{3}{\sqrt{5}}$$

$$\frac{\sqrt{y}}{y} = \frac{\sqrt{y}}{y}\left(\frac{\sqrt{y}}{\sqrt{y}}\right) = \frac{\cancel{y}}{\cancel{y}\sqrt{y}} = \frac{1}{\sqrt{y}}$$

Rationalize each numerator and then reduce to lowest terms.

a) $\dfrac{\sqrt{2}}{2} =$ _____

b) $\dfrac{a\sqrt{b}}{b} =$ _____

a) $\dfrac{10y}{3\sqrt{2y}}$

b) $\dfrac{7x}{\sqrt{xy}}$

99. After simplifying a radical containing a fraction, we can frequently rationalize the numerator of the new fraction. For example:

$$\sqrt{\frac{a^2p}{b^2q}} = \frac{a\sqrt{p}}{b\sqrt{q}} = \frac{a\sqrt{p}}{b\sqrt{q}}\left(\frac{\sqrt{p}}{\sqrt{p}}\right) = \frac{ap}{b\sqrt{pq}}$$

Simplify and then rationalize the numerator of the new fraction.

a) $\sqrt{\dfrac{7}{20}} =$ _____

b) $\sqrt{\dfrac{ax^2}{y^6}} =$ _____

a) $\dfrac{1}{\sqrt{2}}$

b) $\dfrac{a}{\sqrt{b}}$

134   Square Root Radicals

100. Even when we cannot simplify a radical containing a fraction, we can still rationalize its numerator. For example:

$$\sqrt{\frac{2}{3}} = \frac{\sqrt{2}}{\sqrt{3}} = \frac{\sqrt{2}}{\sqrt{3}}\left(\frac{\sqrt{2}}{\sqrt{2}}\right) = \frac{2}{\sqrt{6}}$$

$$\sqrt{\frac{ax}{y}} = \frac{\sqrt{ax}}{\sqrt{y}} = \frac{\sqrt{ax}}{\sqrt{y}}\left(\frac{\sqrt{ax}}{\sqrt{ax}}\right) = \frac{ax}{\sqrt{axy}}$$

Following the examples, rationalize each numerator.

a) $\sqrt{\dfrac{3m}{2}} =$ _____

b) $\sqrt{\dfrac{b}{cd}} =$ _____

a) $\dfrac{7}{2\sqrt{35}}$

b) $\dfrac{ax}{y^3\sqrt{a}}$

a) $\dfrac{3m}{\sqrt{6m}}$ b) $\dfrac{b}{\sqrt{bcd}}$

3-12 ADDING AND SUBTRACTING RADICALS

In this section, we will discuss the procedure for adding and subtracting square root radicals.

101. Two radical terms are called "like" terms if they contain the same radical. For example:

$2\sqrt{5}$ and $7\sqrt{5}$ are like terms because both contain $\sqrt{5}$.

$3\sqrt{x}$ and $4\sqrt{x}$ are like terms because both contain $\sqrt{x}$.

We use the distributive principle to add like radical terms. For example:

$2\sqrt{5} + 7\sqrt{5} = (2 + 7)\sqrt{5} = 9\sqrt{5}$

$3\sqrt{x} + 4\sqrt{x} = (3 + 4)\sqrt{x} = 7\sqrt{x}$

Notice that using the distributive principle is the same as adding the coefficients of the radicals. Using that method, do these:

a) $5\sqrt{y} + 6\sqrt{y} =$ _____   b) $8\sqrt{2} + 2\sqrt{2} =$ _____

102. If the coefficient of a radical is not explicitly shown, its coefficient is "1". For example:

$\sqrt{7} = 1\sqrt{7}$ $\sqrt{x} = 1\sqrt{x}$

Therefore: a) $\sqrt{7} + 3\sqrt{7} = 1\sqrt{7} + 3\sqrt{7} =$ _____

b) $5\sqrt{x} + \sqrt{x} = 5\sqrt{x} + 1\sqrt{x} =$ _____

c) $\sqrt{3} + \sqrt{3} = 1\sqrt{3} + 1\sqrt{3} =$ _____

a) $11\sqrt{y}$

b) $10\sqrt{2}$

103. Complete each addition.

   a) $9\sqrt{t} + \sqrt{t} =$ _____   c) $\sqrt{15} + 6\sqrt{15} =$ _____

   b) $\sqrt{y} + \sqrt{y} =$ _____   d) $\sqrt{23} + \sqrt{23} =$ _____

   a) $4\sqrt{7}$
   b) $6\sqrt{x}$
   c) $2\sqrt{3}$

---

104. We use the distributive principle to subtract like radical terms. For example:

   $$10\sqrt{x} - 7\sqrt{x} = (10 - 7)\sqrt{x} = 3\sqrt{x}$$
   $$2\sqrt{5} - 6\sqrt{5} = (2 - 6)\sqrt{5} = -4\sqrt{5}$$

   Notice that using the distributive principle is the same as subtracting the coefficients of the radicals. Using that method, do these:

   a) $15\sqrt{y} - 5\sqrt{y} =$ _____   b) $4\sqrt{3} - 9\sqrt{3} =$ _____

   a) $10\sqrt{t}$ c) $7\sqrt{15}$
   b) $2\sqrt{y}$ d) $2\sqrt{23}$

---

105. To perform the subtractions below, it is helpful to write the "1" coefficients explicitly. That is:

   a) $4\sqrt{7} - \sqrt{7} = 4\sqrt{7} - 1\sqrt{7} =$ _____

   b) $\sqrt{x} - 3\sqrt{x} = 1\sqrt{x} - 3\sqrt{x} =$ _____

   c) $\sqrt{2} - \sqrt{2} = 1\sqrt{2} - 1\sqrt{2} =$ _____

   a) $10\sqrt{y}$ b) $-5\sqrt{3}$

---

106. Two radical terms cannot be combined into one term by addition or subtraction if they do not contain the same radical. For example:

   $4\sqrt{3} + 5\sqrt{7}$ cannot be combined into one term.

   $8\sqrt{x} + 3\sqrt{y}$ cannot be combined into one term.

   Perform these additions and subtractions if possible.

   a) $\sqrt{y} + 2\sqrt{y} =$ _____   c) $2\sqrt{7} - \sqrt{7} =$ _____

   b) $\sqrt{3} + 5\sqrt{2} =$ _____   d) $4\sqrt{t} - \sqrt{5} =$ _____

   a) $3\sqrt{7}$
   b) $-2\sqrt{x}$
   c) $0\sqrt{2}$ or 0

---

107. The distributive principle is used to add or subtract radicals with letters as coefficients. For example:

   $$c\sqrt{y} + d\sqrt{y} = (c + d)\sqrt{y}$$
   $$a\sqrt{x} - b\sqrt{x} = (a - b)\sqrt{x}$$

   Complete these.

   a) $p\sqrt{t} + q\sqrt{t} =$ _____   b) $F\sqrt{V} - D\sqrt{V} =$ _____

   a) $3\sqrt{y}$
   b) Not possible
   c) $1\sqrt{7}$ or $\sqrt{7}$
   d) Not possible

---

a) $(p + q)\sqrt{t}$
b) $(F - D)\sqrt{V}$

108. Each expression below contains one radical term and one non-radical term.

$$3 + \sqrt{2} \qquad\qquad y - \sqrt{x}$$

Since the terms are "unlike", they cannot be combined into one. That is:

$$3 + \sqrt{2} \text{ does not equal } 3\sqrt{2}$$
$$y - \sqrt{x} \text{ does not equal } -y\sqrt{x}$$

Perform the following additions and subtractions if possible.

a) $2\sqrt{t} - \sqrt{t} = $ _____   c) $4 - \sqrt{m} = $ _____

b) $7 + \sqrt{R} = $ _____   d) $a\sqrt{x} + \sqrt{x} = $ _____

---

109. To multiply radicals, we simply multiply their radicands. That is:

$$\sqrt{9} \cdot \sqrt{4} = \sqrt{(9)(4)} = \sqrt{36}$$

To add or subtract radicals, however, we do not add or subtract their radicands. That is:

$$\sqrt{9} + \sqrt{4} \text{ does not equal } \sqrt{9+4}$$
$$\sqrt{9} - \sqrt{4} \text{ does not equal } \sqrt{9-4}$$

Which of the following are true? _____

a) $\sqrt{2} + \sqrt{5} = \sqrt{2+5}$ c) $\sqrt{a} - \sqrt{b} = \sqrt{a-b}$

b) $\sqrt{2} \cdot \sqrt{5} = \sqrt{(2)(5)}$ d) $\sqrt{a} \cdot \sqrt{b} = \sqrt{ab}$

---

a) $1\sqrt{t}$ or $\sqrt{t}$

b) Not possible

c) Not possible

d) $(a+1)\sqrt{x}$

Only (b) and (d)

SELF-TEST 9 (pages 124-137)

Simplify each radical by factoring out perfect squares.

1. $\sqrt{\dfrac{4d^4}{25b^2}}$

2. $\sqrt{\dfrac{w^3}{at^2}}$

After reducing each fraction to lowest terms, factor out perfect squares.

3. $\sqrt{\dfrac{3p^2s^3}{12s^2}}$

4. $\sqrt{\dfrac{6xy^2}{2x^3y^3}}$

5. Simplify $\dfrac{\sqrt{ab}}{r\sqrt{2a}}$

Continued on following page.

SELF-TEST 9 (pages 124-137) - Continued	
Simplify each radical. 6. $\sqrt{(4m)^2}$ 7. $\sqrt{\left(\dfrac{9h}{v}\right)^2}$	Simplify each expression. 8. $\left(\sqrt{\dfrac{k}{4d}}\right)^2$ 9. $\sqrt{3w} \cdot \sqrt{3w}$
10. Which of the following are equal to "6w"? a) $3\sqrt{2w}$ b) $\sqrt{6w} \cdot \sqrt{6w}$ c) $\sqrt{(6w)^2}$ d) $(6\sqrt{w})^2$ e) $(\sqrt{6w})^2$	
Rationalize each <u>denominator</u>. Report each answer in lowest terms. 11. $\dfrac{2\sqrt{3}}{\sqrt{6}}$ 12. $\dfrac{rt}{k\sqrt{r}}$	Rationalize each <u>numerator</u>. Report each answer in lowest terms. 13. $\dfrac{5\sqrt{2}}{4}$ 14. $\dfrac{K\sqrt{F}}{F}$
15. Add. $\sqrt{H} + 2\sqrt{H}$	16. Subtract. $a\sqrt{t} - p\sqrt{t}$

ANSWERS:
1. $\dfrac{2d^2}{5b}$ 5. $\dfrac{1}{r}\sqrt{\dfrac{b}{2}}$ 9. $3w$ 13. $\dfrac{5}{2\sqrt{2}}$

2. $\dfrac{w}{t}\sqrt{\dfrac{w}{a}}$ 6. $4m$ 10. b, c, e   14. $\dfrac{K}{\sqrt{F}}$

3. $\dfrac{p}{2}\sqrt{s}$ 7. $\dfrac{9h}{v}$ 11. $\sqrt{2}$ 15. $3\sqrt{H}$

4. $\dfrac{1}{x}\sqrt{\dfrac{3}{y}}$ 8. $\dfrac{k}{4d}$ 12. $\dfrac{t\sqrt{r}}{k}$ 16. $(a-p)\sqrt{t}$

3-13   RADICALS CONTAINING BINOMIALS

In this section, we will discuss radicals containing binomials and contrast them with radicals containing two-factor multiplications and radicals containing the indicated squares of binomials.

110. We have seen that an addition of two radicals with different radicands <u>cannot be combined into one radical</u>. That is:

$$\sqrt{x} + \sqrt{9} \text{ does } \underline{not} \text{ equal } \sqrt{x+9}$$

Similarly, a radical containing an addition <u>cannot be broken up into an addition of two radicals</u>. That is:

$$\sqrt{x+9} \text{ does } \underline{not} \text{ equal } \sqrt{x} + \sqrt{9}$$

To show that the statement above is true, we can substitute a number for "x".

If $x = 16$:   a) $\sqrt{x+9} = \sqrt{16+9} = \sqrt{25} =$ _____

               b) $\sqrt{x} + \sqrt{9} = \sqrt{16} + \sqrt{9} = 4 + 3 =$ _____

               c) Does $\sqrt{x+9} = \sqrt{x} + \sqrt{9}$? _____

111. We have seen that a subtraction of two radicals with different radicands cannot be combined into one radical. That is:

$\sqrt{y} - \sqrt{36}$ does not equal $\sqrt{y - 36}$

Similarly, a radical containing a subtraction cannot be broken up into a subtraction of two radicals. That is:

$\sqrt{y - 36}$ does not equal $\sqrt{y} - \sqrt{36}$

To show that the statement above is true, we can substitute a number for "y".

If y = 100: a) $\sqrt{y - 36} = \sqrt{100 - 36} = \sqrt{64} = $ _____

b) $\sqrt{y} - \sqrt{36} = \sqrt{100} - \sqrt{36} = 10 - 6 = $ _____

c) Does $\sqrt{y - 36} = \sqrt{y} - \sqrt{36}$? _____

a) 5
b) 7
c) No

112. If the radicand is a two-factor multiplication, a radical can be factored into a multiplication of two radicals. For example:

$\sqrt{5t} = \sqrt{5} \cdot \sqrt{t}$ $\sqrt{xy} = \sqrt{x} \cdot \sqrt{y}$

If the radicand is a binomial, the radicand cannot be broken up into an addition or subtraction of radicals. For example:

$\sqrt{a + 4} \neq \sqrt{a} + \sqrt{4}$ $\sqrt{c - d} \neq \sqrt{c} - \sqrt{d}$

Answer "true" or "false" for these:

a) $\sqrt{10m} = \sqrt{10} \cdot \sqrt{m}$ _____   c) $\sqrt{R - S} = \sqrt{R} - \sqrt{S}$ _____

b) $\sqrt{10 + m} = \sqrt{10} + \sqrt{m}$ _____   d) $\sqrt{RS} = \sqrt{R} \cdot \sqrt{S}$ _____

a) 8
b) 4
c) No

113. Even when both terms are perfect squares, a radical containing a binomial is in its simplest form. It cannot be broken up into an addition or subtraction of two radicals. For example:

$\sqrt{y^2 + 16} \neq \sqrt{y^2} + \sqrt{16}$ $\sqrt{c^2 - a^2} \neq \sqrt{c^2} - \sqrt{a^2}$

To show that the statements above are true, we can substitute numbers for the variables.

If y = 3: a) $\sqrt{y^2 + 16} = \sqrt{3^2 + 16} = \sqrt{9 + 16} = \sqrt{25} = $ _____
         b) $\sqrt{y^2} + \sqrt{16} = \sqrt{3^2} + \sqrt{16} = \sqrt{9} + \sqrt{16} = 3 + 4 = $ _____

If c = 10 and a = 8:

c) $\sqrt{c^2 - a^2} = \sqrt{10^2 - 8^2} = \sqrt{100 - 64} = \sqrt{36} = $ _____
d) $\sqrt{c^2} - \sqrt{a^2} = \sqrt{10^2} - \sqrt{8^2} = \sqrt{100} - \sqrt{64} = 10 - 8 = $ _____

a) true   c) false
b) false d) true

114. If the radicand is a <u>multiplication</u> with two perfect-square factors, it <u>can</u> be simplified. For example:

$$\sqrt{9x^2} = \sqrt{9} \cdot \sqrt{x^2} = 3x \qquad \sqrt{a^2b^2} = \sqrt{a^2} \cdot \sqrt{b^2} = ab$$

If the radical is a <u>binomial</u> with two perfect-square terms, it <u>cannot</u> be simplified. For example:

$$\sqrt{x^2 + 9} \neq \sqrt{x^2} + \sqrt{9} \text{ or } x + 3$$
$$\sqrt{a^2 - b^2} \neq \sqrt{a^2} - \sqrt{b^2} \text{ or } a - b$$

Answer "true" or "false" for these:

a) $\sqrt{y^2 - 25} = y - 5$ _____   c) $\sqrt{F^2 + T^2} = F + T$ _____

b) $\sqrt{25y^2} = 5y$ _____   d) $\sqrt{F^2T^2} = FT$ _____

a) 5   c) 6
b) 7   d) 2

115. A radical containing an indicated square of a binomial can be simplified. For example:

$$\sqrt{(t + 4)^2} = t + 4 \qquad \sqrt{(x - y)^2} = x - y$$

A radical containing a binomial with two perfect-square terms cannot be simplified. For example:

$$\sqrt{t^2 + 4} \neq t + 2 \qquad \sqrt{x^2 - y^2} \neq x - y$$

Answer "true" or "false" for these:

a) $\sqrt{(y - 9)^2} = y - 9$ _____   c) $\sqrt{B^2 + D^2} = B + D$ _____

b) $\sqrt{y^2 - 9} = y - 3$ _____   d) $\sqrt{(B + D)^2} = B + D$ _____

a) false
b) true
c) false
d) true

116. Simplify if possible.

a) $\sqrt{36p^2} =$ _____   d) $\sqrt{(q - t)^2} =$ _____

b) $\sqrt{(p + 36)^2} =$ _____   e) $\sqrt{q^2 - t^2} =$ _____

c) $\sqrt{p^2 + 36} =$ _____   f) $\sqrt{q^2t^2} =$ _____

a) true
b) false
c) false
d) true

a) 6p   b) p + 36   c) Not possible   d) q - t   e) Not possible   f) qt

---

3-14   OPERATIONS WITH RADICALS CONTAINING BINOMIALS

In this section, we will show that the same procedures are used for operations with radicals containing binomials.

Square Root Radicals

117. In the multiplication below, one radicand is a binomial.

$$\sqrt{5} \cdot \sqrt{y+3} = \sqrt{5(y+3)} = \sqrt{5y+15}$$

Following the example, complete these:

a) $\sqrt{2} \cdot \sqrt{x+7} = $ _____

b) $\sqrt{a} \cdot \sqrt{b+c} = $ _____

---

118. The multiplications below involve non-radicals and radicals.

$$7\sqrt{m} \cdot \sqrt{m+2} = 7\sqrt{m(m+2)} = 7\sqrt{m^2+2m}$$
$$a\sqrt{x} \cdot b\sqrt{x+1} = ab\sqrt{x(x+1)} = ab\sqrt{x^2+x}$$

a) $\sqrt{2x+14}$
b) $\sqrt{ab+ac}$

Following the examples, complete these:

a) $c\sqrt{d} \cdot \sqrt{d-8} = $ _____

b) $2\sqrt{a+b} \cdot p\sqrt{q} = $ _____

---

119. To convert each multiplication below to a single radical, we substituted $\sqrt{9}$ for 3 and $\sqrt{4x^2}$ for 2x.

$$3\sqrt{a+5} = \sqrt{9} \cdot \sqrt{a+5} = \sqrt{9a+45}$$
$$2x\sqrt{y-1} = \sqrt{4x^2} \cdot \sqrt{y-1} = \sqrt{4x^2y-4x^2}$$

a) $c\sqrt{d^2-8d}$
b) $2p\sqrt{aq+bq}$

Convert each of these to a single radical.

a) $4\sqrt{m+3} = $ _____

b) $c\sqrt{d-t} = $ _____

c) $3R\sqrt{S+V} = $ _____

---

120. When the radicand is a binomial, we can sometimes use the distributive principle to factor out perfect squares. For example:

$$\sqrt{9x+9y} = \sqrt{9(x+y)} = 3\sqrt{x+y}$$
$$\sqrt{4a^2b - 12a^2c} = \sqrt{4a^2(b-3c)} = 2a\sqrt{b-3c}$$

a) $\sqrt{16m+48}$
b) $\sqrt{c^2d-c^2t}$
c) $\sqrt{9R^2S+9R^2V}$

Factor out the perfect squares from these:

a) $\sqrt{9p+18q} = $ _____

b) $\sqrt{c^2t-c^2v} = $ _____

c) $\sqrt{16x^2y+16x^2z} = $ _____

Square Root Radicals 141

121. A radical containing a binomial can be simplified only if we can use the distributive principle to factor out a perfect square.

$\sqrt{a^2x^2 + a^2y^2}$ can be simplified to $a\sqrt{x^2 + y^2}$

$\sqrt{x^2 + y^2}$ cannot be simplified

Simplify if possible.

a) $\sqrt{25c^2 + 100d^2}$ = _____

b) $\sqrt{S^2 + T^2}$ = _____

c) $\sqrt{p^2 - q^2}$ = _____

d) $\sqrt{d^2f^2 - d^2t^2}$ = _____

a) $3\sqrt{p + 2q}$

b) $c\sqrt{t - v}$

c) $4x\sqrt{y + z}$

---

122. We simplified the radical below by factoring out perfect squares.

$\sqrt{\dfrac{4x + 4y}{b^2}} = \dfrac{2\sqrt{x + y}}{b}$ or $\dfrac{2}{b}\sqrt{x + y}$

Simplify these by factoring out perfect squares.

a) $\sqrt{\dfrac{9a - 9b}{16}}$ = _____

b) $\sqrt{\dfrac{1}{c^2y^2 + d^2y^2}}$ = _____

a) $5\sqrt{c^2 + 4d^2}$

b) Not possible

c) Not possible

d) $d\sqrt{f^2 - t^2}$

---

123. After simplifying below, we converted the division of radicals to a single radical.

$\sqrt{\dfrac{x^2y}{a^2b - a^2d}} = \dfrac{x\sqrt{y}}{a\sqrt{b - d}} = \dfrac{x}{a}\sqrt{\dfrac{y}{b - d}}$

Simplify these and convert the division of radicals to a single radical.

a) $\sqrt{\dfrac{9p - 9q}{t}}$ = _____

b) $\sqrt{\dfrac{50x + 25y}{cx^2 - dx^2}}$ = _____

a) $\dfrac{3\sqrt{a - b}}{4}$ or $\dfrac{3}{4}\sqrt{a - b}$

b) $\dfrac{1}{y\sqrt{c^2 + d^2}}$

---

124. After simplifying below, we were able to reduce the fraction to lowest terms.

$\dfrac{bx}{\sqrt{ax^2 + cx^2}} = \dfrac{b\cancel{x}}{\cancel{x}\sqrt{a + c}} = \dfrac{b}{\sqrt{a + c}}$

Simplify and reduce to lowest terms.

a) $\dfrac{\sqrt{9p - 18q}}{3q}$ = _____

b) $\dfrac{ady}{\sqrt{a^2y^2 + d^2y^2}}$ = _____

a) $3\sqrt{\dfrac{p - q}{t}}$

b) $\dfrac{5}{x}\sqrt{\dfrac{2x + y}{c - d}}$

142 Square Root Radicals

125. If we multiply two radicals containing the same binomials, the product is the binomial. For example:

$$\sqrt{y-5} \cdot \sqrt{y-5} = \sqrt{(y-5)^2} = y - 5$$

If we square a radical containing a binomial, its square is the binomial. For example:

$$(\sqrt{x+1})^2 = x + 1 \qquad (\sqrt{p-q})^2 = p - q$$

Complete these:

a) $\sqrt{t+7} \cdot \sqrt{t+7} = $ _____

b) $(\sqrt{R-4})^2 = $ _____

c) $\sqrt{F-V} \cdot \sqrt{F-V} = $ _____

d) $(\sqrt{a+b})^2 = $ _____

a) $\dfrac{\sqrt{p-2q}}{q}$

b) $\dfrac{ad}{\sqrt{a^2+d^2}}$

---

126. If we square a radical whose radicand is a fraction containing a binomial term, its square is the fraction. That is:

$$\left(\sqrt{\dfrac{x-7}{5}}\right)^2 = \dfrac{x-7}{5} \qquad \left(\sqrt{\dfrac{a+b}{p-q}}\right)^2 = \underline{\quad\quad}$$

a) $t + 7$

b) $R - 4$

c) $F - V$

d) $a + b$

---

127. To rationalize the denominator below, we multiplied by a fraction whose terms are identical to the radical in the denominator.

$$\dfrac{t}{c\sqrt{a-b}} = \dfrac{t}{c\sqrt{a-b}}\left(\dfrac{\sqrt{a-b}}{\sqrt{a-b}}\right) = \dfrac{t\sqrt{a-b}}{c(a-b)} \text{ or } \dfrac{t\sqrt{a-b}}{ac-bc}$$

Rationalize each denominator.

a) $\dfrac{1}{\sqrt{x+y}} = $ _____

b) $\dfrac{R}{2\sqrt{S-T}} = $ _____

$\dfrac{a+b}{p-q}$

---

128. To rationalize the numerator below, we multiplied by a fraction whose terms are identical to the radical in the numerator.

$$\dfrac{3\sqrt{x-1}}{5} = \dfrac{3\sqrt{x-1}}{5}\left(\dfrac{\sqrt{x-1}}{\sqrt{x-1}}\right) = \dfrac{3(x-1)}{5\sqrt{x-1}} \text{ or } \dfrac{3x-3}{5\sqrt{x-1}}$$

Rationalize each numerator.

a) $\dfrac{\sqrt{a+b}}{7} = $ _____

b) $\dfrac{R\sqrt{S-T}}{V} = $ _____

a) $\dfrac{\sqrt{x+y}}{x+y}$

b) $\dfrac{R\sqrt{S-T}}{2(S-T)}$ or $\dfrac{R\sqrt{S-T}}{2S-2T}$

---

a) $\dfrac{a+b}{7\sqrt{a+b}}$ b) $\dfrac{R(S-T)}{V\sqrt{S-T}}$ or $\dfrac{RS-RT}{V\sqrt{S-T}}$

129. To do the addition and subtraction below, we added and subtracted the coefficients.

$$5\sqrt{x-1} + 4\sqrt{x-1} = 9\sqrt{x-1}$$
$$8\sqrt{a+b} - 3\sqrt{a+b} = 5\sqrt{a+b}$$

Complete these:

a) $10\sqrt{y-8} + 20\sqrt{y-8} =$ _____

b) $12\sqrt{p+q} - 6\sqrt{p+q} =$ _____

130. If the coefficient of a radical is not shown, its coefficient is "1". That is:

$$\sqrt{x-5} = 1\sqrt{x-5} \qquad \sqrt{R+S} = 1\sqrt{R+S}$$

Using the fact above, complete these:

a) $7\sqrt{y+3} - \sqrt{y+3} =$ _____

b) $\sqrt{a+b} + \sqrt{a+b} =$ _____

c) $\sqrt{c-d} - \sqrt{c-d} =$ _____

a) $30\sqrt{y-8}$

b) $6\sqrt{p+q}$

a) $6\sqrt{y+3}$ b) $2\sqrt{a+b}$ c) 0, from $0\sqrt{c-d}$

3-15   SQUARING TERMS CONTAINING A RADICAL

In this section, we will discuss the procedure for squaring terms containing a radical.

131. To square a term containing a radical, we multiply the term by itself. That is:

$$(2\sqrt{3})^2 = (2\sqrt{3})(2\sqrt{3}) = 2 \cdot 2 \cdot \sqrt{3} \cdot \sqrt{3} = 4(3) = 12$$
$$(a\sqrt{bc})^2 = (a\sqrt{bc})(a\sqrt{bc}) = a \cdot a \cdot \sqrt{bc} \cdot \sqrt{bc} = a^2bc$$

Complete each squaring.

a) $(7\sqrt{2})^2 = (7\sqrt{2})(7\sqrt{2}) = 7 \cdot 7 \cdot \sqrt{2} \cdot \sqrt{2} =$ _____

b) $(R\sqrt{3S})^2 = (R\sqrt{3S})(R\sqrt{3S}) = R \cdot R \cdot \sqrt{3S} \cdot \sqrt{3S} =$ _____

a) 98, from 49(2)

b) $3R^2S$

144 Square Root Radicals

132. Squaring a non-fractional term containing a radical is equivalent to squaring each factor in the term. For example:

$$(6\sqrt{y})^2 = 6^2 \cdot (\sqrt{y})^2 = 36y$$

$$(b\sqrt{ct})^2 = b^2 \cdot (\sqrt{ct})^2 = b^2ct$$

Square these by squaring each factor in the term.

a) $(10\sqrt{5})^2 = $ _____

b) $(3x\sqrt{y})^2 = $ _____

c) $(p\sqrt{2q})^2 = $ _____

d) $(R\sqrt{ST})^2 = $ _____

---

133. A common error is shown below. The <u>common</u> <u>error</u> <u>is</u> <u>forgetting to</u> <u>square the non-radical factor.</u>

$$(3\sqrt{t})^2 = 3t \qquad \text{(error)}$$

$$(4x\sqrt{y})^2 = 4xy \qquad \text{(error)}$$

If the terms above are squared correctly, we get:

$$(3\sqrt{t})^2 = 9t \qquad (4x\sqrt{y})^2 = 16x^2y$$

Avoid the common error in squaring these:

a) $(8\sqrt{q})^2 = $ _____

b) $(5a\sqrt{b})^2 = $ _____

c) $(p\sqrt{3t})^2 = $ _____

d) $(4x\sqrt{2y})^2 = $ _____

a) 500, from: 100(5)

b) $9x^2y$

c) $2p^2q$

d) R^2ST

---

134. To square the terms below, we also squared each factor.

$$(2\sqrt{y+3})^2 = 2^2 \cdot (\sqrt{y+3})^2 = 4(y+3) \text{ or } 4y + 12$$

$$(m\sqrt{a-b})^2 = m^2 \cdot (\sqrt{a-b})^2 = m^2(a-b) \text{ or } am^2 - bm^2$$

Square each term.

a) $(5\sqrt{x-1})^2 = $ _____

b) $(c\sqrt{p^2+q^2})^2 = $ _____

a) $64q$

b) $25a^2b$

c) $3p^2t$

d) $32x^2y$

---

135. To square a fraction containing a radical, we multiply the fraction by itself. For example:

a) $\left(\dfrac{\sqrt{3}}{4}\right)^2 = \left(\dfrac{\sqrt{3}}{4}\right)\left(\dfrac{\sqrt{3}}{4}\right) = \dfrac{\sqrt{3} \cdot \sqrt{3}}{4 \cdot 4} = $ _____

b) $\left(\dfrac{c}{2\sqrt{b}}\right)^2 = \left(\dfrac{c}{2\sqrt{b}}\right)\left(\dfrac{c}{2\sqrt{b}}\right) = \dfrac{c \cdot c}{2 \cdot 2 \cdot \sqrt{b} \cdot \sqrt{b}} = $ _____

a) $25(x-1)$ or $25x - 25$

b) $c^2(p^2 + q^2)$ or $c^2p^2 + c^2q^2$

---

a) $\dfrac{3}{16}$ \qquad b) $\dfrac{c^2}{4b}$

Square Root Radicals 145

136. Squaring a fraction containing a radical is equivalent to squaring both of its terms. That is:

$$\left(\frac{3x}{\sqrt{7}}\right)^2 = \frac{(3x)^2}{(\sqrt{7})^2} = \frac{9x^2}{7}$$

$$\left(\frac{p\sqrt{q}}{t}\right)^2 = \frac{(p\sqrt{q})^2}{(t)^2} = \frac{p^2 q}{t^2}$$

Following the examples, square these:

a) $\left(\dfrac{5\sqrt{2}}{9}\right)^2 = $ _____

b) $\left(\dfrac{a}{\sqrt{b}}\right)^2 = $ _____

c) $\left(\dfrac{\sqrt{2x}}{3y}\right)^2 = $ _____

137. Square these:

a) $\left(\dfrac{2\sqrt{5x}}{d}\right)^2 = $ _____

b) $\left(\dfrac{1}{\sqrt{F}}\right)^2 = $ _____

c) $\left(\dfrac{1}{3\sqrt{5t}}\right)^2 = $ _____

a) $\dfrac{50}{81}$

b) $\dfrac{a^2}{b}$

c) $\dfrac{2x}{9y^2}$

138. After squaring a fraction, we can sometimes reduce the new fraction to lowest terms. For example:

$$\left(\frac{\sqrt{2}}{2}\right)^2 = \frac{2}{4} = \frac{1}{2} \qquad \left(\frac{ab}{\sqrt{a}}\right)^2 = \frac{a^2 b^2}{a} = ab^2$$

Square these and reduce to lowest terms.

a) $\left(\dfrac{\sqrt{6}}{3}\right)^2 = $ _____

b) $\left(\dfrac{t}{\sqrt{ct}}\right)^2 = $ _____

a) $\dfrac{20x}{d^2}$

b) $\dfrac{1}{F}$

c) $\dfrac{1}{45t}$

139. In each example below, we squared and then reduced to lowest terms.

$$\left(\frac{3\sqrt{5}}{5}\right)^2 = \frac{9(5)}{25} = \frac{9}{5} \qquad \left(\frac{R}{d\sqrt{R}}\right)^2 = \frac{R^2}{d^2 R} = \frac{R}{d^2}$$

Square these and reduce to lowest terms.

a) $\left(\dfrac{3}{2\sqrt{3}}\right)^2 = $ _____

b) $\left(\dfrac{b\sqrt{t}}{t}\right)^2 = $ _____

a) $\dfrac{2}{3}$, from: $\dfrac{6}{9}$

b) $\dfrac{t}{c}$, from: $\dfrac{t^2}{ct}$

140. We squared the term below in which the radical contains a fraction.

$$\left(3b\sqrt{\frac{c}{d}}\right)^2 = (3b)^2 \cdot \left(\sqrt{\frac{c}{d}}\right)^2 = 9b^2\left(\frac{c}{d}\right) = \frac{9b^2c}{d}$$

Following the example, square these.

a) $\left(2y\sqrt{\frac{a}{x}}\right)^2 = $ _____

b) $\left(\frac{m}{2}\sqrt{\frac{t}{2p}}\right)^2 = $ _____

a) $\dfrac{3}{4}$

b) $\dfrac{b^2}{t}$

141. To square the fraction below, we squared both terms.

$$\left(\frac{2\sqrt{x-1}}{3}\right)^2 = \frac{2^2 \cdot (\sqrt{x-1})^2}{3^2} = \frac{4(x-1)}{9} \text{ or } \frac{4x-4}{9}$$

Following the example, square these.

a) $\left(\dfrac{1}{3\sqrt{y+4}}\right)^2 = $ _____

b) $\left(\dfrac{\sqrt{p-3}}{4}\right)^2 = $ _____

a) $\dfrac{4ay^2}{x}$

b) $\dfrac{m^2t}{8p}$

a) $\dfrac{1}{9(y+4)}$ or $\dfrac{1}{9y+36}$ b) $\dfrac{p-3}{16}$

3-16 SQUARING BINOMIALS CONTAINING A RADICAL

In this section, we will discuss the procedure for squaring binomials when one of the terms contains a radical.

142. To square a binomial containing a radical, we multiply the binomial by itself. That is:

$$(3 + \sqrt{y})^2 = (3 + \sqrt{y})(3 + \sqrt{y})$$

Therefore, squaring a binomial is the same as multiplying two identical binomials. We can use the FOIL method to do so. We get:

$$(3 + \sqrt{y})^2 = (3 + \sqrt{y})(3 + \sqrt{y}) = 3(3) + 3\sqrt{y} + 3\sqrt{y} + \sqrt{y} \cdot \sqrt{y}$$
$$= 9 + 6\sqrt{y} + y$$

We can use the usual shortcut to square the binomial above. To see that fact, let's examine $9 + 6\sqrt{y} + y$.

1) The <u>first</u> <u>term</u> (9) is the square of 3.
2) The <u>second</u> <u>term</u> ($6\sqrt{y}$) is double the product of the two terms of the binomial. That is: $6\sqrt{y} = 2(3)(\sqrt{y})$.
3) The <u>third</u> <u>term</u> (y) is the square of $\sqrt{y}$.

Square Root Radicals   147

143. Let's use the shortcut to square $(\sqrt{x} + 5)$.

   a) The square of $\sqrt{x}$ is _____.
   b) Double the product of $\sqrt{x}$ and 5 is _____.
   c) The square of 5 is _____.
   d) Therefore, $(\sqrt{x} + 5)^2 =$ _____.

144. Use the shortcut to square each binomial below.

   a) $(8 + \sqrt{R})^2 =$ _____
   b) $(\sqrt{a} + 10)^2 =$ _____

a) x
b) $10\sqrt{x}$
c) 25
d) $x + 10\sqrt{x} + 25$

145. We used the FOIL method to square $(7 - \sqrt{t})$ below.

$(7 - \sqrt{t})^2 = (7 - \sqrt{t})(7 - \sqrt{t}) = 7(7) - 7\sqrt{t} - 7\sqrt{t} + \sqrt{t} \cdot \sqrt{t}$
$= 49 - 14\sqrt{t} + t$

Let's use the shortcut to show that we get the same result.

   a) The square of 7 is _____.
   b) Double the product of 7 and $\sqrt{t}$ is _____.
   c) The square of $\sqrt{t}$ is _____.
   d) Therefore, $(7 - \sqrt{t})^2 =$ _____.

a) $64 + 16\sqrt{R} + R$
b) $a + 20\sqrt{a} + 100$

146. Let's use the shortcut to square $(\sqrt{x} - 4)$.

   a) The square of $\sqrt{x}$ is _____.
   b) Double the product of $\sqrt{x}$ and 4 is _____.
   c) The square of 4 is _____.
   d) Therefore, $(\sqrt{x} - 4)^2 =$ _____.

a) 49
b) $14\sqrt{t}$
c) t
d) $49 - 14\sqrt{t} + t$

147. When using the shortcut to square a difference, <u>remember to subtract the second term</u>. Use the shortcut for these.

   a) $(9 - \sqrt{m})^2 =$ _____
   b) $(\sqrt{S} - 1)^2 =$ _____

a) x
b) $8\sqrt{x}$
c) 16
d) $x - 8\sqrt{x} + 16$

148. Use the shortcut for these.

   a) $(x + \sqrt{y})^2 =$ _____
   b) $(\sqrt{V} - T)^2 =$ _____

a) $81 - 18\sqrt{m} + m$
b) $S - 2\sqrt{S} + 1$

148 Square Root Radicals

149. We used the FOIL method to square $(4 + 3\sqrt{x})$ below.

$$(4 + 3\sqrt{x})^2 = (4 + 3\sqrt{x})(4 + 3\sqrt{x}) = 4(4) + 4(3\sqrt{x}) + 4(3\sqrt{x}) + 3\sqrt{x} \cdot 3\sqrt{x}$$
$$= 16 + 24\sqrt{x} + 9x$$

Let's use the shortcut to show that we get the same results.

a) The square of 4 is _____.
b) Double the product of 4 and $3\sqrt{x}$ is _____.
c) The square of $3\sqrt{x}$ is _____.
d) Therefore, $(4 + 3\sqrt{x})^2 =$ _____.

a) $x^2 + 2x\sqrt{y} + y$

b) $V - 2T\sqrt{V} + T^2$

---

150. Let's use the shortcut to square $(a\sqrt{b} - c)$.

a) The square of $a\sqrt{b}$ is _____.
b) Double the product of $a\sqrt{b}$ and "c" is _____.
c) The square of "c" is _____.
d) Therefore, $(a\sqrt{b} - c)^2 =$ _____.

a) 16
b) $24\sqrt{x}$
c) $9x$
d) $16 + 24\sqrt{x} + 9x$

---

151. Use the shortcut for these.

a) $(3\sqrt{t} + 1)^2 =$ _____.
b) $(y - d\sqrt{x})^2 =$ _____.

a) $a^2 b$
b) $2ac\sqrt{b}$
c) c^2
d) $a^2 b - 2ac\sqrt{b} + c^2$

---

152. We squared the fraction below by squaring each term of the fraction.

$$\left(\frac{2 + \sqrt{x}}{3}\right)^2 = \frac{(2 + \sqrt{x})^2}{3^2} = \frac{4 + 4\sqrt{x} + x}{9}$$

Use the same method to square these.

a) $\left(\dfrac{\sqrt{d} - 1}{5}\right)^2 =$ _____

b) $\left(\dfrac{c}{a + p\sqrt{y}}\right)^2 =$ _____

a) $9t + 6\sqrt{t} + 1$

b) $y^2 - 2dy\sqrt{x} + d^2 x$

---

153. We used the FOIL method to square $(3 + \sqrt{x + 2})$ below. Notice how we combined terms to get the final trinomial.

$$(3 + \sqrt{x+2})^2 = (3 + \sqrt{x+2})(3 + \sqrt{x+2})$$
$$= 3(3) + 3\sqrt{x+2} + 3\sqrt{x+2} + \sqrt{x+2} \cdot \sqrt{x+2}$$
$$= 9 + 6\sqrt{x+2} + x + 2$$
$$= 11 + 6\sqrt{x+2} + x$$

a) $\dfrac{d - 2\sqrt{d} + 1}{25}$

b) $\dfrac{c^2}{a^2 + 2ap\sqrt{y} + p^2 y}$

Continued on following page.

Square Root Radicals   149

153. Continued

Using the shortcut to square $(3 + \sqrt{x+2})$, we get:

$$(3 + \sqrt{x+2})^2 = 3^2 + 2(3)\sqrt{x+2} + (\sqrt{x+2})^2$$
$$= 9 + 6\sqrt{x+2} + x + 2$$
$$= 11 + 6\sqrt{x+2} + x$$

Let's use the shortcut to square $(4 - \sqrt{y-1})$.

a) The square of 4 is _____.

b) Double the product of 4 and $\sqrt{y-1}$ is _____.

c) The square of $\sqrt{y-1}$ is _____.

d) The four-term expression is _____.

e) The final trinomial is _____.

---

154. Use the shortcut method to square these. Combine terms to express each square as a trinomial.

a) $(6 + \sqrt{V-9})^2 =$ _____
   = _____

b) $(10 - \sqrt{t+1})^2 =$ _____
   = _____

a) 16

b) $8\sqrt{y-1}$

c) $y - 1$

d) $16 - 8\sqrt{y-1} + y - 1$

e) $15 - 8\sqrt{y-1} + y$

---

155. Square this fraction and write the denominator as a trinomial.

$$\left(\frac{1}{8 + \sqrt{2x-5}}\right)^2 =$$ _____

= _____

a) $36 + 12\sqrt{V-9} + V - 9$
   $27 + 12\sqrt{V-9} + V$

b) $100 - 20\sqrt{t+1} + t + 1$
   $101 - 20\sqrt{t+1} + t$

$$\frac{1}{64 + 16\sqrt{2x-5} + 2x - 5} = \frac{1}{59 + 16\sqrt{2x-5} + 2x}$$

3-17 CONJUGATES AND RATIONALIZING

In this section, we will show what is meant by the "conjugate" of a binomial containing a radical and then use conjugates to rationalize denominators and numerators.

156. When multiplying the sum and difference of the same terms, the product is the first term squared minus the second term squared. For example:

$$(x + 3)(x - 3) = x^2 - 3^2 \text{ or } x^2 - 9$$
$$(a + b)(a - b) = a^2 - b^2$$

The same pattern is used when one of the terms is a square root radical. Two examples are shown. Notice that neither product contains a radical.

$$(2 + \sqrt{x})(2 - \sqrt{x}) = 2^2 - (\sqrt{x})^2 \text{ or } 4 - x$$
$$(\sqrt{y} + 5)(\sqrt{y} - 5) = (\sqrt{y})^2 - 5^2 \text{ or } y - 25$$

Following the examples, complete these.

a) $(3 + \sqrt{t})(3 - \sqrt{t}) = $ _____

b) $(\sqrt{a} + 4)(\sqrt{a} - 4) = $ _____

---

157. Notice how the product below simplifies to a single number.

$$(3 + \sqrt{2})(3 - \sqrt{2}) = 3^2 - (\sqrt{2})^2 = 9 - 2 = 7$$

Following the example, complete these.

a) $(5 + \sqrt{11})(5 - \sqrt{11}) = $ _____

b) $(\sqrt{3} + 1)(\sqrt{3} - 1) = $ _____

a) $9 - t$

b) $a - 16$

---

158. When two binomials are the sum and difference of the same terms and one term is a square root radical, they are called "conjugates". Two pairs of "conjugates" are shown below.

$(4 + \sqrt{x})$ and $(4 - \sqrt{x})$ $(\sqrt{5} + 2)$ and $(\sqrt{5} - 2)$

We say: $4 - \sqrt{x}$ is the conjugate of $4 + \sqrt{x}$.

$\sqrt{5} + 2$ is the conjugate of $\sqrt{5} - 2$.

Write the conjugate of each binomial.

a) $2 + \sqrt{t}$ _____   c) $\sqrt{10} - 3$ _____

b) $\sqrt{R} - 1$ _____   d) $4 - \sqrt{3}$ _____

a) 14, from $25 - 11$

b) 2, from $3 - 1$

---

159. Since we always get a non-radical expression when multiplying conjugates, we can use conjugates to rationalize denominators. For example, to rationalize the denominator below, we multiplied by a fraction whose terms are the conjugate of $2 + \sqrt{x}$.

$$\frac{5}{2 + \sqrt{x}} = \frac{5}{2 + \sqrt{x}}\left(\frac{2 - \sqrt{x}}{2 - \sqrt{x}}\right) = \frac{5(2 - \sqrt{x})}{4 - x} \text{ or } \frac{10 - 5\sqrt{x}}{4 - x}$$

Continued on following page.

a) $2 - \sqrt{t}$

b) $\sqrt{R} + 1$

c) $\sqrt{10} + 3$

d) $4 + \sqrt{3}$

Square Root Radicals 151

159. Continued

Following the example, multiply by $\dfrac{\sqrt{y}+1}{\sqrt{y}+1}$ to rationalize the denominator below.

$$\dfrac{a}{\sqrt{y}-1} = \dfrac{a}{\sqrt{y}-1}\left(\dfrac{\sqrt{y}+1}{\sqrt{y}+1}\right) = \underline{\hspace{2cm}}$$

---

160. To rationalize the denominator below, we multiplied by a fraction whose terms are the <u>conjugate</u> of $7 - \sqrt{2}$.

$$\dfrac{1}{7-\sqrt{2}} = \dfrac{1}{7-\sqrt{2}}\left(\dfrac{7+\sqrt{2}}{7+\sqrt{2}}\right) = \dfrac{1(7+\sqrt{2})}{7-2} = \dfrac{7+\sqrt{2}}{5}$$

Following the example, multiply by $\dfrac{\sqrt{11}-3}{\sqrt{11}-3}$ to rationalize the denominator below.

$$\dfrac{5}{\sqrt{11}+3} = \dfrac{5}{\sqrt{11}+3}\left(\dfrac{\sqrt{11}-3}{\sqrt{11}-3}\right) = \underline{\hspace{2cm}}$$

$\dfrac{a(\sqrt{y}+1)}{y-1}$ or $\dfrac{a\sqrt{y}+a}{y-1}$

---

161. Use conjugates to rationalize each denominator below.

a) $\dfrac{x}{3+\sqrt{2}} = \underline{\hspace{3cm}}$

b) $\dfrac{11}{\sqrt{R}-2} = \underline{\hspace{3cm}}$

$\dfrac{5(\sqrt{11}-3)}{2}$ or $\dfrac{5\sqrt{11}-15}{2}$

---

162. We used the conjugate of $\sqrt{y} - 3$ to rationalize the numerator below.

$$\dfrac{\sqrt{y}-3}{5} = \dfrac{\sqrt{y}-3}{5}\left(\dfrac{\sqrt{y}+3}{\sqrt{y}+3}\right) = \dfrac{y-9}{5(\sqrt{y}+3)} \text{ or } \dfrac{y-9}{5\sqrt{y}+15}$$

Following the example, rationalize this numerator.

$$\dfrac{2+\sqrt{x}}{7} = \underline{\hspace{3cm}}$$

a) $\dfrac{x(3-\sqrt{2})}{7}$ or $\dfrac{3x-x\sqrt{2}}{7}$

b) $\dfrac{11(\sqrt{R}+2)}{R-4}$ or $\dfrac{11\sqrt{R}+22}{R-4}$

---

163. We used the conjugate of $2 + \sqrt{3}$ to rationalize the numerator below.

$$\dfrac{2+\sqrt{3}}{5} = \dfrac{2+\sqrt{3}}{5}\left(\dfrac{2-\sqrt{3}}{2-\sqrt{3}}\right) = \dfrac{4-3}{5(2-\sqrt{3})} = \dfrac{1}{5(2-\sqrt{3})}$$

$$\text{or } \dfrac{1}{10-5\sqrt{3}}$$

Following the example, rationalize this numerator.

$$\dfrac{\sqrt{5}-1}{x} = \underline{\hspace{3cm}}$$

$\dfrac{4-x}{7(2-\sqrt{x})}$ or $\dfrac{4-x}{14-7\sqrt{x}}$

$\dfrac{4}{x(\sqrt{5}+1)}$ or $\dfrac{4}{x\sqrt{5}+x}$

SELF-TEST 10 (pages 137-152)

1. Which of the following are true?

   a) $\sqrt{PR} = \sqrt{P} \cdot \sqrt{R}$ b) $\sqrt{P + R} = \sqrt{P} + \sqrt{R}$ c) $\sqrt{x^2 - y^2} = x - y$

   d) $\sqrt{x^2 y^2} = xy$ e) $\sqrt{(t + 4)^2} = t + 4$

Multiply.

2. $a\sqrt{p + t} \cdot h\sqrt{b}$

3. $\sqrt{2r - d} \cdot \sqrt{2r - d}$

Simplify by factoring out perfect squares.

4. $\sqrt{8a^2 b + 4a^2 p}$

5. $\sqrt{\dfrac{9r - 36t}{25}}$

---

6. Rationalize the denominator.

   $\dfrac{x + 1}{2\sqrt{x + 1}}$

7. Subtract.

   $5\sqrt{h + k} - \sqrt{h + k}$

Square each fraction. Report answers in lowest terms.

8. $\left(\dfrac{5}{2\sqrt{10}}\right)^2$ 9. $\left(\dfrac{b\sqrt{6t}}{2t}\right)^2$ 10. $\left(\dfrac{3\sqrt{2x - 8}}{2}\right)^2$

Square each binomial.

11. $(d\sqrt{p} + w)^2$

12. $(3 - \sqrt{y + 5})^2$

Multiply.

13. $(\sqrt{5} + 2)(\sqrt{5} - 2)$

14. $(4 + \sqrt{a})(4 - \sqrt{a})$

15. Rationalize the numerator.

   $\dfrac{\sqrt{x} - 1}{2}$

ANSWERS:
1. a, d, e
2. $ah\sqrt{b(p + t)}$
3. $2r - d$
4. $2a\sqrt{2b + p}$
5. $\dfrac{3\sqrt{r - 4t}}{5}$
6. $\dfrac{\sqrt{x + 1}}{2}$
7. $4\sqrt{h + k}$
8. $\dfrac{5}{8}$
9. $\dfrac{3b^2}{2t}$
10. $\dfrac{9(x - 4)}{2}$
11. $d^2 p + 2dw\sqrt{p} + w^2$
12. $14 - 6\sqrt{y + 5} + y$
13. 1
14. $16 - a$
15. $\dfrac{x - 1}{2(\sqrt{x} + 1)}$

Square Root Radicals 153

3-18 REAL AND IMAGINARY NUMBERS

In this section, we will discuss the difference between "real" and "imaginary" numbers. Some operations with "imaginary" numbers are shown.

164. The set of "real" numbers includes the number "0" and all positive and negative whole numbers, decimal numbers, and fractions. For example, each number below is a "real" number.

   $-8 \quad\quad -5.3 \quad\quad -\dfrac{1}{2} \quad\quad 0 \quad\quad \dfrac{5}{3} \quad\quad 8.7 \quad\quad 29$

   Earlier in the chapter, we saw that the square root of a negative number is not a "real" number. That is:

   $\sqrt{-25} \neq 5$, since $5^2 = 25$ $\quad\quad\quad \sqrt{-25} \neq -5$, since $(-5)^2 = 25$

   Mathematicians have developed another type of number to represent the square roots of negative numbers. To contrast them with "real" numbers, they call them "imaginary" numbers. That is:

   $\sqrt{-1}, \sqrt{-4}, \sqrt{-33}$ are called "imaginary" numbers.

   The choice of the word "imaginary" for some numbers was probably not too good because it suggests that the numbers do not exist or that they are useless. However:

   1) The word "imaginary" is merely used as a contrast to the word "real".

   2) "Imaginary" numbers do exist and they are useful.

165. Any imaginary number can be written as a real number times $\sqrt{-1}$. For example:

   $\sqrt{-4} = \sqrt{(4)(-1)} = \sqrt{4} \cdot \sqrt{-1} = 2\sqrt{-1}$
   $\sqrt{-9} = \sqrt{(9)(-1)} = \sqrt{9} \cdot \sqrt{-1} = 3\sqrt{-1}$

   Write each of these as a real number times $\sqrt{-1}$.

   a) $\sqrt{-16} =$ _____ b) $\sqrt{-25} =$ _____ c) $\sqrt{-100} =$ _____

166. The opposite of any imaginary number can also be written as a real number times $\sqrt{-1}$. For example:

   $-\sqrt{-4} = -\sqrt{(4)(-1)} = -\sqrt{4} \cdot \sqrt{-1} = -2\sqrt{-1}$
   $-\sqrt{-9} = -\sqrt{(9)(-1)} = -\sqrt{9} \cdot \sqrt{-1} = -3\sqrt{-1}$

   Write each of these as a real number times $\sqrt{-1}$.

   a) $-\sqrt{-36} =$ _____ b) $-\sqrt{-49} =$ _____ c) $-\sqrt{-81} =$ _____

   a) $4\sqrt{-1}$
   b) $5\sqrt{-1}$
   c) $10\sqrt{-1}$

154 Square Root Radicals

167. Mathematicians use the letter "i" for $\sqrt{-1}$. Therefore:

Instead of $5\sqrt{-1}$, they write 5i.
Instead of $-9\sqrt{-1}$, they write -9i.

Convert each of these to "i" notation.

a) $2\sqrt{-1}$ = _____   b) $10\sqrt{-1}$ = _____   c) $-8\sqrt{-1}$ = _____

a) $-6\sqrt{-1}$
b) $-7\sqrt{-1}$
c) $-9\sqrt{-1}$

---

168. To add or subtract imaginary numbers, we add or subtract their coefficients. For example:

4i + 3i = 7i   10i - 8i = 2i

Complete these:

a) 10i + 5i = _____   b) 3i - 7i = _____

a) 2i
b) 10i
c) -8i

---

169. To perform the addition and subtraction below, we wrote the "1" coefficient explicitly.

i + 5i = 1i + 5i = 6i
9i - i = 9i - 1i = 8i

Complete these:

a) i + i = _____   b) i - 2i = _____

a) 15i
b) -4i

---

170. When an imaginary number is subtracted from itself, the difference is "0". That is:

7i - 7i = 0i = 0   i - i = _____

a) 2i
b) -1i or -i

---

171. Since $i = \sqrt{-1}$, $i^2 = \sqrt{-1} \cdot \sqrt{-1}$ or $(\sqrt{-1})^2$ or -1. Therefore, "i^2" can be expressed as the real number "-1". That is:

Just as $\sqrt{3} \cdot \sqrt{3} = (\sqrt{3})^2 = 3$, $\sqrt{-1} \cdot \sqrt{-1} = (\sqrt{-1})^2 = -1$

Since $i^2 = -1$, we can express "i^2" terms as real numbers by substituting -1 for "i^2". For example:

$5i^2 = 5(-1) = -5$
$-3i^2 = -3(-1) = 3$

Express each of these as a real number.

a) $10i^2$ = _____   b) $-8i^2$ = _____   c) $-i^2$ = _____

0i or 0

Square Root Radicals   155

172. To multiply an imaginary number by a real number, we multiply its coefficient by the real number. That is:

$$5(3i) = 15i \qquad 4(-2i) = -8i$$

Complete these:

a) $-9(6i) = $ _____   c) $8(-i) = $ _____

b) $-2(-10i) = $ _____   d) $-4(i) = $ _____

a) -10
b) 8
c) 1

173. When multiplying two imaginary numbers, we get an "i^2" factor. Therefore, the product can be expressed as a real number. For example:

$$(5i)(2i) = 10i^2 = 10(-1) = -10$$
$$(-3i)(6i) = -18i^2 = -18(-1) = 18$$

Express each product as a real number.

a) $(10i)(7i) = $ _____   c) $(i)(-9i) = $ _____

b) $(-4i)(-8i) = $ _____   d) $(-i)(-i) = $ _____

a) -54i
b) 20i
c) -8i
d) -4i

a) -70   b) -32   c) 9   d) -1

3-19 ADDING AND SUBTRACTING COMPLEX NUMBERS

In this section we will show what is meant by a "complex" number and then discuss the procedure for adding and subtracting complex numbers.

174. A "complex" number is the sum of a real number and an imaginary number. The general form is $\boxed{a + bi}$, where "a" is the real part and "bi" is the imaginary part. Some examples are shown.

$$2 + 5i \qquad -1 + 3i \qquad 10 + i$$

When the coefficient of the imaginary number is negative, we usually write the complex number in subtraction form. That is:

Instead of $8 + (-9i)$, we write $8 - 9i$

Instead of $1 + (-i)$, we write _____

$1 - i$

156 Square Root Radicals

175. Any real number can be written as a complex number in which the coefficient of the imaginary number is "0". For example:

$$6 = 6 + 0i \qquad -4 = -4 + 0i$$

Any imaginary number can be written as a complex number in which the real number is "0". For example:

$$5i = 0 + 5i \qquad -9i = 0 - 9i$$

Write each of these as a complex number.

a) 10 = _____   c) i = _____

b) -1 = _____   d) -3i = _____

---

176. When either "a" or "b" is "0" in a complex number, we frequently write it without the "0" term. For example:

Instead of 0 + 3i, we write 3i.

Instead of 2 + 0i, we write 2.

A "complete" complex number is one in which both terms are explicitly written.

0 + 3i and 2 + 0i are "complete" complex numbers.

An "incomplete" complex number is one in which the "0" term is not explicitly written.

3i and 2 are "incomplete" complex numbers.

Which of the following are "incomplete" complex numbers?

a) -5 + 0i   b) -4   c) -10i   d) 0 - 8i

a) 10 + 0i
b) -1 + 0i
c) 0 + i
d) 0 - 3i

---

177. The procedure for adding complex numbers is the same as the procedure for adding binomials.

$$(5 + 2i) + (3 + 9i) = 8 + 11i$$
$$(-4 + 3i) + (1 - 7i) = -3 - 4i$$

Find each sum.

a) (3 + 10i) + (1 - i) = _____

b) (2 - 5i) + (-6 - i) = _____

(b) and (c)

---

a) 4 + 9i

b) -4 - 6i

178. In each addition below, we got an incomplete complex number as the sum.

$$(6 + 5i) + (-6 + 3i) = 0 + 8i = 8i$$
$$(-1 - 7i) + (-3 + 7i) = -4 + 0i = -4$$

Find each sum.

a) $(10 + 8i) + (-7 - 8i) =$ _____

b) $(5 + i) + (-5 - 2i) =$ _____

---

179. In each addition below, we added an incomplete and a complete complex number.

$$(5 + 3i) + 6i = 5 + 9i$$
$$10 + (-5 + i) = 5 + i$$

Find each sum.

a) $(4 - 7i) + 1 =$ _____   b) $-9i + (-1 - i) =$ _____

a) 3

b) -1i or -i

---

180. We added the three complex numbers below in vertical columns. Use the same method for the other addition.

Add $3 + 2i$, $5 - i$, and $-6 - 4i$.

Add $-1 + i$, $-5 - 3i$, and $2 + 6i$.

$$\begin{array}{r} 3 + 2i \\ 5 - i \\ \underline{-6 - 4i} \\ 2 - 3i \end{array}$$

a) $5 - 7i$

b) $-1 - 10i$

---

181. To get the opposite of a complex number, we change the sign of each term. That is:

The opposite of $3 + 5i$ is $-3 - 5i$.

The opposite of $-1 - 9i$ is $1 + 9i$.

Write the opposite of each complex number.

a) $10 - 7i$ _____   b) $-6 + i$ _____

$$\begin{array}{r} -1 + i \\ -5 - 3i \\ \underline{2 + 6i} \\ -4 + 4i \end{array}$$

---

a) $-10 + 7i$

b) $6 - i$

158   Square Root Radicals

182. To subtract complex numbers, we add the opposite of the second complex number. That is:

$$\text{Opposite}$$
$$(3 - 2i) - (5 + i) = (3 - 2i) + \overbrace{(-5 - i)} = -2 - 3i$$
$$(8 + i) - (-3 - 6i) = (8 + i) + (3 + 6i) = 11 + 7i$$

Complete these subtractions.

a) (1 - i) - (7 - 4i) = (1 - i) + (   ) = _____

b) (-2 + 3i) - (-1 + 4i) = (-2 + 3i) + (   ) = _____

183. In each subtraction below, we got an incomplete complex number as the difference.

$$(5 - 3i) - (5 + 2i) = (5 - 3i) + (-5 - 2i) = 0 - 5i = -5i$$
$$(4 + 7i) - (1 + 7i) = (4 + 7i) + (-1 - 7i) = 3 + 0i = 3$$

Complete these subtractions.

a) (-6 + i) - (-6 + 3i) = (-6 + i) + (   ) = _____

b) (3 - i) - (7 - i) = (3 - i) + (   ) = _____

a) + (-7 + 4i) = -6 + 3i

b) + (1 - 4i) = -1 - i

184. Each subtraction below involves an incomplete and a complete complex number.

$$10 - (3 - 8i) = 10 + (-3 + 8i) = 7 + 8i$$
$$(5 + 6i) - 4i = (5 + 6i) + (-4i) = 5 + 2i$$

Complete each subtraction.

a) (-2 + 7i) - 6 = _____   b) i - (-5 + 4i) = _____

a) + (6 - 3i) = -2i

b) + (-7 + i) = -4

a) -8 + 7i   b) 5 - 3i

3-20 MULTIPLYING COMPLEX NUMBERS AND CONJUGATES

In this section, we will discuss the procedure for multiplying complex numbers. The meaning of the "conjugate" of a complex number is shown.

185. To multiply a complete complex number by a incomplete complex number in which "b" is "0", we use the distributive principle. That is:

$$4(3 + 2i) = 4(3) + 4(2i) = 12 + 8i$$
$$5(1 - 3i) = 5(1) - 5(3i) = 5 - 15i$$

Continued on following page.

Square Root Radical   159

185. Continued

Complete these multiplications.

a) $2(-1 + 4i) = $ _____

b) $-3(-2 - i) = $ _____

---

186. In the multiplication below, we used the distributive principle to multiply a complete complex number by an incomplete complex number in which "a" is "0". Notice how we substituted -1 for i^2 and wrote the product as a complex number.

$$2i(3 + 4i) = 6i + 8i^2 = 6i + 8(-1) = -8 + 6i$$

Write each product as a complex number.

a) $5i(1 + 2i) = $ _____

b) $-3i(4 - i) = $ _____

a) $-2 + 8i$

b) $6 + 3i$

---

187. We converted the product below to a complex number.

$$-3i(5 + 2i) = -15i - 6i^2 = -15i - 6(-1) = -15i + 6 = 6 - 15i$$

Write each product as a complex number.

a) $5i(7 - i) = $ _____

b) $-i(2 + 8i) = $ _____

a) $-10 + 5i$, from: $5i + 10i^2$

b) $-3 - 12i$, from: $-12i + 3i^2$

---

188. We used the FOIL method to multiply two complete complex numbers below. Notice how we substituted -1 for i^2 and then wrote the product as a complex number.

$$(3 + i)(2 + 4i) = 6 + 12i + 2i + 4i^2$$
$$= 6 + 14i + 4(-1)$$
$$= 6 + 14i - 4 = 2 + 14i$$

Write each product as a complex number.

a) $(1 + 2i)(3 + 5i) = $

b) $(2 + i)(5 + 9i) = $

a) $5 + 35i$, from: $35i - 5i^2$

b) $8 - 2i$, from: $-2i - 8i^2$

189. We converted the product below to a complex number.

$$(4 - 5i)(3 + 2i) = 12 + 8i - 15i - 10i^2$$
$$= 12 - 7i - 10(-1)$$
$$= 12 - 7i + 10 = 22 - 7i$$

Write each product as a complex number.

a) $(1 + i)(6 - 4i) =$

b) $(2 - i)(7 + 3i) =$

a) $-7 + 11i$, from: $3 + 11i + 10i^2$

b) $1 + 23i$, from: $10 + 23i + 9i^2$

190. Write each product as a complex number.

a) $(-4 - 3i)(1 - 2i) =$

b) $(-2 + 7i)(-1 - i) =$

a) $10 + 2i$, from: $6 + 2i - 4i^2$

b) $17 - i$, from: $14 - i - 3i^2$

191. In the multiplication below, the complex numbers are the sum and difference of the same two terms. Notice how the product simplifies to an incomplete complex number with only a real number part.

$$(5 + 3i)(5 - 3i) = 5^2 - (3i)^2 = 25 - 9i^2$$
$$= 25 - 9(-1)$$
$$= 25 + 9 = 34$$

Simplify each product to a real number.

a) $(2 + 7i)(2 - 7i) =$

b) $(4 - i)(4 + i) =$

a) $-10 + 5i$, from: $-4 + 5i + 6i^2$

b) $9 - 5i$, from: $2 - 5i - 7i^2$

a) 53, from: $4 - 49i^2$

b) 17, from: $16 - i^2$

192. We simplified the product below to a real number.

$$(-4 + 5i)(-4 - 5i) = (-4)^2 - (5i)^2 = 16 - 25i^2$$
$$= 16 - 25(-1)$$
$$= 16 + 25 = 41$$

Simplify each product to a real number.

a) $(-10 + 9i)(-10 - 9i) =$

b) $(-1 - i)(-1 + i) =$

a) 181, from: $100 - 81i^2$
b) 2, from: $1 - i^2$

193. When two complex numbers are the sum and difference of the same two terms, they are called "<u>conjugates</u>". That is:

$(4 + 9i)$ and $(4 - 9i)$ are "conjugates"

$(-8 + i)$ and $(-8 - i)$ are "conjugates"

Write the conjugate of each complex number.

a) $6 - i$ _____   b) $-2 + 7i$ _____   c) $-12 - 5i$ _____

a) $6 + i$ b) $-2 - 7i$ c) $-12 + 5i$

3-21 DIVIDING COMPLEX NUMBERS

In this section, we will discuss the procedure for dividing complex numbers.

194. When dividing complex numbers, we always write them as a fraction. For example:

$(8 + 2i) \div (4 + 7i)$ is written: $\dfrac{8 + 2i}{4 + 7i}$

When a complex number is divided by itself, the quotient is "1". That is:

$\dfrac{4 + 3i}{4 + 3i} = 1 \qquad \dfrac{5 - i}{5 - i} = 1 \qquad \dfrac{-8 - 9i}{-8 - 9i} =$ _____

1

Square Root Radicals 161

162 Square Root Radicals

195. The procedure for dividing complex numbers is similar to the procedure for rationalizing a denominator like $5-\sqrt{2}$. For example, to perform the division below, we multiplied by a fraction <u>whose terms are identical to the conjugate of the denominator</u>. Notice how we simplified the quotient.

$$\frac{4 + 3i}{5 - 2i} = \left(\frac{4 + 3i}{5 - 2i}\right)\left(\frac{5 + 2i}{5 + 2i}\right) = \frac{20 + 23i + 6i^2}{25 - 4i^2}$$

$$= \frac{20 + 23i + 6(-1)}{25 - 4(-1)}$$

$$= \frac{14 + 23i}{29}$$

To perform the division below, we multiply by a fraction whose terms are the conjugate of (4 + i). Complete the division.

$$\frac{1 + 2i}{4 + i} = \left(\frac{1 + 2i}{4 + i}\right)\left(\frac{4 - i}{4 - i}\right) = $$

---

196. In the last frame, we got $\frac{14 + 23i}{29}$ and $\frac{6 + 7i}{17}$ as the quotients. Quotients of that type can be written as complex numbers in which "a" and "b" are fractions. That is:

$$\frac{14 + 23i}{29} = \frac{14}{29} + \frac{23i}{29} \qquad \frac{6 + 7i}{17} = \underline{\hspace{2cm}}$$

$\frac{6 + 7i}{17}$, from:

$$\frac{4 + 7i - 2i^2}{16 - i^2}$$

---

197. When converting quotients to complex numbers, be sure to reduce to lowest terms. For example:

$$\frac{8 - 11i}{6} = \frac{8}{6} - \frac{11i}{6} = \frac{4}{3} - \frac{11i}{6} \qquad \frac{7 + 5i}{10} = \underline{\hspace{2cm}}$$

$\frac{6}{17} + \frac{7i}{17}$

---

198. Some quotients can be converted to complex numbers in which "a" and "b" are whole numbers. For example:

$$\frac{6 - 8i}{2} = \frac{6}{2} - \frac{8i}{2} = 3 - 4i \qquad \frac{20 + 4i}{4} = \underline{\hspace{2cm}}$$

$\frac{7}{10} + \frac{i}{2}$

---

199. Some quotients can be converted to complex numbers in which "a" and "b" are decimal numbers. For example:

$$\frac{3 + 6i}{5} = \frac{3}{5} + \frac{6i}{5} = 0.6 + 1.2i \qquad \frac{9 - 3i}{2} = \underline{\hspace{2cm}}$$

$5 + i$

$4.5 - 1.5i$

200. Write each quotient as a complex number in which "a" and "b" are either fractions, whole numbers, or decimal numbers.

   a) $\dfrac{-5 + 9i}{1 - i} =$

   b) $\dfrac{4 + 2i}{1 - 3i} =$

---

201. To divide a complex number by an incomplete complex number in which "a" is "0", we can multiply by $\left(\dfrac{i}{i}\right)$ to get a real-number denominator. For example:

$$\dfrac{6 + 3i}{2i} = \left(\dfrac{6 + 3i}{2i}\right)\left(\dfrac{i}{i}\right) = \dfrac{6i + 3i^2}{2i^2} = \dfrac{6i + 3(-1)}{2(-1)} = \dfrac{-3 + 6i}{-2} = 1.5 - 3i$$

$$\dfrac{5 - i}{3i} = \left(\dfrac{5 - i}{3i}\right)\left(\dfrac{i}{i}\right) = \dfrac{5i - i^2}{3i^2} = \dfrac{5i - (-1)}{3(-1)} = \dfrac{1 + 5i}{-3} = -\dfrac{1}{3} - \dfrac{5i}{3}$$

Write each quotient as a complex number.

   a) $\dfrac{4 + 7i}{i} =$

   b) $\dfrac{-9 - 5i}{-4i} =$

Answers:

a) $-7 + 2i$, from: $\dfrac{-14 + 4i}{2}$

b) $-\dfrac{1}{5} + \dfrac{71}{5}$ or or $-0.2 + 1.4i$ from: $\dfrac{-2 + 14i}{10}$

---

a) $7 - 4i$, from: $\dfrac{-7 + 4i}{-1}$ b) $\dfrac{5}{4} - \dfrac{9i}{4}$ or $1.25 - 2.25i$, from: $\dfrac{5 - 9i}{4}$

3-22 IMAGINARY NUMBERS AND COMPLEX NUMBERS AS SOLUTIONS OF EQUATIONS

In this section, we will show how the solution of a quadratic equation can be either an imaginary number or a complex number.

202. To solve $x^2 = 9$ below, we found both square roots of 9. The solutions are 3 and -3. To solve $y^2 = -4$ below, we found both square roots of -4. The solutions are $2i$ and $-2i$.

$$x^2 = 9 \qquad\qquad y^2 = -4$$
$$x = \pm\sqrt{9} \qquad\qquad y = \pm\sqrt{-4}$$
$$x = \pm 3 \qquad\qquad y = \pm\sqrt{4} \cdot \sqrt{-1}$$
$$\qquad\qquad\qquad y = \pm 2i$$

Continued on following page.

164 Square Root Radicals

202. Continued

We checked 2i as a solution of $y^2 = -4$ below. Check -2i as a solution of the same equation.

$$y^2 = -4$$
$$(2i)^2 = -4$$
$$4i^2 = -4$$
$$4(-1) = -4$$
$$-4 = -4$$

$$y^2 = -4$$

203. We solved the equation at the left below. Solve the other equation.

$$x^2 + 25 = 0 \qquad\qquad t^2 + 1 = 0$$
$$x^2 = -25$$
$$x = \pm\sqrt{-25}$$
$$x = \pm 5i$$

$(-2i)^2 = -4$
$4i^2 = -4$
$4(-1) = -4$
$-4 = -4$

204. As we saw earlier, the standard form for quadratic equations is $ax^2 + bx + c = 0$. The quadratic formula used to solve such equations is:

$$\boxed{\text{Two solutions} = \frac{(-b) \pm \sqrt{b^2 - 4ac}}{2a}}$$

We used the quadratic fromula to solve $x^2 - 4x + 5 = 0$ below. From the equation: $a = 1$, $b = -4$, and $c = 5$. Notice that the two solutions are $2 + i$ and $2 - i$.

$$\text{Two solutions} = \frac{4 \pm \sqrt{(-4)^2 - 4(1)(5)}}{2(1)}$$
$$= \frac{4 \pm \sqrt{16 - 20}}{2}$$
$$= \frac{4 \pm \sqrt{-4}}{2}$$
$$= \frac{4 \pm 2i}{2}$$
$$= 2 \pm i$$

We checked $2 + i$ as a solution below. Check $2 - i$ as a solution.

$$x^2 - 4x + 5 = 0 \qquad\qquad x^2 - 4x + 5 = 0$$
$$(2 + i)^2 - 4(2 + i) + 5 = 0$$
$$4 + 4i + i^2 - 8 - 4i + 5 = 0$$
$$4 + 4i - 1 - 8 - 4i + 5 = 0$$
$$9 - 9 + 4i - 4i = 0$$
$$0 = 0$$

$t = \pm 1i$ or $\pm i$

205. Use the quadratic formula to solve the equation below.

$$x^2 - 2x + 2 = 0$$

$(2 - i)^2 - 4(2 - i) + 5 = 0$

$4 - 4i + i^2 - 8 + 4i + 5 = 0$

$4 - 4i - 1 - 8 + 4i + 5 = 0$

$9 - 9 - 4i + 4i = 0$

$0 = 0$

$x = 1 + i$ and $1 - i$

SELF-TEST 11 (pages 153-165)

Find each sum.

1. $(-3 + 7i) + (5 - i)$

2. $(1 + 4i) + (-6 - 4i)$

3. $8i + (1 - 5i)$

Find each difference.

4. $(9 - 5i) - (-1 + 2i)$

5. $(2 - 3i) - (2 + i)$

6. $(2 + 7i) - 6$

Find each product.

7. $-2i(7 - 3i)$

8. $(-5 + 4i)(1 - i)$

9. $(-3 - 6i)(-3 + 6i)$

Find each quotient.

10. $\dfrac{5 + 5i}{1 + 2i}$

11. $\dfrac{10 + 7i}{2 - 2i}$

12. $\dfrac{1 - 4i}{3i}$

13. Find the two solutions of: $y^2 + 36 = 0$

Using the quadratic formula, solve each equation below.

14. $x^2 - 6x + 10 = 0$

15. $w^2 - 2w + 5 = 0$

ANSWERS:
1. $2 + 6i$
2. -5
3. $1 + 3i$
4. $10 - 7i$
5. $-4i$
6. $-4 + 7i$
7. $-6 - 14i$
8. $-1 + 9i$
9. 45
10. $3 - i$
11. $\dfrac{3}{4} + \dfrac{17i}{4}$
12. $-\dfrac{4}{3} - \dfrac{i}{3}$
13. $y = 6i$ and $-6i$
14. $x = 3 + i$ and $3 - i$
15. $w = 1 + 2i$ and $1 - 2i$

SUPPLEMENTARY PROBLEMS - CHAPTER 3

Assignment 8

1. Which of the following are perfect squares?

   a) 25   b) 1   c) 8   d) 90   e) 100

Find the two square roots of each of the following.

   2. 9   3. 36   4. 81   5. 400

Using a calculator, find the principal square root of each of the following.

   6. 289   7. 144,400   8. 234.09   9. .000576

10. In $W = \sqrt{G}$, find W when G = 100.

11. In $h = r\sqrt{kt}$, find "h" when r = 3, k = 32, and t = 2.

12. In $A = K\sqrt{\dfrac{P}{R}}$, find A when K = 10, P = 48, and R = 12.

13. In $p = \sqrt{d^2 - b^2}$, find "p" when d = 10 and b = 8.

14. In $F = \dfrac{B}{\sqrt{AN}}$, find F when B = 80, A = 2, and N = 50.

15. In $s = \sqrt{\dfrac{a+t}{t}}$, find "s" when a = 120 and t = 5.

Do the following multiplications.

16. $\sqrt{5x} \cdot \sqrt{3y}$ 17. $p\sqrt{2t} \cdot 4r\sqrt{d}$ 18. $\sqrt{a} \cdot \sqrt{3w} \cdot \sqrt{7}$

Convert each multiplication to a single radical.

19. $5\sqrt{3}$ 20. $d\sqrt{b}$ 21. $2\sqrt{3x}$ 22. $r\sqrt{8}$

Simplify each radical by factoring out all perfect squares.

23. $\sqrt{45}$ 24. $\sqrt{300x}$ 25. $\sqrt{28r^2w}$ 26. $\sqrt{4h^5p^3}$

Multiply. Simplify each product by factoring out all perfect squares.

27. $\sqrt{2} \cdot \sqrt{18y}$ 28. $\sqrt{6a^3} \cdot \sqrt{8ab}$ 29. $\sqrt{10tw} \cdot \sqrt{20tw^2}$ 30. $\sqrt{h^5k} \cdot \sqrt{12k^3}$

Divide. Simplify each quotient by factoring out all perfect squares.

31. $\dfrac{\sqrt{40}}{\sqrt{5}}$ 32. $\dfrac{\sqrt{27x^2}}{\sqrt{3x}}$ 33. $\dfrac{\sqrt{d^2p^5}}{\sqrt{dp}}$ 34. $\dfrac{\sqrt{75a^5b^3}}{\sqrt{3a^2b}}$

Assignment 9

Simplify each radical by factoring out all perfect squares.

1. $\sqrt{\dfrac{4}{9}}$ 2. $\sqrt{\dfrac{p^6}{a^4}}$ 3. $\sqrt{\dfrac{8t^3}{25d^2}}$ 4. $\sqrt{\dfrac{16x^2y}{9w^2}}$

Reduce each fraction to lowest terms; then factor out all perfect squares.

5. $\sqrt{\dfrac{a^2b^2}{2a}}$ 6. $\sqrt{\dfrac{27r^3v}{3r^2}}$ 7. $\sqrt{\dfrac{5a^2b^3}{15b^2d}}$ 8. $\sqrt{\dfrac{2p}{8h^2p}}$

Assignment 9 (continued)

Simplify. Each answer should contain only one radical, in lowest terms, with all perfect squares factored out.

9. $\dfrac{\sqrt{6k}}{\sqrt{2r}}$
10. $\dfrac{\sqrt{mp}}{t\sqrt{5m}}$
11. $\dfrac{\sqrt{8b^2h^3}}{\sqrt{4b}}$
12. $\dfrac{\sqrt{rt}}{\sqrt{r^3t^2}}$

Simplify each radical.

13. $\sqrt{(4x)^2}$
14. $\sqrt{4x^2}$
15. $\sqrt{\left(\dfrac{9p}{4r}\right)^2}$
16. $\sqrt{\dfrac{9p^2}{4r^2}}$

Simplify each expression.

17. $\sqrt{8} \cdot \sqrt{8}$
18. $(\sqrt{3})^2$
19. $\left(\sqrt{\dfrac{a}{d}}\right)^2$
20. $\sqrt{\dfrac{9s}{w}} \cdot \sqrt{\dfrac{9s}{w}}$

21. Which of the following are true?

   a) $\sqrt{9y} \cdot \sqrt{9y} = 3y$
   b) $2\sqrt{x} \cdot 2\sqrt{x} = 4x$
   c) $(\sqrt{9t})^2 = 3t$
   d) $\sqrt{(9t)^2} = 9t$

Rationalize each denominator. Report answers in lowest terms.

22. $\dfrac{3}{2\sqrt{3}}$
23. $\dfrac{a}{d\sqrt{4k}}$
24. $\sqrt{\dfrac{mv^2}{pt^2}}$
25. $\sqrt{\dfrac{2r}{w}}$

Rationalize each numerator. Report answers in lowest terms.

26. $\dfrac{\sqrt{20}}{5}$
27. $\dfrac{x\sqrt{y}}{y}$
28. $\sqrt{\dfrac{a^2d}{hp^2}}$
29. $\sqrt{\dfrac{v}{4s}}$

Find each sum.

30. $7\sqrt{3} + 2\sqrt{3}$
31. $\sqrt{F} + 3\sqrt{F}$
32. $p\sqrt{y} + r\sqrt{y}$

Find each difference.

33. $2\sqrt{10} - \sqrt{10}$
34. $\sqrt{x} - 5\sqrt{x}$
35. $K\sqrt{A} - W\sqrt{A}$

36. Which of the following are true?

   a) $\sqrt{d} \cdot \sqrt{k} = \sqrt{dk}$
   b) $\sqrt{d} + \sqrt{k} = \sqrt{d+k}$
   c) $\sqrt{7} - \sqrt{3} = \sqrt{7-3}$
   d) $a\sqrt{h} - b\sqrt{h} = a - b\sqrt{h}$

Assignment 10

1. Which of the following are true?

   a) $\sqrt{(p-r)^2} = p - r$
   b) $\sqrt{x^2 + 9} = x + 3$
   c) $\sqrt{b-d} = \sqrt{b} - \sqrt{d}$
   d) $\sqrt{at} = \sqrt{a} \cdot \sqrt{t}$

Simplify each of the following if possible.

2. $\sqrt{a^2 + b^2}$
3. $\sqrt{(a+b)^2}$
4. $\sqrt{c^2 - d^2}$
5. $\sqrt{c^2d^2}$

Do these multiplications.

6. $\sqrt{2r} \cdot \sqrt{r-1}$
7. $p\sqrt{w+2} \cdot \sqrt{d}$
8. $2\sqrt{a} \cdot b\sqrt{a+4}$
9. $\sqrt{h-p} \cdot \sqrt{h-p}$

Simplify by factoring out all perfect squares. Report answers in lowest terms.

10. $\sqrt{9x^2 + 36y^2}$
11. $\sqrt{a^2t - a^2w}$
12. $\dfrac{\sqrt{8h^2p - 4h^2r}}{6p}$
13. $\sqrt{\dfrac{m^2v^2 + r^2v^2}{16v^2}}$

Rationalize each denominator.

14. $\dfrac{2}{\sqrt{r-1}}$
15. $\dfrac{3(y+2)}{\sqrt{y+2}}$

Rationalize each numerator.

16. $\dfrac{\sqrt{a+b}}{c}$
17. $\dfrac{k\sqrt{t-1}}{2t}$

Assignment 10 (continued)

Find each sum or difference.

18. $8\sqrt{x+y} + 3\sqrt{x+y}$
19. $4\sqrt{m-1} - \sqrt{m-1}$
20. $\sqrt{h+4} + \sqrt{h+4}$

Complete each squaring. Report answers in lowest terms.

21. $\left(\dfrac{2\sqrt{6}}{3}\right)^2$
22. $\left(\dfrac{4b}{a\sqrt{2b}}\right)^2$
23. $\left(\dfrac{\sqrt{rw}}{w}\right)^2$
24. $\left(2x\sqrt{\dfrac{x-y}{x}}\right)$

Do these squarings involving binomials. Combine like terms where possible.

25. $(3 - \sqrt{F})^2$
26. $\left(\dfrac{\sqrt{x}+4}{4}\right)^2$
27. $(2 + \sqrt{t+5})^2$
28. $(3 - \sqrt{1-R})^2$

Do these multiplications involving conjugates.

29. $(5 + \sqrt{3})(5 - \sqrt{3})$
30. $(\sqrt{a} - 2)(\sqrt{a} + 2)$
31. $(1 + \sqrt{w})(1 - \sqrt{w})$

Rationalize each denominator.   Rationalize each numerator.

32. $\dfrac{2}{5 - \sqrt{3}}$
33. $\dfrac{3}{\sqrt{x}+1}$
34. $\dfrac{2 - \sqrt{2}}{4}$
35. $\dfrac{\sqrt{r}+5}{2}$

Assignment 11

Write each of the following radicals in "i" notation, where $i = \sqrt{-1}$.

1. $\sqrt{-16}$
2. $\sqrt{-9}$
3. $\sqrt{-100}$
4. $-\sqrt{-49}$
5. $-\sqrt{-1}$

Do the following problems involving operations with "i".

6. $3i + 9i$
7. $5i - 5i$
8. $4i^2$
9. $7(-6i)$
10. $(-2i)(5i)$

Add. Write each sum as a complex number.

11. $(-3 + 5i) + (1 - 4i)$
12. $(7 - i) + (-7 - 3i)$
13. $(-2 + 6i) + (7 - 6i)$

Subtract. Write each difference as a complex number.

14. $(1 + 3i) - (5 - 2i)$
15. $(-8 - i) - (4 - i)$
16. $(7 + 5i) - (7 - 5i)$

Multiply. Write each product as a complex number.

17. $(2 - 3i)(3 + 4i)$
18. $(-7 + 2i)(-1 - i)$
19. $(0 - 6i)(4 - 5i)$

Multiply these conjugate pairs of complex numbers.

20. $(5 + 2i)(5 - 2i)$
21. $(-3 + i)(-3 - i)$
22. $(-8 - 6i)(-8 + 6i)$

Divide. Write each quotient as a complex number in lowest terms.

23. $\dfrac{3 - i}{1 + i}$
24. $\dfrac{6 + 17i}{3 - 4i}$
25. $\dfrac{-3 - 5i}{-1 + 3i}$
26. $\dfrac{-2 + 4i}{-5 - i}$

Find the two solutions of each quadratic equation.

27. $x^2 = -9$
28. $y^2 + 49 = 0$
29. $w^2 + 4 = 0$

Using the quadratic formula, find the two solutions of each equation.

30. $x^2 - 4x + 8 = 0$
31. $t^2 - 2t + 10 = 0$
32. $y^2 - 4y + 13 = 0$

Chapter 4 RADICAL EQUATIONS AND FORMULAS

In this chapter, we will discuss a method for solving radical equations and for rearranging radical formulas. Methods for simplifying radical solutions are included. We will also show that either the equivalence method or the substitution method can be used to eliminate a variable from a system of formulas involving square root radicals.

4-1 THE SQUARING PRINCIPLE FOR RADICAL EQUATIONS

In this section, we will show how the "squaring principle" can be used to solve radical equations when the radical is isolated on one side.

1. We saw earlier that the square of a square root radical is the radicand. That is:

    $\left(\sqrt{x}\right)^2 = x$ $\left(\sqrt{y-4}\right)^2 = y - 4$ $\left(\sqrt{\dfrac{t}{3}}\right)^2 = \dfrac{t}{3}$

    Write each square.

    a) $\left(\sqrt{3m}\right)^2 = $ _____   b) $\left(\sqrt{\dfrac{1}{d}}\right)^2 = $ _____   c) $\left(\sqrt{\dfrac{p+3}{2}}\right)^2 = $ _____

2. The **SQUARING PRINCIPLE FOR EQUATIONS** says this:

    > If we square both sides of an equation, the new equation is equivalent to the original equation.

    By using the "squaring principle", we can eliminate the radical in a radical equation. Two examples are shown.

    $\sqrt{y} = 4$ $\sqrt{x-1} = 9$
    $\left(\sqrt{y}\right)^2 = 4^2$ $\left(\sqrt{x-1}\right)^2 = 9^2$
    $y = 16$ $x - 1 = 81$

    Write the non-radical equation obtained by squaring both sides of these.

    a) $\sqrt{5t} = 1$ b) $6 = \sqrt{4a-3}$ c) $\sqrt{\dfrac{4x}{7}} = 2$

Answers:

a) $3m$

b) $\dfrac{1}{d}$

c) $\dfrac{p+3}{2}$

3. When using the squaring principle, both sides of the equation must be squared. The common error is forgetting to square the non-radical side.

   Write the non-radical equation obtained by squaring both sides of these.

   a) $\sqrt{x} = \dfrac{4}{3}$ b) $3 = \sqrt{\dfrac{100}{y+1}}$ c) $\sqrt{\dfrac{5m}{7}} = \dfrac{1}{2}$

   a) $5t = 1$
   b) $36 = 4a - 3$
   c) $\dfrac{4x}{7} = 4$

4. When the radical is isolated on one side, only two steps are needed to solve a radical equation. They are:

   1. Eliminate the radical by squaring both sides.
   2. Then solve the resulting non-radical equation.

   We used the two steps to solve the equation below and then checked the solution. Solve and check the other equation.

   $\sqrt{x-3} = 2$ $\sqrt{3y} = 6$

   $\left(\sqrt{x-3}\right)^2 = 2^2$

   $x - 3 = 4$

   $x = 7$

   Check

   $\sqrt{x-3} = 2$

   $\sqrt{7-3} = 2$

   $\sqrt{4} = 2$

   $2 = 2$

   a) $x = \dfrac{16}{9}$
   b) $9 = \dfrac{100}{y+1}$
   c) $\dfrac{5m}{7} = \dfrac{1}{4}$

5. Following the example, solve the other equations.

   $\sqrt{\dfrac{3m}{4}} = 6$ a) $\sqrt{\dfrac{t}{5}} = 3$ b) $1 = \sqrt{\dfrac{7}{2x}}$

   $\dfrac{3m}{4} = 36$

   $3m = 144$

   $m = 48$

   $y = 12$

   Check:

   $\sqrt{3(12)} = 6$

   $\sqrt{36} = 6$

   $6 = 6$

   a) $t = 45$
   b) $x = \dfrac{7}{2}$

6. Following the example, solve the other equation.

$$\sqrt{\frac{3x-1}{2}} = 4 \qquad \sqrt{\frac{200}{y+1}} = 5$$

$$\frac{3x-1}{2} = 16$$

$$3x - 1 = 32$$

$$3x = 33$$

$$x = 11$$

$y = 7$, from:

$$\frac{200}{y+1} = 25$$

$$200 = 25(y+1)$$

$$200 = 25y + 25$$

---

7. Following the example, solve the other equations.

$$\sqrt{5x} = \frac{3}{4} \qquad a) \sqrt{3p} = \frac{2}{3} \qquad b) \frac{1}{2} = \sqrt{7m}$$

$$5x = \frac{9}{16}$$

$$80x = 9$$

$$x = \frac{9}{80}$$

a) $p = \frac{4}{27}$

b) $m = \frac{1}{28}$

---

8. Following the example, solve the other equation.

$$\frac{1}{3} = \sqrt{\frac{1}{d}} \qquad \sqrt{\frac{p+1}{2}} = \frac{3}{2}$$

$$\frac{1}{9} = \frac{1}{d}$$

$$\cancel{9}d\left(\frac{1}{\cancel{9}}\right) = 9\cancel{d}\left(\frac{1}{\cancel{d}}\right)$$

$$d = 9$$

$p = \frac{7}{2}$, from:

$$\frac{p+1}{2} = \frac{9}{4}$$

$$\cancel{4}^2\left(\frac{p+1}{\cancel{2}}\right) = \cancel{4}\left(\frac{9}{\cancel{4}}\right)$$

$$2p + 2 = 9$$

9. If we square both sides of $\sqrt{y+1} = \frac{3}{2}$, we get: $y + 1 = \frac{9}{4}$. There are two ways to solve the non-radical equation.

Clearing the fraction and then solving the non-fractional equation.	Using the addition axiom with the equation as it stands.
$y + 1 = \frac{9}{4}$	$y + 1 = \frac{9}{4}$
$4y + 4 = 9$	$y = \frac{9}{4} - 1$
$4y = 5$	$y = \frac{9}{4} - \frac{4}{4}$
$y = \frac{5}{4}$	$y = \frac{5}{4}$

Using either method, solve each of these.

a) $\sqrt{x - 1} = \frac{3}{4}$ b) $\sqrt{t - \frac{1}{3}} = 3$

---

10. Following the example, solve the other equation.

$\sqrt{x^2 + 9} = 5$ $\sqrt{y^2 - 36} = 8$

$x^2 + 9 = 25$

$x^2 = 16$

$x = \pm 4$

a) $x = \frac{25}{16}$ or $1\frac{9}{16}$

b) $t = \frac{28}{3}$ or $9\frac{1}{3}$

---

11. Following the example, solve the other equation.

$\sqrt{2x^2 - 18} = 0$ $\sqrt{16y^2 - 9} = 4$

$2x^2 - 18 = 0$

$2x^2 = 18$

$x^2 = 9$

$x = \pm 3$

$y = \pm 10$

$y = \pm \frac{5}{4}$

Radical Equations and Formulas   173

4-2   ISOLATING RADICALS IN RADICAL EQUATIONS

Before applying the squaring principle to a radical equation, we usually isolate the radical first. We will discuss the procedure for isolating radicals in this section.

---

12. We used the addition axiom to isolate the radical below. Isolate the radical in the other equation.

    $$3 + \sqrt{x} = 7 \qquad\qquad 10 = \sqrt{\frac{5x-1}{3}} - 30$$

    $$(-3) + 3 + \sqrt{x} = 7 + (-3)$$

    $$\sqrt{x} = 4$$

---

13. We used the multiplication axiom to isolate the radical below. Isolate the radical in the other equation.

    $$5\sqrt{y+3} = 1 \qquad\qquad 10 = 7\sqrt{\frac{t}{4}}$$

    $$\sqrt{y+3} = \frac{1}{5}$$

    $40 = \sqrt{\dfrac{5x-1}{3}}$

---

14. We used both axioms to isolate the radical below. Isolate the radical in the other equation.

    $$2 + 3\sqrt{R} = 7 \qquad\qquad 15 = 10\sqrt{\frac{y}{y-1}} - 25$$

    $$3\sqrt{R} = 5$$

    $$\sqrt{R} = \frac{5}{3}$$

    $\dfrac{10}{7} = \sqrt{\dfrac{t}{4}}$

---

15. We isolated the radical below. Notice how we multiplied both sides by 5 to clear the fraction. Isolate the radical in the other equation.

    $$\frac{3\sqrt{y}}{5} = 2 \qquad\qquad 1 = \frac{7\sqrt{x-3}}{6}$$

    $$\cancel{5}\left(\frac{3\sqrt{y}}{\cancel{5}}\right) = 5(2)$$

    $$3\sqrt{y} = 10$$

    $$\sqrt{y} = \frac{10}{3}$$

    $4 = \sqrt{\dfrac{y}{y-1}}$

$\sqrt{x-3} = \dfrac{6}{7}$, from:

$6 = 7\sqrt{x-3}$

16. To clear the fraction below, we multiplied both sides by $\sqrt{t}$. Isolate the radical in each of the other equations.

$$\frac{3}{\sqrt{t}} = 7 \qquad \text{a) } \frac{5}{\sqrt{6x}} = 1 \qquad \text{b) } 4 = \frac{9}{\sqrt{y^2 - 25}}$$

$$\sqrt{t}\left(\frac{3}{\sqrt{t}}\right) = 7\left(\sqrt{t}\right)$$

$$3 = 7\sqrt{t}$$

$$\sqrt{t} = \frac{3}{7}$$

---

17. To clear the fraction below, we multiplied both sides by $3\sqrt{b}$. Isolate the radical in the other equation.

$$\frac{5}{3\sqrt{b}} = 2 \qquad\qquad 3 = \frac{1}{4\sqrt{V-5}}$$

$$3\sqrt{b}\left(\frac{5}{3\sqrt{b}}\right) = 2\left(3\sqrt{b}\right)$$

$$5 = 6\sqrt{b}$$

$$\sqrt{b} = \frac{5}{6}$$

a) $\sqrt{6x} = 5$

b) $\sqrt{y^2 - 25} = \frac{9}{4}$, from:

$4\sqrt{y^2 - 25} = 9$

---

18. To clear the fractions below, we multiplied both sides by 6. Isolate the radical in the other equation.

$$\frac{5\sqrt{x}}{6} = \frac{1}{3} \qquad\qquad \frac{3\sqrt{y-7}}{4} = \frac{9}{8}$$

$$6\left(\frac{5\sqrt{x}}{6}\right) = 6\left(\frac{1}{3}\right)$$

$$5\sqrt{x} = 2$$

$$\sqrt{x} = \frac{2}{5}$$

$\sqrt{V-5} = \frac{1}{12}$, from:

$12\sqrt{V-5} = 1$

---

19. To clear the fractions below, we multiplied both sides by 6. Isolate the radical in the other equation.

$$\frac{4\sqrt{x}}{3} = \frac{1}{2} \qquad\qquad \frac{1}{5} = \frac{3\sqrt{2y}}{4}$$

$$6\left(\frac{4\sqrt{x}}{3}\right) = 6\left(\frac{1}{2}\right)$$

$$8\sqrt{x} = 3$$

$$\sqrt{x} = \frac{3}{8}$$

$\sqrt{y-7} = \frac{3}{2}$, from:

$6\sqrt{y-7} = 9$

$\sqrt{2y} = \frac{4}{15}$, from:

$4 = 15\sqrt{2y}$

20. To clear the fraction below, we multiplied both sides by $12\sqrt{y}$. Isolate the radical in the other equation.

$$\frac{2}{3\sqrt{y}} = \frac{1}{4} \qquad\qquad \frac{1}{3} = \frac{5}{2\sqrt{7d}}$$

$$\overset{4}{\cancel{12}\cancel{\sqrt{y}}}\left(\frac{2}{\cancel{3}\cancel{\sqrt{y}}}\right) = \overset{3}{\cancel{12}}\sqrt{y}\left(\frac{1}{\cancel{4}}\right)$$

$$8 = 3\sqrt{y}$$

$$\sqrt{y} = \frac{8}{3}$$

---

21. To clear the fraction below, we multiplied both sides by 5. Isolate the radical in the other equation.

$$\frac{4 + \sqrt{x}}{5} = 2 \qquad\qquad 1 = \frac{3 + 2\sqrt{y}}{7}$$

$$\cancel{5}\left(\frac{4 + \sqrt{x}}{\cancel{5}}\right) = 5(2)$$

$$4 + \sqrt{x} = 10$$

$$\sqrt{x} = 6$$

$\sqrt{7d} = \frac{15}{2}$, from:

$2\sqrt{7d} = 15$

---

22. To clear the fraction below, we multiplied both sides by $3 + \sqrt{x}$. Notice that $2(3 + \sqrt{x})$ is an instance of the distributive principle.

$$\frac{7}{3 + \sqrt{x}} = 2$$

$$\cancel{(3+\sqrt{x})}\left(\frac{7}{\cancel{3+\sqrt{x}}}\right) = 2\left(3 + \sqrt{x}\right)$$

$$7 = 6 + 2\sqrt{x}$$

$$1 = 2\sqrt{x}$$

$$\sqrt{x} = \frac{1}{2}$$

Following the example, isolate the radical in the equation below.

$$4 = \frac{15}{2 + 3\sqrt{y}}$$

$\sqrt{y} = 2$, from:

$7 = 3 + 2\sqrt{y}$

$\sqrt{y} = \frac{7}{12}$, from:

$15 = 8 + 12\sqrt{y}$

176 Radical Equations and Formulas

4-3 ISOLATING RADICALS AND SOLVING RADICAL EQUATIONS

In this section, we will discuss a method for solving radical equations containing a radical that is not isolated.

23. The method for solving radical equations containing a radical that is not isolated involves three steps:

    1. Isolating the radical.
    2. Applying the squaring principle.
    3. Solving the resulting non-radical equation.

An example is shown. Use the same steps to solve the other equation.

$$2 + \sqrt{3x} = 8 \qquad\qquad 5 + \sqrt{T + 10} = 10$$
$$\sqrt{3x} = 6$$
$$3x = 36$$
$$x = 12$$

24. Following the example, solve the other equation.

$$4\sqrt{y} = 12 \qquad\qquad 3\sqrt{2F} = 18$$
$$\sqrt{y} = 3$$
$$y = 9$$

$T = 15$, since:
$$\sqrt{T + 10} = 5$$
$$T + 10 = 25$$

25. Following the example, solve the other equation.

$$2\sqrt{\frac{m}{3}} = 8 \qquad\qquad 30 = 10\sqrt{\frac{3t}{5}}$$
$$\sqrt{\frac{m}{3}} = 4$$
$$\frac{m}{3} = 16$$
$$m = 48$$

$F = 18$, since:
$$\sqrt{2F} = 6$$
$$2F = 36$$

$t = 15$, since: $3 = \sqrt{\frac{3t}{5}}$
$$9 = \frac{3t}{5}$$
$$45 = 3t$$

26. Following the example, solve the other equation.

$\sqrt{\dfrac{5x}{x-1}} - 3 = 0$ $\sqrt{\dfrac{V}{2V-3}} + 7 = 8$

$\sqrt{\dfrac{5x}{x-1}} = 3$

$\dfrac{5x}{x-1} = 9$

$5x = 9(x-1)$

$5x = 9x - 9$

$-4x = -9$

$x = \dfrac{9}{4}$ or $2\dfrac{1}{4}$

---

27. Following the example, solve the other equation.

$\dfrac{3\sqrt{2x}}{4} = 6$ $\dfrac{1}{2\sqrt{P}} = 4$

$3\sqrt{2x} = 24$

$\sqrt{2x} = 8$

$2x = 64$

$x = 32$

$V = 3$, since:

$\sqrt{\dfrac{V}{2V-3}} = 1$

$\dfrac{V}{2V-3} = 1$

$V = 2V - 3$

$-V = -3$

---

28. Following the example, solve the other equation.

$\dfrac{3}{\sqrt{x+1}} = 2$ $4 = \dfrac{1}{\sqrt{D-1}}$

$3 = 2\sqrt{x+1}$

$\dfrac{3}{2} = \sqrt{x+1}$

$\dfrac{9}{4} = x + 1$

$x = \dfrac{9}{4} - 1$

$x = \dfrac{5}{4}$ or $1\dfrac{1}{4}$

$P = \dfrac{1}{64}$, since:

$1 = 8\sqrt{P}$

$\dfrac{1}{8} = \sqrt{P}$

$D = 1\dfrac{1}{16}$ or $\dfrac{17}{16}$, since: $4\sqrt{D-1} = 1$

$\sqrt{D-1} = \dfrac{1}{4}$

$D - 1 = \dfrac{1}{16}$

29. Following the example, solve the other equation.

$$\frac{20}{3 + 2\sqrt{x}} = 5 \qquad\qquad \frac{4}{3 + \sqrt{4T - 3}} = 1$$

$$20 = 5(3 + 2\sqrt{x})$$
$$20 = 15 + 10\sqrt{x}$$
$$5 = 10\sqrt{x}$$
$$\sqrt{x} = \frac{1}{2}$$
$$x = \frac{1}{4}$$

---

30. To clear the fractions below, we multiplied both sides by 4. Solve the other equation.

$$\frac{\sqrt{3x}}{4} = \frac{1}{2} \qquad\qquad \frac{2}{3} = \frac{\sqrt{2y}}{9}$$

$$\cancel{4}\left(\frac{\sqrt{3x}}{\cancel{4}}\right) = \cancel{4}^{2}\left(\frac{1}{\cancel{2}}\right)$$
$$\sqrt{3x} = 2$$
$$3x = 4$$
$$x = \frac{4}{3}$$

$T = 1$, since:
$$4 = 3 + \sqrt{4T - 3}$$
$$1 = \sqrt{4T - 3}$$
$$1 = 4T - 3$$
$$4 = 4T$$

---

31. To clear the fractions below, we multiplied both sides by 20. Solve the other equation.

$$\frac{\sqrt{m + 1}}{5} = \frac{1}{4} \qquad\qquad \frac{1}{3} = \frac{\sqrt{R - 1}}{2}$$

$$\cancel{20}^{4}\left(\frac{\sqrt{m + 1}}{\cancel{5}}\right) = \cancel{20}^{5}\left(\frac{1}{\cancel{4}}\right)$$
$$4\sqrt{m + 1} = 5$$
$$\sqrt{m + 1} = \frac{5}{4}$$
$$m + 1 = \frac{25}{16}$$
$$m = \frac{9}{16}$$

$y = 18$, since:
$$\cancel{9}^{3}\left(\frac{2}{\cancel{3}}\right) = \cancel{9}\left(\frac{\sqrt{2y}}{\cancel{9}}\right)$$
$$6 = \sqrt{2y}$$
$$36 = 2y$$

32. To clear the fractions below, we multiplied both sides by $15\sqrt{y}$. Solve the other equation.

$$\frac{2}{5\sqrt{y}} = \frac{1}{3}$$

$$\overset{3}{\cancel{15}}\sqrt{\cancel{y}}\left(\frac{2}{\cancel{5}\sqrt{\cancel{y}}}\right) = \overset{5}{\cancel{15}}\sqrt{y}\left(\frac{1}{\cancel{3}}\right)$$

$$6 = 5\sqrt{y}$$

$$\frac{6}{5} = \sqrt{y}$$

$$y = \frac{36}{25}$$

$$\frac{5}{6} = \frac{1}{3\sqrt{H-1}}$$

$R = 1\frac{4}{9}$ or $\frac{13}{9}$, since:

$$\overset{2}{\cancel{6}}\left(\frac{1}{\cancel{3}}\right) = \overset{3}{\cancel{6}}\left(\frac{\sqrt{R-1}}{\cancel{2}}\right)$$

$$2 = 3\sqrt{R-1}$$

$$\frac{2}{3} = \sqrt{R-1}$$

$$\frac{4}{9} = R - 1$$

$H = 1\frac{4}{25}$ or $\frac{29}{25}$

4-4 FORMULA EVALUATIONS REQUIRING EQUATION-SOLVING

In this section, we will discuss some formula evaluations that require solving a radical equation.

33. To perform the evaluation below, we had to solve a radical equation. Complete the other evaluation.

In the formula below, find A when $s = 10$.

$$\boxed{s = \sqrt{A}}$$

$$10 = \sqrt{A}$$

$$A = 100$$

In the formula below, find "s" when $v = 8$ and $a = 2$.

$$\boxed{v = \sqrt{2as}}$$

$s = 16$, from:

$$8 = \sqrt{4s}$$

$$64 = 4s$$

34. Following the example, complete the other evaluation.

In the formula below, find R when I = 6 and P = 72.

$$\boxed{I = \sqrt{\frac{P}{R}}}$$

$$6 = \sqrt{\frac{72}{R}}$$

$$36 = \frac{72}{R}$$

$$36R = 72$$

$$R = 2$$

In the formula below, find "s" when t = 10 and g = 3.

$$\boxed{t = \sqrt{\frac{2s}{g}}}$$

35. In the evaluation below, we reported only the positive root because the negative root does not make sense. Complete the other evaluation.

In the formula below, find X when Z = 5 and R = 3.

$$\boxed{Z = \sqrt{R^2 + X^2}}$$

$$5 = \sqrt{3^2 + X^2}$$

$$5 = \sqrt{9 + X^2}$$

$$25 = 9 + X^2$$

$$X^2 = 16$$

$$X = 4$$

In the formula below, find "v_f" when v_o = 8, a = 6, and s = 3.

$$\boxed{v_o = \sqrt{v_f^2 - 2as}}$$

36. a) In the formula below, find I_c when d = 5, I_a = 100, and m = 2.

$$\boxed{d = \sqrt{\frac{I_a - I_c}{m}}}$$

b) In the formula below, find Q when R = 100 and P = 95.

$$\boxed{R = P + \sqrt{Q}}$$

s = 150, from:

$$100 = \frac{2s}{3}$$

v_f = 10, from:

$$8 = \sqrt{v_f^2 - 36}$$

$$64 = v_f^2 - 36$$

$$v_f^2 = 100$$

Radical Equations and Formulas   181

37. a) In the formula below, find T when $a = 100$, $R = 4$, and $V = 50$.

$$\boxed{\dfrac{a\sqrt{T}}{R} = V}$$

b) In the formula below, find L_2 when $K = 10$, $M = 100$, and $L_1 = 5$.

$$\boxed{K = \dfrac{M}{\sqrt{L_1 L_2}}}$$

a) $I_c = 50$, from:

$$25 = \dfrac{100 - I_c}{2}$$

b) $Q = 25$, from:

$$5 = \sqrt{Q}$$

---

38. Following the example, complete the other evaluation.

In the formula below, find "g_p" when $g_o = 4$, $r_o = 60$, and $r_p = 12$.

$$\boxed{\sqrt{\dfrac{g_p}{g_o}} = \dfrac{r_o}{r_p}}$$

$$\sqrt{\dfrac{g_p}{4}} = \dfrac{60}{12}$$

$$\sqrt{\dfrac{g_p}{4}} = 5$$

$$\dfrac{g_p}{4} = 25$$

$$g_p = 100$$

In the formula below, find S when $T = 45$, $V = 30$, and $D = 10$.

$$\boxed{\sqrt{\dfrac{T}{S}} = \dfrac{V}{D}}$$

a) $T = 4$, from:

$$\sqrt{T} = 2$$

b) $L_2 = 20$, from:

$$\sqrt{5L_2} = 10$$

---

39. Following the example, complete the other evaluation.

In the formula below, find "a" when $w = 640$, $g = 32$, and $r = 4$.

$$\boxed{w = 2g\sqrt{\dfrac{a}{r}}}$$

$$640 = 2(32)\sqrt{\dfrac{a}{4}}$$

$$640 = 64\sqrt{\dfrac{a}{4}}$$

$$10 = \sqrt{\dfrac{a}{4}}$$

$$100 = \dfrac{a}{4}$$

$$a = 400$$

In the formula below, find "g_o" when $r_o = 400$, $r_p = 100$, and $g_p = 160$.

$$\boxed{\dfrac{r_o}{r_p} = \sqrt{\dfrac{g_p}{g_o}}}$$

$S = 5$, from:

$$\sqrt{\dfrac{45}{S}} = 3$$

$g_o = 10$, from:

$$4 = \sqrt{\dfrac{160}{g_o}}$$

182 Radical Equations and Formulas

4-5 EXTRANEOUS SOLUTIONS

To solve a radical equation, we occasionally have to solve a complete quadratic equation. Sometimes one of the two solutions of the quadratic equation does not satisfy the original radical equation. A solution of that type is called an "extraneous" solution. We will discuss "extraneous" solutions in this section.

40. After squaring both sides to eliminate the radical in the equation below, we get a quadratic equation.

$$\sqrt{x+3} = x+1$$
$$\left(\sqrt{x+3}\right)^2 = (x+1)^2$$
$$x+3 = x^2 + 2x + 1$$

or

$$x^2 + x - 2 = 0$$
$$(x-1)(x+2) = 0$$

The two solutions of the quadratic equation are "1" and "-2". We checked the two solutions in the original radical equation below.

Checking "1"	Checking "-2"
$\sqrt{1+3} = 1+1$	$\sqrt{(-2)+3} = (-2)+1$
$\sqrt{4} = 2$	$\sqrt{1} = -1$
$2 = 2$	$1 \neq -1$

Of the two solutions, only "1" satisfies the original equation. Therefore, the other solution (-2) is called an "extraneous" solution. That is, <u>-2 is not a solution of the original equation</u>.

41. When we get a complete quadratic equation by squaring both sides of a radical equation, we do not always get an extraneous solution. Sometimes both solutions satisfy the original radical equation. For example:

$$y - 1 = \sqrt{4y - 7}$$
$$(y-1)^2 = \left(\sqrt{4y-7}\right)^2$$
$$y^2 - 2y + 1 = 4y - 7$$
$$y^2 - 6y + 8 = 0$$
$$(y-2)(y-4) = 0$$

The two solutions of the quadratic equations are "2" and "4". Show that both solutions satisfy the original radical equation.

Checking "2"	Checking "4"
$y - 1 = \sqrt{4y - 7}$	$y - 1 = \sqrt{4y - 7}$

Radical Equations and Formulas 183

42. When solving a radical equation leads to solving a complete quadratic equation, <u>the only way to detect an extraneous solution is to check both solutions in the original radical equation</u>.

Solve this equation and check both solutions.

$$m - 5 = \sqrt{3m - 5}$$

$2 - 1 = \sqrt{4(2) - 7}$
$1 = \sqrt{1}$
$1 = 1$
$4 - 1 = \sqrt{4(4) - 7}$
$3 = \sqrt{9}$
$3 = 3$

a) The two solutions of the quadratic equation are _____ and _____.

b) Is either solution an extraneous solution? _____

When solving a radical equation leads to solving a quadratic equation, always check for an extraneous solution. In an applied problem, you can usually detect an extraneous solution because the value generally makes no sense in the context of the problem.

a) 3 and 10, from
$m^2 - 13m + 30 = 0$

b) Yes, 3 is an extraneous solution.

SELF-TEST 12 (pages 169-184)

Solve each radical equation.

1. $\sqrt{2x - 1} = 3$

2. $\dfrac{\sqrt{5t}}{2} = 10$

3. $\sqrt{\dfrac{2}{3y + 2}} = \dfrac{1}{2}$

4. $\dfrac{3\sqrt{2w}}{2} = 6$

5. $\dfrac{4}{3\sqrt{V - 1}} = 1$

6. $\dfrac{3}{5\sqrt{x + 2}} = \dfrac{1}{4}$

Continued on following page.

184 Radical Equations and Formulas

SELF-TEST 12 (Continued)

7. In $\boxed{N = G - K\sqrt{P}}$, find P when N = 5, G = 20, and K = 3.	8. In $\boxed{h = \sqrt{\dfrac{2d}{a}}}$, find "d" when h = 6 and a = 3.
9. In $\boxed{R = \dfrac{B\sqrt{V}}{H}}$, find V when R = 20, B = 10, and H = 2.	10. In $\boxed{b = \sqrt{c^2 - a^2}}$, find "a" when b = 8 and c = 10.
Solve each radical equation. 11. $\sqrt{x+7} = x + 1$	12. $y - 2 = \sqrt{2y - 1}$

ANSWERS:
1. x = 5
2. t = 80
3. y = 2
4. w = 8
5. $V = \dfrac{25}{9}$
6. x = 4
7. P = 25
8. d = 54
9. V = 16
10. a = 6
11. x = 2 (not x = -3)
12. y = 5 (not y = 1)

4-6 THE SQUARING PRINCIPLE AND RADICAL FORMULAS

To solve for a variable that is part of the radicand in a radical formula, we must use the squaring principle. We will discuss solutions of that type in this section.

43. We saw earlier that the square of any square root radical is the radicand. That is:

$$\left(\sqrt{\dfrac{a}{b}}\right)^2 = \dfrac{a}{b} \qquad \left(\sqrt{c + 2h}\right)^2 = c + 2h$$

Continued on following page.

43. Continued

Therefore, if a radical is isolated on one side of a formula, we can use the squaring principle to eliminate the radical. For example:

$$I = \sqrt{\frac{P}{R}} \qquad v_o = \sqrt{v_f^2 - 2gs}$$

$$I^2 = \frac{P}{R} \qquad v_o^2 = v_f^2 - 2gs$$

Write the non-radical formula obtained by squaring both sides of these.

a) $v = \sqrt{2gs}$ b) $c = \sqrt{a^2 + b^2}$ c) $h = \sqrt{\frac{R_A - R_B}{t}}$

---

44. To solve for "s" below, we applied the squaring principle and then rearranged the non-radical formula. Solve for R in the other formula.

$$t = \sqrt{\frac{2s}{g}} \qquad d = \sqrt{\frac{pL}{R}}$$

$$t^2 = \frac{2s}{g}$$

$$gt^2 = 2s$$

$$s = \frac{gt^2}{2}$$

a) $v^2 = 2gs$

b) $c^2 = a^2 + b^2$

c) $h^2 = \frac{R_A - R_B}{t}$

---

45. Using the same method, solve for the indicated letter in each formula.

a) Solve for "s".   b) Solve for A.   c) Solve for "h".

$v = \sqrt{2gs}$ $d = \sqrt{\frac{A}{0.7854}}$ $c = \sqrt{d + h}$

$R = \frac{pL}{d^2}$, from:

$$d^2 = \frac{pL}{R}$$

$$d^2 R = pL$$

a) $s = \frac{v^2}{2g}$

b) $A = 0.7854 d^2$

c) $h = c^2 - d$

46. When the non-radical side of a formula is a monomial containing more than one factor, we square it by squaring each factor. For example:

$$(bc)^2 = b^2c^2 \qquad (3dt)^2 = 9d^2t^2$$

Apply the squaring principle to each formula below.

 a) $\sqrt{h} = 2b$ b) $\sqrt{a + 3c} = mf$ c) $\sqrt{Z^2 - R^2} = 2\pi fL$

---

47. We solved for Q below. Solve for "d" in the other formula.

$$\sqrt{\frac{K}{Q}} = bt \qquad\qquad \sqrt{a + 2d} = hp$$

$$\frac{K}{Q} = b^2t^2$$

$$K = b^2t^2Q$$

$$Q = \frac{K}{b^2t^2}$$

a) $h = 4b^2$

b) $a + 3c = m^2f^2$

c) $Z^2 - R^2 = 4\pi^2f^2L^2$

---

48. When the non-radical side of a formula is a fraction, we square it by squaring both terms. For example:

$$\left(\frac{mt}{a}\right) = \frac{m^2t^2}{a^2} \qquad\qquad \left(\frac{1}{2\pi f}\right) = \frac{1}{4\pi^2 f^2}$$

Apply the squaring principle to each formula below.

 a) $\sqrt{D_1 D_2} = \frac{B}{A}$ b) $\sqrt{\frac{a}{b}} = \frac{c}{3d}$ c) $\sqrt{\frac{Fr}{m}} = \frac{2\pi r}{t}$

$d = \dfrac{h^2p^2 - a}{2}$

---

49. Solve for the indicated letter in each formula.

 a) Solve for "b".   b) Solve for "m".   c) Solve for R.

$$\sqrt{b} = \frac{x}{y} \qquad\qquad \sqrt{m} = \frac{1}{2t} \qquad\qquad 4Q = \sqrt{R}$$

a) $D_1 D_2 = \dfrac{B^2}{A^2}$

b) $\dfrac{a}{b} = \dfrac{c^2}{9d^2}$

c) $\dfrac{Fr}{m} = \dfrac{4\pi^2 r^2}{t^2}$

Radical Equations and Formulas   187

50. We solved for L_1 below. Solve for C in the other formula.

$$\sqrt{L_1 L_2} = \frac{M}{K} \qquad \sqrt{LC} = \frac{1}{2\pi f}$$

$$L_1 L_2 = \frac{M^2}{K^2}$$

$$K^2 L_1 L_2 = M^2$$

$$L_1 = \frac{M^2}{K^2 L_2}$$

a) $b = \dfrac{x^2}{y^2}$

b) $m = \dfrac{1}{4t^2}$

c) $R = 16Q^2$

---

51. We solved for "b_o" below. Solve for "r" in the other formula.

$$\sqrt{\frac{b_p}{b_o}} = \frac{d_o}{d_p} \qquad \frac{t}{2\pi} = \sqrt{\frac{r}{a}}$$

$$\frac{b_p}{b_o} = \frac{d_o^2}{d_p^2}$$

$$b_p d_p^2 = b_o d_o^2$$

$$b_o = \frac{b_p d_p^2}{d_o^2}$$

$C = \dfrac{1}{4\pi^2 f^2 L}$

---

52. Sometimes a solution is a fraction that can be reduced to lowest terms. For example:

$$b = \frac{c^2 d^2}{ad} = \frac{c^2 d}{a}\left(\frac{d}{d}\right) = \frac{c^2 d}{a}$$

$$H = \frac{mP}{P^2 T^2} = \frac{m}{PT^2}\left(\frac{P}{P}\right) = \frac{m}{PT^2}$$

Following the examples, reduce each solution below to lowest terms.

a) $k = \dfrac{p^2 q^2}{q} =$ _____   b) $V = \dfrac{bM}{M^2 t^2} =$ _____

$r = \dfrac{at^2}{4\pi^2}$

---

53. We solved for "b" below. Notice how we reduced the solution to lowest terms. Solve for "k" in the other formula.

$$\sqrt{bm} = 3cm \qquad \sqrt{\frac{dk}{2}} = \frac{d}{P}$$

$$bm = 9c^2 m^2$$

$$b = \frac{9c^2 m^2}{m}$$

$$b = 9c^2 m \left(\frac{m}{m}\right)$$

$$b = 9c^2 m$$

a) $k = p^2 q$

b) $V = \dfrac{b}{Mt^2}$

54. We solved for "t" below. Notice how we reduced the solution to lowest terms. Solve for "m" in the other formula.

$$\sqrt{\frac{bD}{t}} = Dh \qquad\qquad \sqrt{\frac{Fr}{m}} = \frac{2\pi r}{t}$$

$$\frac{bD}{t} = D^2h^2$$

$$bD = D^2h^2 t$$

$$t = \frac{bD}{D^2h^2}$$

$$t = \frac{b}{Dh^2}\left(\frac{D}{D}\right)$$

$$t = \frac{b}{Dh^2}$$

$k = \dfrac{2d}{P^2}$, from:

$k = \dfrac{2d^2}{dP^2}$

---

55. We used the opposing principle to solve for "d" below. Notice that we got the opposite of the binomial "h - t" by interchanging the two terms.

$$-d = h - t$$
$$d = t - h$$

Notice how we used the opposing principle in the final step to solve for F_2 below. Solve for R in the other formula.

$$p = \sqrt{\frac{F_1 - F_2}{k}} \qquad\qquad b = \sqrt{v - R}$$

$$p^2 = \frac{F_1 - F_2}{k}$$

$$p^2 k = F_1 - F_2$$

$$p^2 k - F_1 = -F_2$$

$$F_2 = F_1 - p^2 k$$

$m = \dfrac{Ft^2}{4\pi^2 r}$, from:

$m = \dfrac{Frt^2}{4\pi^2 r^2}$

---

56. We solved for "s" below. Notice how we used the opposing principle with $v_0^2 - v_f^2 = -2gs$. Solve for P in the other formula.

$$v_0 = \sqrt{v_f^2 - 2gs} \qquad\qquad Q = \sqrt{T^2 - mP}$$

$$v_0^2 = v_f^2 - 2gs$$

$$v_0^2 - v_f^2 = -2gs$$

$$v_f^2 - v_0^2 = 2gs$$

$$s = \frac{v_f^2 - v_0^2}{2g}$$

$R = v - b^2$, from:

$b^2 - v = -R$

Radical Equations and Formulas   189

57. We solved for "F" below. Notice how we wrote the solution in two different forms. Solve for "m" in the other formula.

$$\frac{t}{2b} = \sqrt{a - F} \qquad\qquad \sqrt{c - m} = \frac{K}{R}$$

$$\frac{t^2}{4b^2} = a - F$$

$$\frac{t^2}{4b^2} - a = -F$$

$$F = a - \frac{t^2}{4b^2}$$

or

$$F = \frac{4ab^2 - t^2}{4b^2}$$

$P = \dfrac{T^2 - Q^2}{m}$, from:

$Q^2 - T^2 = -mP$

$T^2 - Q^2 = mP$

---

58. We solved for "t" below. Notice that we factored "t" out of two terms. Using the same steps, solve for "m" in the other formula.

$$\sqrt{1 - \frac{T}{t}} = P \qquad\qquad V = \sqrt{1 + \frac{M}{m}}$$

$$1 - \frac{T}{t} = P^2$$

$$t\left(1 - \frac{T}{t}\right) = t(P^2)$$

$$t - T = tP^2$$

$$t - tP^2 = T$$

$$t(1 - P^2) = T$$

$$t = \frac{T}{1 - P^2}$$

$m = c - \dfrac{K^2}{R^2}$

or

$m = \dfrac{cR^2 - K^2}{R^2}$

---

59. To clear the fractions below, we multiplied both sides by T. Clear the fractions in the other formula.

$$\frac{R}{T} = \frac{K^2}{T^2} \qquad\qquad \frac{2s}{a} = \frac{V^2}{a^2}$$

$$T^2\left(\frac{R}{T}\right) = T^2\left(\frac{K^2}{T^2}\right)$$

$$\left(\frac{T^2}{T}\right)(R) = \left(\frac{T^2}{T^2}\right)(K^2)$$

$$RT = K^2$$

$m = \dfrac{M}{V^2 - 1}$

$2as = V^2$

190 Radical Equations and Formulas

60. We solved for P below. Solve for M in the other formula.

$$\sqrt{\frac{T}{P}} = \frac{N}{P} \qquad\qquad \sqrt{\frac{GM}{r_o}} = \frac{v}{r_o}$$

$$\frac{T}{P} = \frac{N^2}{P^2}$$

$$P^2\left(\frac{T}{P}\right) = P^2\left(\frac{N^2}{P^2}\right)$$

$$PT = N^2$$

$$P = \frac{N^2}{T}$$

61. We solved for "d" below. Solve for V in the other formula.

$$\sqrt{\frac{cm}{d}} = \frac{v}{d} \qquad\qquad \frac{H}{V} = \sqrt{\frac{S}{V}}$$

$$\frac{cm}{d} = \frac{v^2}{d^2}$$

$$d^2\left(\frac{cm}{d}\right) = d^2\left(\frac{v^2}{d^2}\right)$$

$$cdm = v^2$$

$$d = \frac{v^2}{cm}$$

$M = \dfrac{v^2}{Gr_o}$

$V = \dfrac{H^2}{S}$

4-7 ISOLATING RADICALS AND REARRANGING FORMULAS

To solve for a variable in the radicand, we sometimes have to isolate the radical before applying the squaring principle. We will discuss the method in this section.

62. To solve for "r" below, we isolated the radical first. Solve for "d" in the other formula.

$$\frac{\sqrt{r}}{b} = t \qquad\qquad w = \frac{\sqrt{d}}{2}$$

$$\sqrt{r} = bt$$

$$r = b^2 t^2$$

$d = 4w^2$

63. To solve for "b" below, we isolated the radical first. Solve for "x" in the other formula.

$$H = \frac{p}{\sqrt{b}} \qquad\qquad \frac{a\sqrt{x}}{c} = h$$

$$H\sqrt{b} = p$$

$$\sqrt{b} = \frac{p}{H}$$

$$b = \frac{p^2}{H^2}$$

---

64. We solved for "y" below. Solve for C in the other formula.

$$\frac{1}{a\sqrt{by}} = v \qquad\qquad f = \frac{1}{2\pi\sqrt{LC}}$$

$$1 = av\sqrt{by}$$

$$\frac{1}{av} = \sqrt{by}$$

$$\frac{1}{a^2v^2} = by$$

$$1 = a^2bv^2y$$

$$y = \frac{1}{a^2bv^2} \quad \text{or} \quad \frac{1}{(av)^2 b}$$

$$x = \frac{c^2h^2}{a^2}$$

or $\quad x = \left(\frac{ch}{a}\right)^2$

---

65. We solved for L_2 below. Solve for "a" in the other formula.

$$K = \frac{M}{\sqrt{L_1 L_2}} \qquad\qquad \frac{D}{\sqrt{ap}} = V$$

$$K\sqrt{L_1 L_2} = M$$

$$\sqrt{L_1 L_2} = \frac{M}{K}$$

$$L_1 L_2 = \frac{M^2}{K^2}$$

$$K^2 L_1 L_2 = M^2$$

$$L_2 = \frac{M^2}{K^2 L_1}$$

$$C = \frac{1}{4\pi^2 f^2 L}$$

or $\quad C = \frac{1}{(2\pi f)^2 L}$

$$a = \frac{D^2}{pV^2}$$

Radical Equations and Formulas 191

192 Radical Equations and Formulas

66. We solved for "h" below. Solve for "t" in the other formula.

$$s = \frac{\sqrt{h+2r}}{d} \qquad \frac{\sqrt{d+t}}{2a} = K$$

$$sd = \sqrt{h+2r}$$

$$s^2d^2 = h + 2r$$

$$h = s^2d^2 - 2r$$

---

67. We solved for "a" below. Solve for "g_o" in the other formula.

$$w = 2g\sqrt{\frac{a}{r}} \qquad r_o = r_p\sqrt{\frac{g_p}{g_o}}$$

$$\frac{w}{2g} = \sqrt{\frac{a}{r}}$$

$$\frac{w^2}{4g^2} = \frac{a}{r}$$

$$rw^2 = 4ag^2$$

$$a = \frac{rw^2}{4g^2}$$

$t = 4a^2K^2 - d$

or $\quad t = (2aK)^2 - d$

---

68. We solved for T below. Notice that we left the solution as $(A-B)^2$ rather than squaring to get $A^2 - 2AB + B^2$. Solve for "m" in the other formula.

$$A = B + \sqrt{T} \qquad c + \sqrt{m} = d$$

$$\sqrt{T} = A - B$$

$$T = (A - B)^2$$

$g_o = \dfrac{g_p r_p^2}{r_o^2}$

---

69. Notice how we used the oppositing principle to solve for "q" below. Solve for V in the other formula.

$$p - \sqrt{q} = r \qquad D = M - \sqrt{V}$$

$$-\sqrt{q} = r - p$$

$$\sqrt{q} = p - r$$

$$q = (p - r)^2$$

$m = (d - c)^2$

---

70. We solved for "x" below. Solve for "p" in the other formula.

$$a = d - m\sqrt{x} \qquad r - a\sqrt{p} = T$$

$$a - d = -m\sqrt{x}$$

$$d - a = m\sqrt{x}$$

$$\sqrt{x} = \frac{d-a}{m}$$

$$x = \left(\frac{d-a}{m}\right)^2$$

$V = (M - D)^2$

$p = \left(\dfrac{r - T}{a}\right)^2$

4-8 SOLVING FOR NON-RADICAND VARIABLES

In this section, we will discuss the method for solving for a variable which is not part of the radicand in a radical formula.

---

71. We solved for "m" below. Since "m" is not part of the radicand, we did not have to use the squaring principle. Solve for M in the other formula.

$$v = \frac{m}{\sqrt{t}} \qquad\qquad K = \frac{M}{\sqrt{L_1 L_2}}$$

$$v\sqrt{t} = m$$

or $\quad m = v\sqrt{t}$

---

72. We solved for "t" below. Solve for "m" in the other formula.

$$b = \frac{\sqrt{ax}}{t} \qquad\qquad \frac{c\sqrt{y}}{m} = R$$

$$bt = \sqrt{ax}$$

$$t = \frac{\sqrt{ax}}{b}$$

$M = K\sqrt{L_1 L_2}$

---

73. We solved for S below. Solve for "a" in the other formula.

$$F = S + \sqrt{T} \qquad\qquad t = a - b\sqrt{d}$$

$$F - \sqrt{T} = S$$

or $\quad S = F - \sqrt{T}$

$m = \dfrac{c\sqrt{y}}{R}$

---

74. We solved for "b" below. Solve for "c" in the other formula.

$$\frac{b\sqrt{P}}{Q} = T \qquad\qquad t = d + c\sqrt{m}$$

$$b\sqrt{P} = QT$$

$$b = \frac{QT}{\sqrt{P}}$$

$a = t + b\sqrt{d}$

$c = \dfrac{t - d}{\sqrt{m}}$

194 Radical Equations and Formulas

75. In the last frame, we got these two solutions.

$$b = \frac{QT}{\sqrt{P}} \qquad c = \frac{t-d}{\sqrt{m}}$$

Though we frequently leave denominators like those above in non-rationalized form, we sometimes rationalize them. We get:

$$b = \frac{QT}{\sqrt{P}} = \frac{QT}{\sqrt{P}}\left(\frac{\sqrt{P}}{\sqrt{P}}\right) = \frac{QT\sqrt{P}}{P}$$

$$c = \frac{t-d}{\sqrt{m}} = \underline{\qquad}$$

76. We solved for "r_o" below. Solve for "a" in the other formula.

$$\sqrt{\frac{g_p}{g_o}} = \frac{r_o}{r_p} \qquad \frac{a}{b} = \sqrt{\frac{c}{d}}$$

$$r_o = r_p\sqrt{\frac{g_p}{g_o}}$$

$c = \dfrac{(t-d)\sqrt{m}}{m}$

77. To get the reciprocal of a radical containing a fraction, we invert the fraction. That is:

The reciprocal of $\sqrt{\dfrac{c}{d}}$ is $\sqrt{\dfrac{d}{c}}$, since:

$$\sqrt{\frac{c}{d}} \cdot \sqrt{\frac{d}{c}} = \sqrt{\frac{cd}{cd}} = \sqrt{1} = 1$$

Write the reciprocal of each radical.

a) $\sqrt{\dfrac{m}{t}} = \underline{\qquad}$ b) $\sqrt{\dfrac{r_o}{r_p}} = \underline{\qquad}$ c) $\sqrt{\dfrac{Fr}{m}} = \underline{\qquad}$

$a = b\sqrt{\dfrac{c}{d}}$

78. The fraction below means: divide "p" by $\sqrt{\dfrac{d}{t}}$. We can simplify it by converting the division to a multiplication.

$$\frac{p}{\sqrt{\dfrac{d}{t}}} = p\left(\text{the reciprocal of } \sqrt{\frac{d}{t}}\right) = p\sqrt{\frac{t}{d}}$$

Notice that the conversion above is equivalent to inverting the fractional radicand and putting the radicand in the numerator.

Convert each division below to a multiplication.

a) $\dfrac{m}{\sqrt{\dfrac{d_1}{d_2}}} = \underline{\qquad}$ b) $\dfrac{c}{\sqrt{\dfrac{ab}{d}}} = \underline{\qquad}$

a) $\sqrt{\dfrac{t}{m}}$

b) $\sqrt{\dfrac{r_p}{r_o}}$

c) $\sqrt{\dfrac{m}{Fr}}$

Radical Equations and Formulas   195

79. We solved for "r_p" below and simplified the solution. Solve for T in the other formula and simplify the solution.

$$\sqrt{\frac{g_p}{g_o}} = \frac{r_o}{r_p} \qquad\qquad \sqrt{\frac{Fr}{m}} = \frac{2\pi r}{T}$$

$$r_p \sqrt{\frac{g_p}{g_o}} = r_o$$

$$r_p = \frac{r_o}{\sqrt{\frac{g_p}{g_o}}}$$

$$r_p = r_o \sqrt{\frac{g_o}{g_p}}$$

a) $m\sqrt{\dfrac{d_2}{d_1}}$

b) $c\sqrt{\dfrac{d}{ab}}$

80. The fraction below means: divide $\sqrt{\dfrac{a}{b}}$ by "c". We converted the division to a multiplication. Notice that we wrote $\dfrac{1}{c}$ in front of the radical.

$$\frac{\sqrt{\frac{a}{b}}}{c} = \sqrt{\frac{a}{b}} \text{ (the reciprocal of "c")} = \sqrt{\frac{a}{b}}\left(\frac{1}{c}\right) = \frac{1}{c}\sqrt{\frac{a}{b}}$$

Convert each fraction below to a multiplication.

a) $\dfrac{\sqrt{\frac{p}{r}}}{t} = $ _____   b) $\dfrac{\sqrt{\frac{2s}{a}}}{mV} = $ _____

$T = 2\pi r\sqrt{\dfrac{m}{Fr}}$

81. We solved for P below and simplified the solution. Solve for "b" in the other formula and simplify the solution.

$$PR = \sqrt{\frac{T}{N}} \qquad\qquad 2ab = \sqrt{\frac{p}{mt}}$$

$$P = \frac{\sqrt{\frac{T}{N}}}{R}$$

$$P = \frac{1}{R}\sqrt{\frac{T}{N}}$$

a) $\dfrac{1}{t}\sqrt{\dfrac{p}{r}}$

b) $\dfrac{1}{mV}\sqrt{\dfrac{2s}{a}}$

$b = \dfrac{1}{2a}\sqrt{\dfrac{p}{mt}}$

82. We solved for R below. Since there is a "t" outside the radical and in the denominator of the radicand, we simplified the solution by moving "t" under the radical. To do so, we substituted $\sqrt{t^2}$ for "t". Solve for F in the other formula.

$$\sqrt{\frac{V}{t}} = \frac{R}{t} \qquad\qquad \sqrt{\frac{A}{B}} = \frac{F}{B}$$

$$R = t\sqrt{\frac{V}{t}}$$

$$R = \sqrt{t^2} \cdot \sqrt{\frac{V}{t}}$$

$$R = \sqrt{\frac{t^2 V}{t}}$$

$$R = \sqrt{tV}$$

$F = \sqrt{AB}$

4-9 RADICAL FORMULAS AND SQUARED VARIABLES

In this section, we will discuss the method for solving for a squared variable that is part of the radicand in a radical formula.

83. To solve for Q below, we used both the squaring principle and the square root principle. Solve for "a" in the other formula.

$$F = \sqrt{P^2 + Q^2} \qquad\qquad c = \sqrt{a^2 + b^2}$$

$$F^2 = P^2 + Q^2$$

$$Q^2 = F^2 - P^2$$

$$Q = \sqrt{F^2 - P^2}$$

$a = \sqrt{c^2 - b^2}$

84. Notice how we used the opposing principle to solve for "d" below. Solve for T in the other formula.

$$m = \sqrt{c^2 - d^2} \qquad\qquad H = \sqrt{S^2 - T^2}$$

$$m^2 = c^2 - d^2$$

$$m^2 - c^2 = -d^2$$

$$c^2 - m^2 = d^2$$

$$d = \sqrt{c^2 - m^2}$$

$T = \sqrt{S^2 - H^2}$

85. a) Solve for v_f.

$$v_o = \sqrt{v_f^2 - 2as}$$

b) Solve for R.

$$\sqrt{Z^2 - R^2} = 2\pi fL$$

---

86. To solve for "v" below, we isolated the radical first. Solve for "m" in the other formula.

$$d = \frac{1}{\sqrt{t^2 + v^2}} \qquad k = \frac{a}{\sqrt{m^2 + 1}}$$

$$d\sqrt{t^2 + v^2} = 1$$

$$\sqrt{t^2 + v^2} = \frac{1}{d}$$

$$t^2 + v^2 = \frac{1}{d^2}$$

$$v^2 = \frac{1}{d^2} - t^2$$

$$v = \sqrt{\frac{1}{d^2} - t^2}$$

$$\text{or } v = \sqrt{\frac{1 - d^2 t^2}{d^2}}$$

a) $v_f = \sqrt{v_o^2 + 2as}$

b) $R = \sqrt{Z^2 - 4\pi^2 f^2 L^2}$
or $R = \sqrt{Z^2 - (2\pi fL)^2}$

$$m = \sqrt{\frac{a^2}{k^2} - 1} \text{ or } m = \sqrt{\frac{a^2 - k^2}{k^2}}$$

SELF-TEST 13 (pages 184-198)

1. Solve for "a".

   $$\sqrt{h - a} = 2t$$

2. Solve for B_2.

   $$\frac{W}{P} = \sqrt{\frac{B_1}{B_2}}$$

3. Solve for "s".

   $$\sqrt{2ks} = dk$$

Continued on following page.

198   Radical Equations and Formulas

SELF-TEST 13 (Continued)

4. Solve for "r". $v = \sqrt{\dfrac{p-r}{r}}$	5. Solve for A. $F = R - G\sqrt{A}$	6. Solve for "t_2". $w = 2p\sqrt{\dfrac{t_1}{t_2}}$
7. Solve for G. $T = \dfrac{P}{\sqrt{GP}}$	8. Solve for "b_1". $\dfrac{b_2}{b_1} = \sqrt{\dfrac{d_1}{d_2}}$	9. Solve for R. $N = \sqrt{H^2 - R^2}$

ANSWERS:

1. $a = h - 4t^2$
2. $B_2 = \dfrac{B_1 P^2}{w^2}$
3. $s = \dfrac{d^2 k}{2}$
4. $r = \dfrac{p}{v^2 + 1}$
5. $A = \left(\dfrac{R - F}{G}\right)^2$
6. $t_2 = \dfrac{4p^2 t_1}{w^2}$
7. $G = \dfrac{P}{T^2}$
8. $b_1 = b_2 \sqrt{\dfrac{d_2}{d_1}}$
9. $R = \sqrt{H^2 - N^2}$

4-10   SIMPLIFYING RADICAL SOLUTIONS BY FACTORING OUT PERFECT SQUARES

When solving for a squared variable in a formula, we can sometimes simplify the solution by factoring out perfect squares. We will discuss that process in this section.

87. We solved for V below. Notice how we were able to factor out a perfect square in the solution. Solve for "p" in the other formula.

$$h = \dfrac{V^2}{64} \qquad\qquad 25k = \dfrac{p^2}{q}$$

$$V^2 = 64h$$

$$V = \sqrt{64h}$$

$$V = 8\sqrt{h}$$

$p = 5\sqrt{kq}$

88. We solved for "c" below. Notice how we factored out a perfect square from the solution. Solve for D_2 in the other formula.

$$\frac{c^2}{d^2} = \frac{k}{t} \qquad \qquad \frac{F_1}{F_2} = \frac{(D_1)^2}{(D_2)^2}$$

$$c^2 t = d^2 k$$

$$c^2 = \frac{d^2 k}{t}$$

$$c = \sqrt{\frac{d^2 k}{t}}$$

$$c = \sqrt{(d^2)\left(\frac{k}{t}\right)}$$

$$c = d\sqrt{\frac{k}{t}}$$

89. To solve for "b" below, we began by squaring the fraction. Solve for H in the other formula.

$$a = \left(\frac{b}{c}\right)^2 \qquad \qquad R = \left(\frac{H}{T}\right)^2$$

$$a = \frac{b^2}{c^2}$$

$$ac^2 = b^2$$

$$b = \sqrt{ac^2}$$

$$b = c\sqrt{a}$$

$D_2 = D_1\sqrt{\dfrac{F_2}{F_1}}$

90. To solve for "c" below, we began by squaring the fraction. Solve for "r_o" in the other formula.

$$\frac{a}{b} = \left(\frac{c}{d}\right)^2 \qquad \qquad \frac{g_o}{g_p} = \left(\frac{r_p}{r_o}\right)^2$$

$$\frac{a}{b} = \frac{c^2}{d^2}$$

$$ad^2 = bc^2$$

$$c^2 = \frac{ad^2}{b}$$

$$c = \sqrt{\frac{ad^2}{b}}$$

$$c = d\sqrt{\frac{a}{b}}$$

$H = T\sqrt{R}$

$r_o = r_p\sqrt{\dfrac{g_p}{g_o}}$

91. To solve for "m" below, we began by squaring the fraction. Solve for T in the other formula.

$$t = 2d\left(\frac{m}{p}\right)^2 \qquad\qquad R = 2a\left(\frac{V}{T}\right)^2$$

$$t = 2d\left(\frac{m^2}{p^2}\right)$$

$$t = \frac{2dm^2}{p^2}$$

$$p^2 t = 2dm^2$$

$$m^2 = \frac{p^2 t}{2d}$$

$$m = \sqrt{\frac{p^2 t}{2d}}$$

$$m = p\sqrt{\frac{t}{2d}}$$

$T = V\sqrt{\dfrac{2a}{R}}$

92. We solved for "c" below. In the final step, notice how we wrote "1" as a factor in the numerator to factor out the "b²". Solve for R in the other formula.

$$a = \left(\frac{b}{c}\right)^2 \qquad\qquad DK = \left(\frac{AT}{R}\right)^2$$

$$a = \frac{b^2}{c^2}$$

$$ac^2 = b^2$$

$$c^2 = \frac{b^2}{a}$$

$$c = \sqrt{\frac{b^2}{a}}$$

$$c = \sqrt{\frac{(b^2)(1)}{a}}$$

$$c = b\sqrt{\frac{1}{a}}$$

$R = AT\sqrt{\dfrac{1}{DK}}$

93. We solved for "t" below. Notice how we reduced the fractional radicand to lowest terms. Solve for R in each of the other formulas.

$$ct^2 = ac^2 \qquad\text{a) } bR^2 = bT^2 X \qquad \text{b) } k^2 R^2 = 3km^2$$

$$t^2 = \frac{ac^2}{c}$$

$$t = \sqrt{\frac{ac^2}{c}}$$

$$t = \sqrt{ac}$$

94. We solved for "t" below. Notice how we factored out perfect squares. Solve for "f" in the other formula.

$$9ct = \frac{1}{cty} \qquad\qquad 2\pi fL = \frac{1}{2\pi fC}$$

$$9c^2t^2y = 1$$

$$t^2 = \frac{1}{9c^2y}$$

$$t = \sqrt{\frac{1}{9c^2y}}$$

$$t = \frac{\sqrt{1}}{\sqrt{9c^2y}}$$

$$t = \frac{1}{3c\sqrt{y}}$$

a) $R = T\sqrt{X}$

b) $R = m\sqrt{\dfrac{3}{k}}$

$$f = \frac{1}{2\pi\sqrt{LC}}$$

4-11 THE EQUIVALENCE METHOD AND FORMULA DERIVATION

In this section, we will show how the equivalence method can be used to derive a new formula from a system of formulas containing squared variables and square root radicals.

95. The two major steps in a formula derivation are:

1. Eliminating a common variable or variables from the system.
2. Solving for one of the variables in the new formula.

The common variables in the system at the right are F and "m". Let's eliminate F and then solve for "v".

$$F = mg$$
$$Fr = mv^2$$

1. Solve for F in each formula.

$$F = mg$$
$$F = \frac{mv^2}{r}$$

2. Use the equivalence principle to eliminate F.

$$mg = \frac{mv^2}{r}$$

3. Solve for "v" in the new formula. (Be sure to reduce to lowest terms.)

Radical Equations and Formulas

96. At the left below, we eliminated G and then solved for T. Eliminate B from the other system and then solve for D.

$$\boxed{\begin{aligned} T &= G^2N \\ G &= PR \end{aligned}} \qquad \boxed{\begin{aligned} V &= B^2D \\ B &= ST \end{aligned}}$$

$$G = \sqrt{\frac{T}{N}} \quad \text{and} \quad G = PR$$

$$\sqrt{\frac{T}{N}} = PR$$

$$\frac{T}{N} = P^2R^2$$

$$T = NP^2R^2$$

$v = \sqrt{gr}$, from:

$v = \sqrt{\dfrac{mgr}{m}}$

---

97. At the left below, we eliminated B and then solved for S. Notice how we wrote the solution in a simpler form by converting the division to a multiplication. Eliminate G from the other system and then solve for R.

$$\boxed{\begin{aligned} V &= B^2D \\ B &= ST \end{aligned}} \qquad \boxed{\begin{aligned} T &= G^2N \\ G &= PR \end{aligned}}$$

$$B = \sqrt{\frac{V}{D}} \quad \text{and} \quad B = ST$$

$$\sqrt{\frac{V}{D}} = ST$$

$$S = \frac{\sqrt{\frac{V}{D}}}{T}$$

$$S = \frac{1}{T}\sqrt{\frac{V}{D}}$$

$D = \dfrac{V}{S^2T^2}$

---

98. At the left below, we eliminated "t" and then solved for "a". Eliminate G from the other system and then solve for P.

$$\boxed{\begin{aligned} v &= at \\ s &= \tfrac{1}{2}at^2 \end{aligned}} \qquad \boxed{\begin{aligned} T &= G^2P \\ P &= \tfrac{N}{G} \end{aligned}}$$

$$t = \frac{v}{a} \quad \text{and} \quad t = \sqrt{\frac{2s}{a}}$$

$$\frac{v}{a} = \sqrt{\frac{2s}{a}}$$

$$\frac{v^2}{a^2} = \frac{2s}{a}$$

$$\cancel{a^2}\left(\frac{v^2}{\cancel{a^2}}\right) = \cancel{a^2}\left(\frac{2s}{\cancel{a}}\right)$$

$$v^2 = 2as$$

$$a = \frac{v^2}{2s}$$

$R = \dfrac{1}{P}\sqrt{\dfrac{T}{N}}$

99. At the left below, we eliminated R and then solved for T. Notice how we simplified the solution by moving the K under the radical as K^2. Eliminate "t" from the other system and then solve for "v".

$$\boxed{\begin{array}{l} H = R^2K \\ K = \dfrac{T}{R} \end{array}}$$

$$\boxed{\begin{array}{l} v = at \\ s = \dfrac{1}{2}at^2 \end{array}}$$

$R = \sqrt{\dfrac{H}{K}}$ and $R = \dfrac{T}{K}$

$\sqrt{\dfrac{H}{K}} = \dfrac{T}{K}$

$T = K\sqrt{\dfrac{H}{K}}$

$T = \sqrt{K^2}\sqrt{\dfrac{H}{K}}$

$T = \sqrt{\dfrac{HK^2}{K}}$

$T = \sqrt{HK}$

$P = \dfrac{N^2}{T}$

---

100. At the left below, we eliminated "v" and then solved for "a". Notice that we reduced the solution to lowest terms. Eliminate "v" from the other system and then solve for "F".

$$\boxed{\begin{array}{l} a = \dfrac{v^2}{r} \\ v = \dfrac{2\pi r}{t} \end{array}}$$

$$\boxed{\begin{array}{l} F = \dfrac{mv^2}{r} \\ v = \dfrac{2\pi r}{t} \end{array}}$$

$v = \sqrt{ar}$ and $v = \dfrac{2\pi r}{t}$

$\sqrt{ar} = \dfrac{2\pi r}{t}$

$ar = \dfrac{4\pi^2 r^2}{t^2}$

$art^2 = 4\pi^2 r^2$

$a = \dfrac{4\pi^2 r^2}{rt^2}$

$a = \dfrac{4\pi^2 r}{t^2}$

$v = \sqrt{2as}$, from:

$v = a\sqrt{\dfrac{2s}{a}}$

$v = \sqrt{\dfrac{2a^2 s}{a}}$

$F = \dfrac{4\pi^2 mr}{t^2}$

(Did you reduce to lowest terms?)

204   Radical Equations and Formulas

101. a) Eliminate "g_o" and then solve for "v".

$$v^2 = g_o r_o$$
$$g_o = \frac{GM}{r_o^2}$$

b) Eliminate "v" and then solve for "t".

$$a = \frac{v^2}{r}$$
$$v = \frac{2\pi r}{t}$$

a) $v = \sqrt{\dfrac{GM}{r_o}}$

b) $t = \dfrac{2\pi r}{\sqrt{ar}}$

or $t = 2\pi\sqrt{\dfrac{r}{a}}$

from: $t = \dfrac{2\pi\sqrt{r^2}}{\sqrt{ar}}$

102. At the left below, we eliminated "v" and then solved for "t". Notice how we simplified the solution by converting the division to a multiplication. Eliminate "v" from the other system and then solve for "t".

$$F = \frac{mv^2}{r}$$
$$v = \frac{2\pi r}{t}$$

$$v^2 = \frac{GM}{r_o}$$
$$v = \frac{2\pi r_o}{t}$$

$v = \sqrt{\dfrac{Fr}{m}}$ and $v = \dfrac{2\pi r}{t}$

$\sqrt{\dfrac{Fr}{m}} = \dfrac{2\pi r}{t}$

$t\sqrt{\dfrac{Fr}{m}} = 2\pi r$

$t = \dfrac{2\pi r}{\sqrt{\dfrac{Fr}{m}}}$

$t = 2\pi r\sqrt{\dfrac{m}{Fr}}$

$t = 2\pi r_o\sqrt{\dfrac{r_o}{GM}}$, from:

$t = \dfrac{2\pi r_o}{\sqrt{\dfrac{GM}{r_o}}}$

103. At the left below, we eliminated X and then solved for L. Eliminate "v" from the other system and then solve for "h".

$$Z = \sqrt{R^2 + X^2}$$
$$X = 2\pi fL$$

$$t = \sqrt{v^2 - b^2}$$
$$h = av$$

$X = \sqrt{Z^2 - R^2}$ and $X = 2\pi fL$

$$\sqrt{Z^2 - R^2} = 2\pi fL$$

$$L = \frac{\sqrt{Z^2 - R^2}}{2\pi f}$$

$h = a\sqrt{b^2 + t^2}$

---

104. a) Eliminate "b" and then solve for "c".

$$c = \sqrt{a^2 + b^2}$$
$$b = 2d$$

b) Eliminate X and then solve for "f".

$$X = 2\pi fL$$
$$X = \frac{1}{2\pi fC}$$

---

105. At the left below, we eliminated "e" and then solved for "w". Notice how we had to factor by the distributive principle to solve for "w". Eliminate K from the other system and then solve for T.

$$v = \sqrt{w - e}$$
$$w = \frac{e}{v}$$

$$H = \sqrt{K + T}$$
$$T = \frac{K}{H}$$

$e = w - v^2$ and $e = vw$

$$w - v^2 = vw$$
$$w - vw = v^2$$
$$w(1 - v) = v^2$$
$$w = \frac{v^2}{1 - v}$$

a) $c = \sqrt{a^2 + 4d^2}$

b) $f = \dfrac{1}{2\pi\sqrt{LC}}$

(Did you factor out the perfect squares?)

$T = \dfrac{H^2}{H + 1}$

206 Radical Equations and Formulas

4-12 THE SUBSTITUTION METHOD AND FORMULA DERIVATION

In this section, we will show how the substitution method can be used to derive a new formula from a system of formulas containing squared variables and square root radicals.

---

106. When the variable we want to eliminate is already solved for in one formula, it is easy to use the substitution method. An example is discussed.

    "g_o" is solved for in the bottom formula at the right.

    $$v^2 = g_o r_o$$
    $$g_o = \frac{GM}{r_o^2}$$

    To eliminate "g_o", we can substitute $\frac{GM}{r_o^2}$ for "g_o" in the top formula.

    $$v^2 = g_o r_o$$
    $$v^2 = \left(\frac{GM}{r_o^2}\right) r_o$$

    Solve for "v" in the new formula.
    (Be sure to reduce to lowest terms.)

---

107. Since "v" is solved for in the top formula at the right, we can eliminate "v" by substituting "wr" for it in the bottom formula. The steps are shown.

    $$v = wr$$
    $$v^2 = ar$$

    $$v^2 = ar$$
    $$(wr)^2 = ar$$
    $$w^2 r^2 = ar$$

    Solve for "w" in the new formula.
    (Be sure to reduce to lowest terms.)

    $v = \sqrt{\dfrac{GM}{r_o}}$

---

108. To eliminate B from this system:

    1. We can solve for B in the bottom formula. We get:

    $$R = B^2 T$$
    $$T = \frac{D}{B}$$

    $$B = \frac{D}{T}$$

    2. We then substitute that value for B in the top formula and simplify. We get:

    $$R = B^2 T = \left(\frac{D}{T}\right)^2 T = \left(\frac{D^2}{T^2}\right) T = \frac{D^2 T}{T^2} = \frac{D^2}{T}$$

    Solve for D in the new formula: $R = \dfrac{D^2}{T}$

    $w = \sqrt{\dfrac{a}{r}}$

109. To eliminate "t" from this system, we can solve for "t" in the top formula. We get:

$$t = \frac{v}{a}$$

$\boxed{\begin{array}{l} v = at \\ s = \dfrac{1}{2}at^2 \end{array}}$

$D = \sqrt{RT}$

a) Substitute that value for "t" in the bottom formula and then simplify.

b) Now solve for "v" in the new formula.

---

110. To eliminate "v" from this system, we can substitute $\dfrac{2\pi r}{t}$ for "v" in the top formula. We get:

$$a = \frac{v^2}{r} = \frac{\left(\dfrac{2\pi r}{t}\right)^2}{r} = \frac{\dfrac{4\pi^2 r^2}{t^2}}{r} = \left(\frac{4\pi^2 r^2}{t^2}\right)\left(\frac{1}{r}\right) = \frac{4\pi^2 r^2}{t^2 r} = \frac{4\pi^2 r}{t^2}$$

$\boxed{\begin{array}{l} a = \dfrac{v^2}{r} \\ v = \dfrac{2\pi r}{t} \end{array}}$

a) $s = \dfrac{v^2}{2a}$, from:

$s = \dfrac{1}{2}a\left(\dfrac{v}{a}\right)^2$

$s = \dfrac{1}{2}a\left(\dfrac{v^2}{a^2}\right)$

$s = \dfrac{av^2}{2a^2} = \dfrac{v^2}{2a}$

b) $v = \sqrt{2as}$

Notice how we squared $\dfrac{2\pi r}{t}$, converted the division to a multiplication, and then reduced to lowest terms.

Solve for "t" in the new formula. $a = \dfrac{4\pi^2 r}{t^2}$

---

111. Let's eliminate "m" from this system and then solve for "v".

a) Substitute $\dfrac{W}{g}$ for "m" in the top formula and then simplify.

$\boxed{\begin{array}{l} F = \dfrac{mv^2}{r} \\ m = \dfrac{W}{g} \end{array}}$

$t = 2\pi\sqrt{\dfrac{r}{a}}$, from:

$t = \sqrt{\dfrac{4\pi^2 r}{a}}$

b) Solve for "v" in the new formula.

208   Radical Equations and Formulas

112. To eliminate "e" from this system, we can:

$v = \sqrt{w - e}$

$w = \dfrac{e}{v}$

a) $F = \dfrac{Wv^2}{gr}$, from:

1. Solve for "e" in the bottom formula.

$e = vw$

$F = \dfrac{\left(\dfrac{W}{g}\right)v^2}{r}$

2. Substitute that value for "e" in the top formula.

$v = \sqrt{w - vw}$

$= \dfrac{\dfrac{Wv^2}{g}}{r}$

Solve for "w" in this new formula.
(You have to factor by the distributive principle.)

$= \left(\dfrac{Wv^2}{g}\right)\left(\dfrac{1}{r}\right)$

b) $v = \sqrt{\dfrac{Fgr}{W}}$

---

113. Substitute to eliminate X from this system and then solve for "f".

$Z = \sqrt{R^2 + X^2}$

$X = 2\pi fL$

$w = \dfrac{v^2}{1 - v}$

$f = \dfrac{\sqrt{Z^2 - R^2}}{2\pi L}$

4-13 FORMULAS IN QUADRATIC FORM

Formulas that contain squared variables are similar to either pure, incomplete, or complete quadratic equations. Therefore, we say that they are "in quadratic form". In this section, we will rearrange formulas of that type.

114. Formulas containing only a squared variable are similar to "pure" quadratic equations. To solve for the squared variable, we isolate it (if necessary) and then apply the square root principle. Two examples are shown. Solve for D in the other formula.

$E^2 = PR$ $a = w^2 r$ $H = \dfrac{D^2 N}{2.5}$

$E = \sqrt{PR}$ $w^2 = \dfrac{a}{r}$

$w = \sqrt{\dfrac{a}{r}}$

Radical Equations and Formulas   209

115. Since the formula below contains a p^2-term and a p-term, it is similar to an "incomplete" quadratic equation with "p" as the variable. Therefore, we can solve for "p" by the factoring method. We get:

$$cp^2 = dp$$
$$cp^2 - dp = 0$$
$$p(cp - d) = 0$$
$$p = 0 \quad \text{and} \quad p = \frac{d}{c}$$

However, p = 0 makes little or no sense as the result of a formula rearrangement. Therefore, we ordinarily solve for "p" by multiplying both sides by $\frac{1}{p}$ to eliminate the "0" solution. We did so below. Using the same method, solve for H in the other formula.

$$\frac{1}{p}(cp^2) = \frac{1}{p}(dp) \qquad\qquad aH^2 = BH$$
$$cp = d$$
$$p = \frac{d}{c}$$

$D = \sqrt{\dfrac{2.5H}{N}}$

---

116. We used the same method to solve for "r" below. Solve for "r" in the other formula.

$$\sqrt{\frac{Fr}{m}} = \frac{2\pi r}{T} \qquad\qquad \sqrt{ar} = \frac{2\pi r}{T}$$

$$\frac{Fr}{m} = \frac{4\pi^2 r^2}{T^2}$$

$$\frac{1}{r}\left(\frac{Fr}{m}\right) = \frac{1}{r}\left(\frac{4\pi^2 r^2}{T^2}\right)$$

$$\frac{F}{m} = \frac{4\pi^2 r}{T^2}$$

$$FT^2 = 4\pi^2 mr$$

$$r = \frac{FT^2}{4\pi^2 m}$$

$H = \dfrac{B}{a}$, from:

$\dfrac{1}{H}(aH^2) = \dfrac{1}{H}(BH)$

$aH = B$

---

117. We used the substitution method at the right to eliminate "v" from the system below.

$$\boxed{\begin{aligned} v &= wr \\ v^2 &= ar \end{aligned}}$$

$$v^2 = ar$$
$$(wr)^2 = ar$$
$$w^2 r^2 = ar$$

Solve for "r" in the new formula.

$r = \dfrac{aT^2}{4\pi^2}$, from:

$a = \dfrac{4\pi^2 r}{T^2}$

118. Since the formula below contains a t^2-term, a t-term, and a term without "t", it is similar to a "complete" quadratic equation.

$$\tfrac{1}{2}gt^2 + vt - s = 0$$

Therefore, using the values below, we can use the quadratic formula to solve for "t" as we have done at the right.

$a = \tfrac{1}{2}g$ or $\tfrac{g}{2}$
$b = v$
$c = -s$

$$t = \frac{(-b) \pm \sqrt{b^2 - 4ac}}{2a}$$

$$t = \frac{(-v) \pm \sqrt{(v)^2 - 4\left(\tfrac{g}{2}\right)(-s)}}{2\left(\tfrac{g}{2}\right)}$$

Simplify the solution.

$r = \dfrac{a}{w^2}$

---

119. Use the quadratic formula to solve for P in the formula below.

$dP^2 - 2hP + d = 0$

$a = d$
$b = -2h$
$c = d$

$$P = \frac{(-b) \pm \sqrt{b^2 - 4ac}}{2a}$$

$t = \dfrac{-v \pm \sqrt{v^2 + 2gs}}{g}$

---

120. Let's use the equivalence method to eliminate "q" from the system at the right.

$q = p - t^2$ and $q = pt$

$p - t^2 = pt$

By rearranging, we can write the new formula as a "complete" quadratic in "t". We get:

$t^2 + pt - p = 0$

$(a = 1, b = p, c = -p)$

Using the quadratic formula, solve for "t".

$\boxed{\begin{array}{l} t = \sqrt{p - q} \\ p = \dfrac{q}{t} \end{array}}$

$P = \dfrac{h \pm \sqrt{h^2 - d^2}}{d}$,

from:

$P = \dfrac{2h \pm \sqrt{4h^2 - 4d^2}}{2d}$

$P = \dfrac{\cancel{2}h \pm \cancel{2}\sqrt{h^2 - d^2}}{\cancel{2}d}$

$t = \dfrac{-p \pm \sqrt{p^2 + 4p}}{2}$

SELF-TEST 14 (pages 198-211)

1. Solve for "t".

    $s^2 = \dfrac{t^2}{w}$

2. Solve for "r_1".

    $\dfrac{A_1}{A_2} = \left(\dfrac{r_1}{r_2}\right)^2$

3. Solve for "m".

    $(km)^2 = d^2 k$

4. Eliminate "r" and then solve for "d".

    $A = \pi r^2$
    $d = 2r$

5. Eliminate A and then solve for V.

    $P = A^2 R$
    $A = \dfrac{V}{R}$

6. Eliminate P and then solve for R.

    $S = KP^2$
    $B = PR$

7. Eliminate "a" and then solve for "t".

    $p = \sqrt{t - a}$
    $a = rt$

8. Eliminate "d" and then solve for "w".

    $w = \sqrt{\dfrac{d}{s}}$
    $d = hw$

9. Use the quadratic formula to solve for V.

    $V^2 - 2kV + h = 0$

ANSWERS:

1. $t = s\sqrt{w}$

2. $r_1 = r_2 \sqrt{\dfrac{A_1}{A_2}}$

3. $m = d\sqrt{\dfrac{1}{k}}$

4. $d = 2\sqrt{\dfrac{A}{\pi}}$

5. $V = \sqrt{PR}$

6. $R = B\sqrt{\dfrac{K}{S}}$

7. $t = \dfrac{p^2}{1 - r}$

8. $w = \dfrac{h}{s}$

9. $V = k \pm \sqrt{k^2 - h}$

Radical Equations and Formulas

SUPPLEMENTARY PROBLEMS - CHAPTER 4

Assignment 12

Solve each equation.

1. $\sqrt{x + 2} = 3$
2. $\sqrt{2t} = 6$
3. $\sqrt{\dfrac{3y}{2}} = 3$
4. $4 = \sqrt{\dfrac{8}{5w}}$
5. $\sqrt{\dfrac{2P - 8}{5}} = 2$
6. $\dfrac{2}{3} = \sqrt{\dfrac{1}{r + 1}}$
7. $\sqrt{x^2 + 16} = 5$
8. $\sqrt{9y^2 - 3} = 1$

Solve each equation. Isolate the radical first.

9. $3\sqrt{2t} = 6$
10. $2\sqrt{5x} = 3$
11. $\sqrt{\dfrac{6s}{s - 1}} - 3 = 0$
12. $\sqrt{\dfrac{5w}{2w - 3}} + 1 = 3$
13. $\dfrac{4\sqrt{3d}}{3} = 8$
14. $\dfrac{2}{5\sqrt{2y}} = 1$
15. $\dfrac{4}{\sqrt{A + 1}} = 2$
16. $3 = \dfrac{1}{\sqrt{m - 2}}$
17. $\dfrac{8}{1 + 4\sqrt{x}} = 4$
18. $\dfrac{10}{5 + \sqrt{2R - 3}} = 1$
19. $\dfrac{\sqrt{4d}}{2} = \dfrac{2}{3}$
20. $\dfrac{3}{4} = \dfrac{\sqrt{2s + 6}}{8}$

Complete the following formula evaluations.

21. In $\boxed{t = k\sqrt{w}}$, find "w" when $t = 12$ and $k = 3$.

22. In $\boxed{F = d\sqrt{\dfrac{a}{r}}}$, find "r" when $F = 15$, $d = 5$, and $a = 18$.

23. In $\boxed{d = p + \sqrt{mv}}$, find "v" when $d = 5$, $p = 3$, and $m = 4$.

24. In $\boxed{i = \sqrt{\dfrac{p_2 - p_1}{r}}}$, find "$p_1$" when $i = 3$, $p_2 = 25$, and $r = 2$.

25. In $\boxed{h = \sqrt{s^2 - k^2}}$, find "k" when $h = 12$ and $s = 13$.

26. In $\boxed{N = \dfrac{A}{\sqrt{B_1 B_2}}}$, find B_1 when $N = 10$, $A = 60$, and $B_2 = 9$.

27. In $\boxed{\sqrt{\dfrac{V_1}{V_2}} = \dfrac{A_1}{A_2}}$, find V_2 when $V_1 = 25$, $A_1 = 10$, and $A_2 = 6$.

28. In $\boxed{W = \dfrac{a\sqrt{2t}}{h}}$, find "t" when $W = 4$, $a = 8$, and $h = 12$.

Solve each equation. Be sure to exclude extraneous solutions.

29. $\sqrt{y - 1} = y - 3$
30. $x + 1 = \sqrt{2x + 5}$
31. $\sqrt{2t + 4} = t - 2$
32. $w + 3 = \sqrt{9w + 7}$

Assignment 13

Rearrange these formulas involving isolated radicals.

1. Solve for A.

   $s = \sqrt{A}$

2. Solve for "r".

   $a = \sqrt{\dfrac{d}{r}}$

3. Solve for "m".

   $\sqrt{2m - w} = p$

4. Solve for N.

   $H = \sqrt{\dfrac{GN}{B}}$

5. Solve for P.

   $E = \sqrt{PR}$

6. Solve for "h".

   $\sqrt{ht} = st$

7. Solve for "a".

   $\sqrt{\dfrac{b}{ap}} = \dfrac{h}{p}$

8. Solve for E.

   $\sqrt{\dfrac{E}{F}} = \dfrac{E}{P}$

9. Solve for "w".

   $b = \sqrt{d^2 - 2w}$

10. Solve for K.

    $\sqrt{AK} = \dfrac{1}{T}$

11. Solve for V.

    $G = \sqrt{\dfrac{V}{1-V}}$

12. Solve for "t".

    $v = \sqrt{\dfrac{t-1}{t}}$

Rearrange these formulas. Isolate the radical first.

13. Solve for R.

    $W = P\sqrt{R}$

14. Solve for "t".

    $d = \dfrac{\sqrt{t}}{2}$

15. Solve for "h".

    $s = \dfrac{b\sqrt{h}}{r}$

16. Solve for "x".

    $p = \dfrac{v}{\sqrt{2x}}$

17. Solve for A.

    $F = \dfrac{1}{K\sqrt{AB}}$

18. Solve for C.

    $H = \dfrac{\sqrt{C+N}}{T}$

19. Solve for "p".

    $t = 2\pi\sqrt{\dfrac{p}{d}}$

20. Solve for "r".

    $w = a\sqrt{\dfrac{s}{br}}$

21. Solve for "d".

    $v = t + \sqrt{d}$

22. Solve for "m".

    $a - \sqrt{m} = b$

23. Solve for P.

    $B\sqrt{P} - H = C$

24. Solve for W.

    $F = G - R\sqrt{W}$

Rearrange these formulas. The variable to be solved for is <u>not</u> in the radicand.

25. Solve for B.

    $P = \dfrac{B}{\sqrt{G}}$

26. Solve for A.

    $F = \dfrac{A\sqrt{R}}{D}$

27. Solve for "C".

    $h = C + k\sqrt{p}$

28. Solve for "m".

    $r = t - m\sqrt{a}$

29. Solve for V_1.

    $\sqrt{\dfrac{A_1}{A_2}} = \dfrac{V_1}{V_2}$

30. Solve for "b".

    $\dfrac{a}{b} = \sqrt{\dfrac{r}{t}}$

31. Solve for "h".

    $\sqrt{\dfrac{s_1}{s_2}} = hp$

32. Solve for K.

    $\dfrac{P}{K} = \sqrt{\dfrac{P}{F}}$

Rearrange these formulas involving radicands with squared variables.

33. Solve for "v".

    $d = \sqrt{v^2 + w^2}$

34. Solve for P.

    $G = \sqrt{B^2 - P^2}$

35. Solve for A.

    $D = \dfrac{R}{\sqrt{1 - A^2}}$

36. Solve for "d".

    $t = \dfrac{1}{\sqrt{d^2 + 1}}$

Assignment 14

Rearrange these formulas. Where possible, factor out perfect squares.

1. Solve for "d".
$$A = \frac{d^2}{4}$$

2. Solve for "r_1".
$$\frac{r_1^2}{r_2^2} = \frac{k}{p}$$

3. Solve for R.
$$T = \left(\frac{R}{V}\right)^2$$

4. Solve for F.
$$\left(\frac{B}{F}\right)^2 = P$$

5. Solve for "t_1".
$$\frac{W_1}{W_2} = \left(\frac{t_1}{t_2}\right)^2$$

6. Solve for "m".
$$s = \frac{am^2}{r^2}$$

7. Solve for P.
$$BG^2 = 2KP^2$$

8. Solve for "w".
$$2tw = \frac{1}{2pw}$$

Solve these formula derivation problems. Use either the equivalence or substitution method.

9. Eliminate G and then solve for V.
$$\boxed{\begin{array}{l} G = PR \\ GK = PV^2 \end{array}}$$

10. Eliminate "d" and then solve for "m".
$$\boxed{\begin{array}{l} d = mp \\ h = d^2 t \end{array}}$$

11. Eliminate N and then solve for T.
$$\boxed{\begin{array}{l} H = N^2 T \\ N = AG \end{array}}$$

12. Eliminate "d" and then solve for "a".
$$\boxed{\begin{array}{l} s = \sqrt{a+d} \\ a = \frac{d}{s} \end{array}}$$

13. Eliminate S and then solve for V.
$$\boxed{\begin{array}{l} W = S^2 V \\ V = \frac{F}{S} \end{array}}$$

14. Eliminate "t" and then solve for "r".
$$\boxed{\begin{array}{l} p = \sqrt{t^2 - b^2} \\ r = kt \end{array}}$$

15. Eliminate A and then solve for N.
$$\boxed{\begin{array}{l} A = FN \\ A^2 = BF \end{array}}$$

16. Eliminate "r" and then solve for "w".
$$\boxed{\begin{array}{l} r^2 = \frac{2t}{k} \\ w = kr \end{array}}$$

17. Eliminate G and then solve for P.
$$\boxed{\begin{array}{l} R = \frac{GP^2}{F} \\ V = GR \end{array}}$$

18. Eliminate "r" and then solve for "w".
$$\boxed{\begin{array}{l} b = \frac{r}{w} \\ p = \frac{1}{rw} \end{array}}$$

19. Eliminate V and then solve for S.
$$\boxed{\begin{array}{l} T = \sqrt{R^2 - V^2} \\ V = KS \end{array}}$$

20. Eliminate "c" and then solve for "a".
$$\boxed{\begin{array}{l} c = \sqrt{a^2 + b^2} \\ a^2 = d^2 - c^2 \end{array}}$$

Do these problems involving formulas in quadratic form.

21. Solve for F.
$$V = AF^2$$

22. Solve for "d".
$$ad^2 = dr$$

23. Solve for "t".
$$\sqrt{ct} = 2tw$$

24. Solve for V.
$$E = \frac{MV^2}{2}$$

25. Solve for P.
$$kP^2 - hP = 0$$

26. Solve for "t".
$$dt^2 - pt + n = 0$$

27. Solve for F.
$$F^2 + 2mF + h = 0$$

28. Eliminate "v" and then solve for "w".
$$\boxed{\begin{array}{l} w = \sqrt{r+v} \\ v = sw \end{array}}$$

29. Eliminate "t" and then solve for V.
$$\boxed{\begin{array}{l} t = kV^2 \\ V = \frac{k-t}{2p} \end{array}}$$

Chapter 5 FUNCTIONS, NON-LINEAR GRAPHS, VARIATION

In this chapter, we will define a "function" and discuss functional notation. We will discuss the following non-linear graphs: parabolas, circles, ellipses, and hyperbolas. We will discuss the general concept of variation and the following specific types of variation: direct, inverse, direct square, inverse square, joint, and combined.

5-1 FUNCTIONS

In this section, we will define and discuss functions. We will also define the domain and range of a function.

1. In equations and formulas, almost all letters are variables and all numbers are constants. For example:
 	
 In $y = 3x + 5$:   the variables are "y" and "x".
                     the constants are 3 and 5.
 	
 In $s = 16t^2$:   a) the variables are _____ and _____.
                    b) the constant is _____.	

2. When one variable is solved-for in a two-variable equation or formula, it is called the dependent variable because its value "depends" on the value substituted for the other variable. The other variable is called the independent variable. For example:
 	
 In $y = 2x^2 - x$:   "y" is the dependent variable because its value depends on the value substituted for "x".
                        "x" is the independent variable.
 	
 In $E = 8I$:   a) the dependent variable is _____.
                b) the independent variable is _____.
	a) s and t
	 b) 16

3. A function is usually expressed as an equation or formula with one variable solved-for. A two-variable equation or formula is a function if for each value substituted for the independent variable, there is only one value of the dependent variable. For example:	a) E b) I

Continued on following page.

215

216   Functions, Non-Linear Graphs, Variation

3. Continued.

$y = x^2 + 2x + 3$ is a <u>function</u> because for each value substituted for "x", there is <u>only one</u> value of "y". That is:

If $x = 3$, $y = 18$.

If $x = 0$, $y = 3$.

If $x = -1$, $y = 2$.

$v = \dfrac{s}{10}$ is a <u>function</u> because for each value substituted for "s", there is <u>only one</u> value of "v". That is:

a) If $s = 80$, $v = $ _____ .   b) If $s = 0$, $v = $ _____

---

4. Not all two-variable equations are functions. For example:

$y = \pm\sqrt{x}$ <u>is not</u> a function because there are two values of "y" for every non-zero value of "x". That is:

If $x = 4$, $y = +2$ and -2.   If $x = 9$, $y = $ _____ and _____ .

a) $v = 8$

b) $v = 0$

---

5. Some functions are not defined for one value of the <u>independent</u> variable because they lead to a division by "0" which is impossible. For example:

$y = \dfrac{5}{x}$ is not defined for $x = 0$, since $y = \dfrac{5}{0}$.

$y = \dfrac{10}{x - 2}$ is not defined for $x = 2$, since $y = \dfrac{10}{2 - 2} = \dfrac{10}{0}$

Each function below is not defined for one value of "x". Name that x-value for each.

a) $y = \dfrac{7}{3x}$ b) $y = \dfrac{4}{x^2}$ c) $y = \dfrac{20}{x - 5}$

x = _____   x = _____   x = _____

+3 and -3

---

6. Some functions involving a radical are not defined for negative values of "x" because those values lead to <u>imaginary</u> values of "y". For example:

$y = \sqrt{x}$ is not defined for negative values of "x" because they lead to imaginary values of "y".

Which functions below are not defined for negative values of "x" ? _____

a) $y = 3x$ b) $y = \sqrt{3x}$ c) $y = 3x^2$

a) $x = 0$

b) $x = 0$

c) $x = 5$

Only (b)

Functions. Non-Linear Graphs. Variation   217

7. We substituted some values for "x" in the function $y = \sqrt{x-3}$ below.

If $x = 4$, $y = \sqrt{x-3} = \sqrt{4-3} = \sqrt{1} = 1$

If $x = 3$, $y = \sqrt{x-3} = \sqrt{3-3} = \sqrt{0} = 0$

If $x = 2$, $y = \sqrt{x-3} = \sqrt{2-3} = \sqrt{-1} = i$

Therefore, $y = \sqrt{x-3}$ is not defined for values of "x" less than 3 because they lead to <u>imaginary</u> values of "y".

To avoid imaginary values of "y", the functions below are not defined for various values of "x".

a) $y = \sqrt{x-1}$ is not defined for values of "x" less than _____.

b) $y = \sqrt{x-9}$ is not defined for values of "x" less than _____.

---

8. When a function is a two-variable formula, it is usually not defined for <u>negative values</u> of the independent variable because they usually do not make sense. For example:

$C = \pi d$ is not defined for negative values of "d" because <u>negative diameters</u> do not make sense.

The function below is not defined for negative values of R because <u>negative resistances</u> do not make sense. It is also not defined for $R = 0$. Why not? _____

$$I = \frac{50}{R}$$

a) 1

b) 9

---

9. For a function, the set of permissible values for the <u>independent</u> variable is called the <u>domain</u>; the set of permissible values for the <u>dependent variable</u> is called the <u>range</u>. That is:

For $y = 3x$: The set of permissible values for "x" is the <u>domain</u>.

The set of permissible values for "y" is the <u>range</u>.

For $I = \frac{E}{50}$: a) The set of permissible values for E is the _____.

b) The set of permissible values for I is the _____.

Because $I = \frac{50}{0}$, and division by "0" is impossible.

---

a) domain

b) range

218 Functions, Non-Linear Graphs, Variation

10. For some functions, the domain and range include all positive numbers, "0", and all negative numbers. Some functions of that type are:

$$y = 5x \qquad y = x^2 - 4x + 1$$

For other functions, the domain and range are limited because the function is not defined for certain values. Some functions of that type are:

$$y = \frac{4}{x - 1} \qquad \text{(Not defined for } x = 1\text{)}$$

$$y = \sqrt{3x} \qquad \text{(Not defined for negative values of "x")}$$

$$y = \sqrt{\frac{20}{t}} \qquad \text{(Not defined for } t = 0 \text{ and negative values of "t")}$$

5-2 GRAPHS OF FUNCTIONS

In this section, we will discuss a "vertical line" test to determine whether a graph represents a function or not.

11. When graphing a two-variable equation or formula, we always put the independent variable on the horizontal axis. For example:

For $y = x^2 - 1$, we put "x" on the horizontal axis.

For $E = 50I$, we put _____ on the horizontal axis.

12. A two-variable equation or formula is a function only if there is only one value of the dependent variable for each value of the independent variable. Therefore, we can use a "vertical line" test to determine whether a graph represents a function or not. To show that fact, two graphs are shown and discussed below.

I

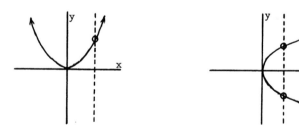

By mentally sliding the dotted vertical line along the horizontal axis, we can see these facts.

At the left, since the dotted line never intersects the graph more than once, there is only one "y" value for each "x" value. Therefore, the graph is a function.

Continued on following page.

12. Continued

    At the right, since the dotted line usually intersects the graph twice, there is usually more than one "y" value for each "x" value. Therefore, the graph is not a function.

    Which of the following graphs are functions? _____

    a)   b)   c)

    Only (b)

13. Which of the following graphs are functions? _____

    a)   b)   c)

    Both (a) and (c)

5-3 FUNCTIONAL NOTATION

In this section, we will discuss functional notation and show how it is used in evaluations.

14. In a function, we say that the dependent variable is a function of the independent variable. That is:

    In $y = 3x + 5$, we say that y is a function of x.

    In $s = 10t^2$, we say that s is a function of t.

    In $I = \dfrac{50}{R}$, we say that I is a function of R.

    Instead of the phrases "function of x", "function of t", and "function of R", we use the symbols f(x), f(t), and f(R). That is:

    In $y = 3x + 5$, we say: $y = f(x)$

    In $s = 10t^2$, we say: $s = f(t)$

    In $I = \dfrac{50}{R}$, we say: _____

220   Functions, Non-Linear Graphs, Variation

15. The symbol f(x) means "function of $\underline{x}$". It <u>does</u> <u>not</u> <u>mean</u> "multiply $\underline{f}$ and $\underline{x}$". Does f(R) mean "function of $\underline{R}$" or "multiply $\underline{f}$ and $\underline{R}$"? _____	I = f(R)

16. For f(x), instead of saying "function of $\underline{x}$", we usually say "$\underline{f}$ of $\underline{x}$". Similarly,
 	
    For f(d), we say "$\underline{f}$ of $\underline{d}$".
 	
    For f(H), we say _____.
	function of R

17. In each equation below, $\underline{y}$ is a function of $\underline{x}$.
 	
    $y = 5x$ $y = x^2 - 1$ $y = 2x^2 + x - 3$
 	
The symbol f(x) is a general symbol for any function of $\underline{x}$. Therefore, we can use it for each function above. We can write.
 	
    $y = f(x) = 5x$
    $y = f(x) = x^2 - 1$
    $y = f(x) = 2x^2 + x - 3$
 	
Similarly, in each formula below, the dependent variable is a function of $\underline{t}$.
 	
    $s = 50t$ $a = \dfrac{100}{t}$ $k = 75t^2$
 	
The symbol f(t) is also a general symbol for any function of $\underline{t}$. Therefore, we can use it for each function above. We can write:
 	
    $s = f(t) = 50t$
    $a = f(t) = \dfrac{100}{t}$
    $k =$ _____
	$\underline{f}$ of $\underline{H}$

18. As you can see from the last frame, f(x) is another way of stating the dependent variable $\underline{y}$. Therefore:
 	
    Instead of $y = f(x) = 5x$, we can write $f(x) = 5x$.
    Instead of $y = f(x) = x^2 - 1$, we can write $f(x) = x^2 - 1$
 	
Similarly, the symbol f(t) is another way of stating the dependent variable. Therefore:
 	
    Instead of $s = f(t) = 50t$, we can write $f(t) = 50t$.
    Instead of $a = f(t) = \dfrac{100}{t}$, we can write _____.
	$f(t) = 75t^2$

	$f(t) = \dfrac{100}{t}$

19. Functional notation can be used in evaluations. For example, we used it below to evaluate $f(x) = 3x + 1$ for $x = 2$ and $x = -1$.

$$f(x) = 3x + 1 \qquad\qquad f(x) = 3x + 1$$
$$f(2) = 3(2) + 1 \qquad\qquad f(-1) = 3(-1) + 1$$
$$f(2) = 7 \qquad\qquad f(-1) = -2$$

a) Find $f(5)$ for the function below.

$$f(x) = 4x^2$$
$$f(5) = \underline{\qquad}$$

b) Find $f(-2)$ for the function below.

$$f(x) = \frac{20}{x}$$
$$f(-2) = \underline{\qquad}$$

20. Functional notation can also be used for formula evaluations. For example, we used it below to evaluate $f(t) = 10t$ for $t = 5$ and $t = 20$.

$$f(t) = 10t \qquad\qquad f(t) = 10t$$
$$f(5) = 10(5) \qquad\qquad f(20) = 10(20)$$
$$f(5) = 50 \qquad\qquad f(20) = 200$$

a) Find $f(10)$ for the function below.

$$f(v) = 5v^2$$
$$f(10) = \underline{\qquad}$$

b) Find $f(2)$ for the function below.

$$f(t) = \frac{80}{t^2}$$
$$f(2) = \underline{\qquad}$$

a) $f(5) = 100$

b) $f(-2) = -10$

21. a) If $f(x) = 2x^2 - x$, find $f(-3)$. b) If $f(p) = p^2 + 6p - 9$, find $f(0)$.

a) $f(10) = 500$

b) $f(2) = 20$

22. We can also evaluate functions for literal expressions. For example, we evaluated the function below for $x = a$ to get $f(a)$.

$$f(x) = x^2 + 3$$
$$f(a) = (a)^2 + 3$$
$$f(a) = a^2 + 3$$

a) Find $f(d)$ for the function below.

$$f(x) = 20x$$
$$f(d) = \underline{\qquad}$$

b) Find $f(b)$ for the function below.

$$f(t) = \frac{100}{t^2}$$
$$f(b) = \underline{\qquad}$$

a) $f(-3) = 21$

b) $f(0) = -9$

23. We evaluated the function below for $x = a+1$ to get $f(a+1)$. Evaluate the other function for $t = a-1$ to get $f(a-1)$.

$$f(x) = x^2 + 5 \qquad\qquad f(t) = t^2 + t$$
$$f(a+1) = (a+1)^2 + 5 \qquad f(a-1) =$$
$$f(a+1) = a^2 + 2a + 1 + 5$$
$$f(a+1) = a^2 + 2a + 6$$

a) $f(d) = 20d$
b) $f(b) = \dfrac{100}{b^2}$

---

24. We evaluated the function below for $x = a+b$ to get $f(a+b)$.

$$f(x) = x + 2x^2$$
$$f(a+b) = (a+b) + 2(a+b)^2$$
$$f(a+b) = a + b + 2(a^2 + 2ab + b^2)$$
$$f(a+b) = a + b + 2a^2 + 4ab + 2b^2$$

Following the example, evaluate the function below for $H = a-b$ to get $f(a-b)$.

$$f(H) = H^2 + 5H$$
$$f(a-b) =$$

$f(a-1) = a^2 - a$
from:
$(a^2 - 2a + 1) + (a - 1)$

---

25. For $y = 2x + 5$, $f(x)$ is equivalent to the dependent variable $\underline{y}$. Therefore, when graphing the function, we sometimes use $f(x)$ instead of $\underline{y}$ on the vertical axis as we have done below.

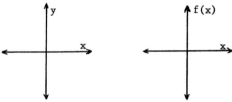

For $s = 10t^2$, $f(t)$ is equivalent to the dependent variable $\underline{s}$. Therefore, at the right below, show how $f(t)$ is sometimes used on one of the axes.

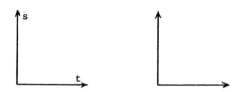

$f(a-b) = (a-b)^2 + 5(a-b)$
$= a^2 - 2ab + b^2 + 5a - 5b$

5-4 FUNCTIONS WITH MORE THAN ONE INDEPENDENT VARIABLE

In this section, we will briefly discuss functions with more than one independent variable.

26. Functions can contain more than one independent variable. For example: In $y = 3x + 2z$, the independent variables are $\underline{x}$ and $\underline{z}$. In $\lambda = \dfrac{L_2 - L_1}{L_1(t_2 - t_1)}$, the independent variables are _____	
27. When a function contains more than one independent variable, we can also use functional notation. For example If $A = LW$, $A = f(L, W)$ If $i = \sqrt{\dfrac{p_2 - p_1}{r}}$, $i = f(\ \ ,\ \ ,\ \ )$	L_2, L_1, t_2, t_1
28. When a function contains more than one independent variable, the functional notation is still equivalent to the dependent variable. That is: Instead of $E = \dfrac{1}{2}mv^2$, we could write: $f(m, v) = \dfrac{1}{2}mv^2$ Instead of $P = \dfrac{Fs}{t}$, we could write: _____	$f(p_2, p_1, r)$
29. When a function contains more than one independent variable, we can also use functional notation for evaluations. For example, we used it below to evaluate $f(x, z) = x^2 + 2z$ for $x = 3$ and $z = -1$. $\quad f(x, z) = x^2 + 2z$ $\quad f(3, -1) = (3)^2 + 2(-1)$ $\quad f(3, -1) = 7$ a) Find $f(10, 5)$ for the function below. $\quad f(m, c) = mc^2$ $\quad f(10, 5) = $ _____ b) Find $f(2, 40, 8)$ for the function below. $\quad f(T_2, V_1, V_2) = \dfrac{T_2 V_1}{V_2}$ $\quad f(2, 40, 8) = $ _____	$f(F, s, t) = \dfrac{Fs}{t}$
	a) $f(10, 5) = 250$ b) $f(2, 40, 8) = 10$

224 Functions, Non-Linear Graphs, Variation

5-5 LINEAR FUNCTIONS

Any equation or formula whose graph is a straight line is a linear function. We will briefly review linear functions in this section.

30. The general equation for a linear function is $y = mx + b$. Two specific equations are given below and graphed at the right.

    $y = -2x + 1$

    $y = x - 2$

    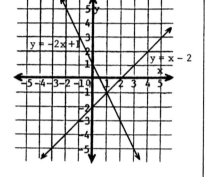

    "m" and "b" are the two constants in the equations. "m" is the slope of the line; "b" is the y-intercept.

    For $y = -2x + 1$, $m = -2$ and $b = 1$.

    For $y = x - 2$, $m = $ _____ and $b = $ _____

31. When a line passes through the origin, the y-intercept is "0". Therefore, $b = 0$ and the general equation for the function is $y = mx$. Two specific equations are given below and graphed at the right.

    $y = 3x$

    $y = -x$

    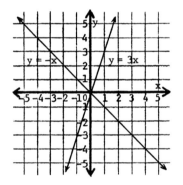

    "m" is the only constant in the equation. "m" is the slope of the line.

    a) For $y = 3x$, the slope is _____.

    b) For $y = -x$, the slope is _____.

    $m = 1$ and $b = -2$

32. Two linear formulas are graphed below. Since negative values do not make sense for the variables, the graphs appear only in the first quadrant.

    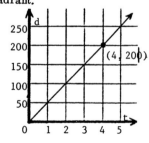
    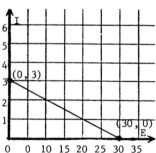

    a) 3
    b) -1

Continued on following page.

32. Continued

Using the plotted points, write the formula for each function.

a) At the left, d = _____   b) At the right, I = _____

| a) d = 50t | b) $I = -\frac{1}{10}E + 3$ or $I = -0.1E + 3$ |

5-6 PARABOLAS (QUADRATIC FUNCTIONS)

In this section, we will discuss parabolas, the graphs of quadratic functions. The following features of a parabola are included: vertex, maximum or minimum value, axis of symmetry, and focus.

33. The general form of a quadratic function is given below.

$$y = ax^2 + bx + c$$

Though <u>a</u> cannot be "0", either <u>b</u> or <u>c</u> or both can be "0". Therefore, the following specific types are possible.

$y = x^2 + 5x + 6$

$y = -3x^2 + x$ (c = 0)

$y = 5x^2 - 9$ (b = 0)

$y = -4x^2$ (b = 0, c = 0)

Which of the following are quadratic functions? _____

a) $y = 25x^2$ b) $y = 2x + 3$ c) $y = 2x^2 - 7x + 5$

34. The graph of any quadratic function is a cup-shaped curve called a "parabola". As examples, we graphed the two quadratic functions below on the next page.

$$y = x^2 - 2x - 3$$

x	y
-2	5
-1	0
0	-3
1	-4
2	-3
3	0
4	5

$$y = -2x^2 + 4x + 1$$

x	y
-1	-5
0	1
1	3
2	1
3	-5

Continued on following page.

Both (a) and (c)

34. Continued

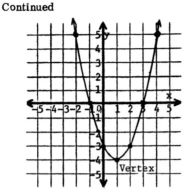
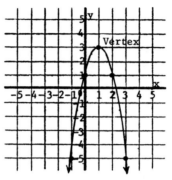

The parabola at the left opens upward; the parabola at the right opens downward. The point where each curve turns is called the "vertex".

a) At the left, the vertex is the lowest point. Its coordinates are (,).

b) At the right, the vertex is the highest point. Its coordinates are (,).

35. Two quadratic functions in which c = 0 are graphed below. Notice that each graph passes through the origin.

$y = x^2 + 4x$		$y = -x^2 + 4x$	
x	y	x	y
-5	5	-1	-5
-4	0	0	0
-3	-3	1	3
-2	-4	2	4
-1	-3	3	3
0	0	4	0
1	5	5	-5

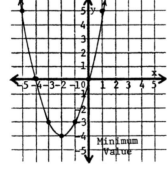
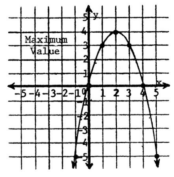

When a parabola opens upward, the vertex gives the minimum value of y. When a parabola opens downward, the vertex gives the maximum value of y.

a) At the left, the vertex is at (-2, -4). Therefore, the minimum value of y is _____.

b) At the right, the vertex is at (2, 4). Therefore, the maximum value of y is _____.

a) (1, -4)
b) (1, 3)

36. Any parabola has an <u>axis of symmetry</u> (see the dotted lines below). That is, if the parabola were folded on that axis, the two halves would match.

a) -4

b) 4

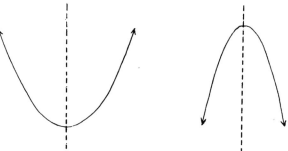

When b = 0 in a quadratic function, the <u>y-axis is the axis of symmetry</u>. As examples, we graphed two functions of that type below.

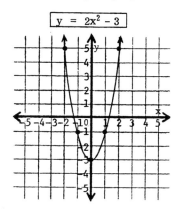
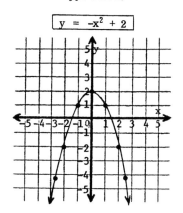

When b = 0, <u>c</u> is the point at which the parabola crosses the y-axis. That is:

For $y = 2x^2 - 3$, c = -3. The parabola crosses the y-axis at -3 and the vertex is at (0, -3).

For $y = -x^2 + 2$, c = 2. The parabola crosses the y-axis at 2 and the vertex is at (,).

37. When both b = 0 and c = 0 in a quadratic function, the y-axis is the axis of symmetry and the vertex is at (0, 0). As examples, we graphed two quadratic functions below.

(0, 2)

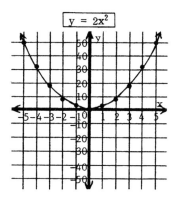
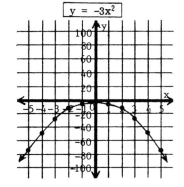

Continued on following page.

228 Functions, Non-Linear Graphs, Variation

37. Continued

   From the graphs, you can see these facts:

   a) For $y = 2x^2$, "0" is the _____ (minimum/maximum) value of y.

   b) For $y = -3x^2$, "0" is the _____ (minimum/maximum) value of y.

   ---

   a) minimum
   b) maximum

38. By reviewing the parabolas in the last four frames, you can see these facts about "a" in a quadratic function.

   When "a" is <u>positive</u>, the parabola opens <u>upward</u>.

   When "a" is <u>negative</u>, the parabola opens <u>downward</u>.

   Which of the following graph as parabolas that open upward? _____

   a) $y = 4x^2$ b) $y = -x^2$ c) $y = -3x^2 + x$ d) $y = x^2 - 16$

   ---

39. There are various formulas of the form $y = ax^2$. Some examples are:

   $P = 5I^2$ $s = 16t^2$ $E = 10v^2$

   Since negative values of the variables usually do not make sense, the graphs appear only in the first quadrant. For each formula, we get half a parabola like the sketch at the right.

   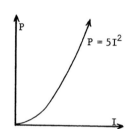

   Both (a) and (d)

40. The parabola has many uses. Projectiles and some space vehicles move in a parabolic path. Parabolic arches are used in construction. Parabolic reflectors are used in spotlights and searchlights, microwave and radar antennas, solar furnaces, and radio and optical telescopes. All reflectors use the "<u>focusing</u>" property of parabolas which is diagrammed below.

   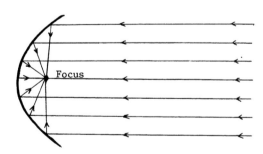

Continued on following page.

40. Continued

The "focusing" property of a parabola is:

1) All incoming rays parallel to the axis of symmetry reflect from the parabola and pass through a single point called the "focus".

2) All rays originating at the focus reflect from the parabola so that they leave parallel to the axis of symmetry.

SELF-TEST 15 (pages 215-229)

1. For the function $y = x^2 - 9x$, the independent variable is _____.
2. For the function $p = 4t + 1$, the dependent variable is _____.
3. For $y = x^2 - 16$, the set of permissible values for "x" is called the _____ (domain/range).
4. When graphing a two-variable equation, we always put the dependent variable on the _____ (horizontal/vertical) axis.

5. If $f(x) = 3x^2 + 5$, $f(10) = $ _____
6. If $f(w) = \dfrac{w}{1 - w^2}$, $f(-2) = $ _____
7. If $f(x) = 8 - 3x$, $f(t - 1) = $ _____
8. If $f(s, v) = \dfrac{s^2}{2v}$, $f(6, 9) = $ _____

For the linear function $y = 3x - 2$:
9. The slope is _____.
10. The y-intercept is _____.

11. Find the equation of the linear function graphed below. _____

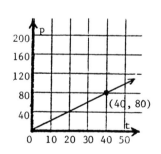

The graph of the quadratic function $y = x^2 - 4x$ shown below is a parabola.

12. The vertex is at (__ , __).
13. The x-intercepts are at (__ , __) and (__ , __).
14. The minimum value of y is _____.

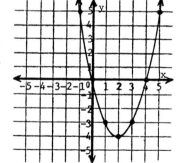

15. Which of the following are quadratic functions? _____
   a) $y = 4x$ b) $y = x^2 - 5$ c) $y = -x^2 + 3x + 8$ d) $w = 2s^2$

16. Which of the following graph as parabolas that open downward? _____
   a) $y = 20x^2$ b) $y = -7x^2 + 4$ c) $y = x^2 - 6x + 5$ d) $y = 3x - 4x^2$

230 Functions, Non-Linear Graphs, Variation

ANSWERS:
1. x
2. p
3. domain
4. vertical
5. 305
6. $\frac{2}{3}$
7. 11 - 3t
8. 2
9. 3
10. -2
11. p = 2t
12. (2, -4)
13. (0, 0) and (4, 0)
14. -4
15. b, c, d
16. b, d

5-7 CIRCLES

In this section, we will discuss the general equation of circles and the graphs of circles.

41. The general equation for a circle whose center is (a, b) and whose radius is r is given below.

$$(x - a)^2 + (y - b)^2 = r^2$$

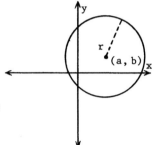

Using the fact that (a, b) is the center and r is the radius, we can identify the coordinates of the center and the length of the radius for each circle below.

$$(x - 2)^2 + (y - 5)^2 = 16$$

   The coordinates of the center are (2, 5).

   The length of the radius is 4, since $r^2 = 16$.

$$(x - 3)^2 + (y - 1)^2 = 25$$

   a) The coordinates of the center are (,).

   b) The length of the radius is _____.

42. Two circles are graphed below. The coordinates of the center and the length of the radius are given for each.

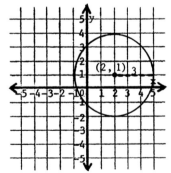 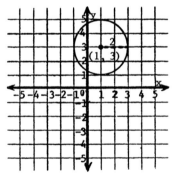

Since $\underline{a}$, $\underline{b}$, and $\underline{r}$ are given for each circle, we can write its equation. Remember that the equation contains $\underline{r^2}$, not $\underline{r}$.

a) (3, 1)

b) 5, since $r^2 = 25$

Continued on following page.

42. For the circle at the left, a = 2, b = 1, and r = 3. Therefore:

$$(x - a)^2 + (y - b)^2 = r^2$$
$$(x - 2)^2 + (y - 1)^2 = 3^2$$
$$(x - 2)^2 + (y - 1)^2 = 9$$

For the circle at the right, a = 1, b = 3, and r = 2. Write its equation.

43. In the general equation for circles, both $(x - a)^2$ and $(y - b)^2$ contain a "−". When either contains a "+", we must convert to a subtraction to identify <u>a</u> or <u>b</u>. For example, we converted to subtraction to find the center of one circle below. Find the center and radius of the other circle.

$(x - 4)^2 + (y + 1)^2 = 36$ $(x + 5)^2 + (y - 2)^2 = 1$
$(x - 4)^2 + [y - (-1)]^2 = 36$

The center is (4, −1).   The center is _____.

The radius is 6.   The radius is _____.

$(x - 1)^2 + (y - 3)^2 = 4$

44. The center and radius are given for each circle below.

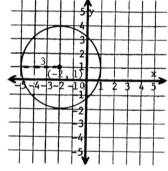 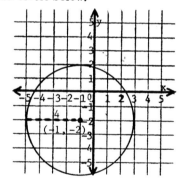

Using (−2, 1) and r = 3, the equation of the circle at the left is written below. Notice how we converted the subtraction to addition. Use (−1, −2) and r = 4 to find the equation of the other circle.

$$[x - (-2)]^2 + (y - 1)^2 = 3^2$$
$$(x + 2)^2 + (y - 1)^2 = 9$$

The center is (−5, 2), from:
$$[x - (-5)]^2 + (y - 2)^2 = 1$$
The radius is 1.

$(x + 1)^2 + (y + 2)^2 = 16$

232   Functions, Non-Linear Graphs, Variation

45. When the center of a circle is at the origin, its coordinates are (0, 0). Therefore, <u>a</u> and <u>b</u> in $(x - a)^2 + (y - b)^2 = r^2$ are "0", and the general equation reduces to:

$$x^2 + y^2 = r^2$$

That is: $x^2 + y^2 = 25$ is the equation of a circle whose center is (0, 0) and whose radius is 5.

$x^2 + y^2 = 64$ is the equation of a circle whose center is (0, 0) and whose radius is _____.

46. Each circle below has its center at the origin. The radius of each is given. Using $x^2 + y^2 = r^2$ and the given radius, write the equation for each.

8

a)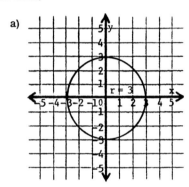

b)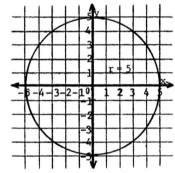

a) $x^2 + y^2 = 9$ b) $x^2 + y^2 = 25$

5-8   ELLIPSES

Orbiting satellites and planets revolving around the sun travel in an oval-shaped curve called an "<u>ellipse</u>". In this section, we will discuss the general equation of ellipses with centers at the origin and their graphs.

47. An ellipse is an oval-shaped curve. The general equation for ellipses with centers at the origin is:

$$\boxed{\dfrac{x^2}{a^2} + \dfrac{y^2}{b^2} = 1}$$

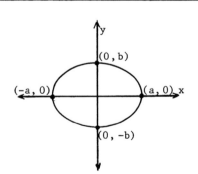

As you can see from the graph, (a, 0) and (-a, 0) are the x-intercepts and (0, b) and (0, -b) are the y-intercepts.

Continued on following page.

47. Continued

   Therefore:

   For $\frac{x^2}{9} + \frac{y^2}{4} = 1$: The x-intercepts are (3, 0) and (-3, 0).
   The y-intercepts are (0, 2) and (0, -2)

   For $\frac{x^2}{16} + \frac{y^2}{25} = 1$: a) The x-intercepts are (,   ) and (,).
   b) The y-intercepts are (,   ) and (,).

   a) (4, 0) and (-4, 0)

   b) (0, 5) and (0, -5)

48. Two ellipses are graphed below. The coordinates of the intercepts for each are given.

   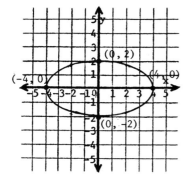
   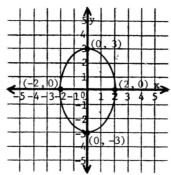

   For the ellipse at the left, a = 4 and b = 2. Therefore, the equation of the ellipse is:

   $$\frac{x^2}{4^2} + \frac{y^2}{2^2} = 1 \quad \text{or} \quad \frac{x^2}{16} + \frac{y^2}{4} = 1$$

   For the ellipse at the right, a = 2 and b = 3. Write the equation of the ellipse.

   $\frac{x^2}{4} + \frac{y^2}{9} = 1$

234 Functions, Non-Linear Graphs, Variation

5-9 HYPERBOLAS

In this section, we will discuss the general equation of hyperbolas and their graphs.

49. A hyperbola is a curve with two parts. The general equation of a hyperbola is:

$$\boxed{\frac{x^2}{a^2} - \frac{y^2}{b^2} = 1}$$

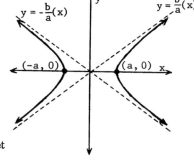

The x-intercepts are $(-a, 0)$ and $(a, 0)$. There are no y-intercepts.

The farther the two parts of the curve are from the origin, the closer they get to two lines called the "asymptotes" without touching them. In the diagram, the dotted lines are the asymptotes. The equations of the asymptotes are:

$$y = \frac{b}{a}(x) \quad \text{and} \quad y = -\frac{b}{a}(x)$$

Let's find the equations of the asymptotes for these.

For $\frac{x^2}{4} - \frac{y^2}{9} = 1$, $a = 2$ and $b = 3$.

The asymptotes are: $y = \frac{3}{2}(x)$ and $y = -\frac{3}{2}(x)$.

For $\frac{x^2}{36} - \frac{y^2}{4} = 1$, $a = 6$ and $b = 2$.

The equations of the asymptotes are: _____

and _____

50. We graphed the hyperbola below at the right.

$$\frac{x^2}{4} - \frac{y^2}{16} = 1$$

The equations of the asymptotes were obtained from:

$$y = \frac{4}{2}(x) \quad \text{and} \quad y = -\frac{4}{2}(x)$$

Do the curves touch the asymptotes? _____

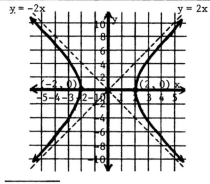

$y = \frac{1}{3}(x)$ and

$y = -\frac{1}{3}(x)$

No. They get closer and closer without touching them.

Functions, Non-Linear Graphs, Variation   235

51. The graph of any equation of the following form is also a hyperbola.

$$xy = k$$

As an example, we graphed the following equation.

$$xy = 5$$

When the equation of a hyperbola is of the form $xy = k$:

a) Which two lines are the asymptotes? _____

b) In which two quadrants do the two parts of the curve appear? _____

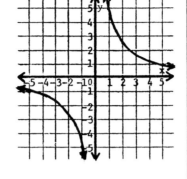

52. There are various formulas of the form $xy = k$. Some examples are:

$$PV = 6 \qquad EI = 40 \qquad vt = 200$$

Since negative values of the variables do not usually make sense, the graphs appear only in the first quadrant. For each formula, we get half a hyperbola like the sketch at the right.

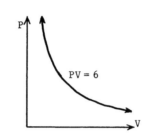

a) the axes

b) Quadrants 1 and 3

5-10   VARIATION

In any two-variable equation, the independent and dependent variables vary together. That is, when the independent variable increases or decreases, the dependent variable either increases or decreases. This process of varying together is called "variation". We will discuss "variation" in this section.

53. At the right, there are various pairs of values for the function below.

$$y = 2x$$

From the table, you can see these facts.

1) As x increases, y increases.
2) As x decreases, y decreases.

x	y
1	2
2	4
3	6
5	10
10	20

a) In $d = 10t$, when t increases, d _____ (increases/decreases).

b) In $F = 50s$, when s decreases, F _____ (increases/decreases).

a) d increases

b) F decreases

54. At the right, there are various pairs of values for the function below.

$$y = \frac{24}{x}$$

x	y
1	24
2	12
3	8
6	4
12	2

From the table, you can see these facts:

  1) As <u>x</u> increases, <u>y</u> decreases.

  2) As <u>x</u> decreases, <u>y</u> increases.

a) In $P = \frac{100}{V}$, as V increases, P _____ (increases/decreases).

b) In $H = \frac{40}{L}$, as L decreases, H _____ (increases/decreases).

---

55. At the right, there are various pairs of values for the function below.

$$y = 2x^2$$

x	y
1	2
2	8
3	18
5	50
10	200

a) P decreases

b) H increases

From the table, you can see these facts:

  1) As <u>x</u> increases, <u>y</u> increases.

  2) As <u>x</u> decreases, <u>y</u> decreases.

a) In $P = 10I^2$, as I increases, P _____ (increases/decreases).

b) In $F = 2.5v^2$, as <u>v</u> decreases, F _____ (increases/decreases).

---

56. At the right, there are various pairs of values for the function below.

$$y = \frac{400}{x^2}$$

x	y
1	400
2	100
5	16
10	4
20	1

a) P increases

b) F decreases

From the table, you can see these facts:

  1) As <u>x</u> increases, <u>y</u> decreases.

  2) As <u>x</u> decreases, <u>y</u> increases.

a) In $F = \frac{100}{d^2}$, as <u>d</u> increases, F _____ (increases/decreases).

b) In $P = \frac{200}{t^2}$, as <u>t</u> decreases, P _____ (increases/decreases).

---

a) F <u>decreases</u>

b) P <u>increases</u>

Functions, Non-Linear Graphs, Variation   237

5-11   DIRECT VARIATION

Any variation of the form $y = kx$ is called "direct" variation. We will discuss "direct" variation in this section.

57. Any equation or formula of the form $\boxed{y = kx}$ is called "direct variation". Some examples are:

$$y = 4x \qquad d = 100t \qquad C = \pi d$$

In $y = kx$, $\underline{k}$ is called the "constant of variation" or the "constant of proportionality". That is:

a) In $y = 4x$, the constant of variation is _____.

b) In $d = 100t$, the constant of proportionality is _____.

58. The following language is used to state a direct variation.

For $y = 4x$, we say: y varies directly as x.
                  or: y is directly proportional to x.

For $d = 100t$, we say: d varies directly as t.
                    or: d is directly proportional to t.

a) In $E = 6R$, _____ varies directly as _____.

b) In $C = \pi d$, _____ is directly proportional to _____.

a) 4
b) 100

59. Using the given constant of variation, we wrote each direct variation below as an equation or formula.

y varies directly as x, and the constant $y = kx$
of variation is k.

P is directly proportional to H, and the $P = 15H$
constant of proportionality is 15.

Write each direct variation below as an equation or formula.

a) m varies directly as t, and the
   constant of proportionality is 50.   _____

b) R is directly proportional to S, and
   the constant of variation is k.   _____

a) E . . . . R
b) C . . . . d

60. Using $\underline{k}$ as the constant of variation, write each of these as a formula.

a) The pressure (P) in a container varies
   directly as the temperature (T).   _____

b) The current (I) in an electric circuit is
   directly proportional to the voltage (E). _____

a) m = 50t
b) R = kS

238 Functions, Non-Linear Graphs, Variation

61. The graph of any direct variation is a straight line through the origin or starting at the origin. As examples, we graphed two direct variations below.

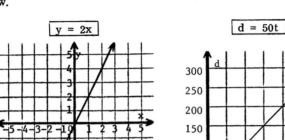

Only straight lines through the origin represent direct variations. Other straight lines that do not pass through the origin, like y = 3x + 2 or P = 5V - 1, are not direct variations.

Which of the following are direct variations? _____

    a) y = 10x   b) R = 4t + 3   c) y = x + 9   d) E = 12I

a) P = kT

b) I = kE

---

62. In a direct variation, whatever happens to the independent variable also happens to the dependent variable. Some specific changes are as follows:

Independent Variable	Dependent Variable
Doubled	Doubled
Tripled	Tripled
Cut in half	Cut in half

As an example, some pairs of values for d = 50t are given at the right.

If t is doubled from 1 to 2, d is doubled from 50 to 100.

If t is tripled from 2 to 6, d is tripled from 100 to 300.

If t is cut in half from 4 to 2, d is cut in half from _____ to _____.

d	t
50	1
100	2
150	3
200	4
250	5
300	6

Only (a) and (d)

---

63. The constant of variation can be a fraction or decimal number. Some examples are:

    $y = \frac{1}{2}x$ $H = \frac{7}{5}T$ F = 1.48s

200 to 100

Continued on following page.

63. Continued

Each equation below is also a direct variation. To show that fact, we can write the constant of variation explicitly. That is:

$$y = \frac{x}{3} \quad \text{can be written:} \quad y = \frac{1}{3}x$$

$$y = \frac{5x}{4} \quad \text{can be written:} \quad y = \frac{5}{4}x$$

Write each formula below so that the constant of variation is explicit.

    a) $V = \dfrac{D}{10}$ _____   b) $F = \dfrac{9R}{7}$ _____

---

64. If we are given one specific pair of values in a direct variation, we can find <u>k</u> by substitution. For example, if <u>y</u> varies directly as <u>x</u> and y = 10 when x = 2, we get:

$$y = kx$$
$$10 = k(2)$$
$$k = \frac{10}{2} = 5$$

Since k = 5, the direct variation is y = 5x. Using that equation, we can find <u>y</u> for other values of <u>x</u>. For example, we found <u>y</u> when x = 3 below. Find <u>y</u> when x = 10.

    y = 5x   y = 5x
    y = 5(3)
    y = 15

a) $V = \dfrac{1}{10}D$

b) $F = \dfrac{9}{7}R$

---

65. Using the two steps from the last frame, we solved the problem below.

If E is directly proportional to R and E = 30 when R = 3, find E when R = 5.

    E = kR   E = 10R
    30 = k(3)   E = 10(5)
    $k = \dfrac{30}{3} = 10$ E = 50

Using the same steps, solve this one.

If <u>d</u> varies directly as <u>t</u> and d = 600 when t = 10, find <u>d</u> when t = 7.

y = 50

240   Functions, Non-Linear Graphs, Variation

66. Using the same two steps, solve each problem.

   a) The amount of current (I) in an electric circuit varies directly as the applied voltage (E). If I = 30 amperes when E = 6 volts, find I when E = 24 volts.

   b) The amount of stretch (s) in a spring is directly proportional to the force (F) applied to it. If s = 6 centimeters when F = 40 kilograms, find s when F = 90 kilograms.

   d = 420, from:
   d = 60t

   a) 120 amperes, from: I = 5E
   b) 13.5 centimeters, from: s = 0.15F

5-12 INVERSE VARIATION

Any variation of the form $y = \dfrac{k}{x}$ is called "inverse" variation. We will discuss "inverse" variation in this section.

67. Any equation or formula of the form $\boxed{y = \dfrac{k}{x}}$ is called "inverse variation". Some examples are:

$$y = \frac{12}{x} \qquad v = \frac{100}{t} \qquad V_1 = \frac{8.5}{P_1}$$

In $y = \dfrac{k}{x}$, k is called the "constant of variation" or the "variation constant". That is:

   a) In $y = \dfrac{12}{x}$, the constant of variation is _____.

   b) In $V_1 = \dfrac{8.5}{P_1}$, the variation constant is _____.

68. The following language is used to state an inverse variation.

   For $y = \dfrac{12}{x}$, we say: y varies inversely as x.
   or: y is inversely proportional to x.

   a) In $v = \dfrac{100}{t}$, _____ varies inversely as _____.

   b) In $V_1 = \dfrac{8.5}{P_1}$, _____ is inversely proportional to _____.

   a) 12
   b) 8.5

   a) v . . . . . . t
   b) V_1 . . . . . P_1

Functions, Non-Linear Graphs, Variation 241

69. Following the example, write each inverse variation as an equation or formula.

y varies inversely as x, and the constant of variation is 50. $y = \dfrac{50}{x}$

a) P is inversely proportional to V, and the variation constant is 4.75. _____

b) a varies inversely as t, and the constant of variation is k. _____

70. Using k as the constant of variation, write each of these as a formula.

a) P = $\dfrac{4.75}{V}$

b) $a = \dfrac{k}{t}$

a) The time (t) required to travel a fixed distance varies inversely as the velocity (v). _____

b) The current (I) in an electric circuit with a fixed voltage is inversely proportional to the resistance (R). _____

71. Any inverse variation can be written in the form $\boxed{xy = k}$. For example:

$y = \dfrac{24}{x}$ can be written $xy = 24$

$P = \dfrac{100}{V}$ can be written $PV = 100$

a) $t = \dfrac{k}{v}$

b) $I = \dfrac{k}{R}$

Since $xy = k$ is a standard form for a hyperbola, any inverse variation graphs as a hyperbola or half a hyperbola. Two examples are given below.

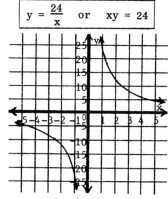
$y = \dfrac{24}{x}$ or $xy = 24$

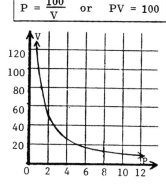
$P = \dfrac{100}{V}$ or $PV = 100$

Which of the following are inverse variations? _____

a) $y = 10x$ b) $y = \dfrac{10}{x}$ c) $E = 40I$ d) $I = \dfrac{24}{R}$

Both (b) and (d)

72. In inverse variations, the change in the dependent variable is the <u>reciprocal</u> of any change in the dependent variable. Some specific changes are as follows:

Independent Variable	Dependent Variable
Doubled	Cut in half
Tripled	Cut to one-third
Cut in half	Doubled

As an example, some pairs of values for $y = \dfrac{24}{x}$ are given at the right.

x	y
1	24
2	12
3	8
4	6
6	4
12	2

If $\underline{x}$ is doubled from 1 to 2, $\underline{y}$ is cut in half from 24 to 12.

If $\underline{x}$ is tripled from 2 to 6, $\underline{y}$ is cut to one-third from 12 to 4.

If $\underline{x}$ is cut in half from 6 to 3, $\underline{y}$ is doubled from _____ to _____.

---

73. If we are given one specific pair of values in an inverse variation, we can find $\underline{k}$ by substitution. For example, if $\underline{y}$ varies inversely as $\underline{x}$ and $y = 50$ when $x = 4$, we get:

$$y = \frac{k}{x}$$

$$50 = \frac{k}{4}$$

$$k = (50)(4) = 200$$

Since $k = 200$, the inverse variation is $y = \dfrac{200}{x}$. We can use that equation to find $\underline{y}$ for other values of $\underline{x}$.

a) Find $\underline{y}$ when $x = 5$.

$$y = \frac{200}{x}$$

b) Find $\underline{y}$ when $x = 50$.

$$y = \frac{200}{x}$$

4 to 8

---

74. Using the two steps from the last frame, we solved the problem below.

If I is inversely proportional to R and $I = 20$ when $R = 2$, find I when $R = 8$.

$$I = \frac{k}{R} \qquad\qquad I = \frac{40}{R}$$

$$20 = \frac{k}{2} \qquad\qquad I = \frac{40}{8}$$

$$k = 40 \qquad\qquad I = 5$$

a) $y = 40$
b) $y = 4$

Continued on following page.

74. Continued

Using the same steps, solve this one.

If t varies inversely as v and $t = 4$ when $v = 100$, find t when $v = 80$.

$t = 5$, from:

$t = \dfrac{400}{v}$

75. Using the same two steps, solve these.

a) In an electric circuit with fixed power, the current (I) is inversely proportional to the voltage (E). If I = 8 amperes when E = 12 volts, find I when E = 10 volts.

b) In a container with a fixed temperature, the pressure (P) of a gas varies inversely as its volume (V). If P = 10 grams per square centimeter when V = 25 cubic centimeters, find P when V = 40 cubic centimeters.

a) I = 9.6 amperes   b) P = 6.25 grams per square centimeter

SELF-TEST 16 (pages 230-244)

1. Write the equation of the circle whose center is at (-3, 2) and whose radius is 4.

2. The ellipse shown below passes through (-3, 0), (3, 0), (0, 2), and (0, -2). Write its equation.

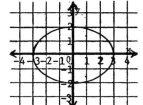

3. Which of the following equations graph as hyperbolas? _____

a) $x^2 + y^2 = 100$

b) $xy = 240$

c) $\dfrac{x^2}{4} + \dfrac{y^2}{16} = 1$

d) $\dfrac{x^2}{16} - \dfrac{y^2}{4} = 1$

Continued on following page.

SELF-TEST 16 (pages 230-244) - Continued

4. In $w = 6s$, when s increases, does w increase or decrease? _____

5. In $d = \dfrac{25}{p}$, when p decreases, does d increase or decrease? _____

6. If A is directly proportional to W and A = 24 when W = 8, find A when W = 5.

   A = _____

7. If power P varies directly as voltage E and P = 5 watts when E = 20 volts, find P when E = 12 volts.

   P = _____

8. If velocity v is directly proportional to time t and v = 20 m/sec when t = 8 sec, find v when t = 50 sec.

   v = _____

9. If r is inversely proportional to h and r = 10 when h = 8, find r when h = 20.

   r = _____

10. If current (I) in a circuit is inversely proportional to resistance (R) and I = 2 amperes when R = 12 ohms, find I when R = 15 ohms.

    I = _____

ANSWERS:
1. $(x+3)^2 + (y-2)^2 = 16$
2. $\dfrac{x^2}{9} + \dfrac{y^2}{4} = 1$
3. (b) and (d)
4. increases
5. increases
6. A = 15
7. P = 3 watts
8. v = 125 m/sec
9. r = 4
10. I = 1.6 amperes

5-13 DIRECT SQUARE VARIATION

Any variation of the form $y = kx^2$ is called "direct square" variation. We will discuss "direct square" variation in this section.

---

76. Any equation or formula of the form $\boxed{y = kx^2}$ is called "direct square variation". Some examples are:

$$y = 3x^2 \qquad P = 40I^2 \qquad A = .7854d^2$$

In $y = kx^2$, k is called the "constant of variation" or the "variation constant". The following language is used to state a direct square variation.

For $y = 3x^2$, we say: y varies directly as the square of x.

For $P = 40I^2$, we say: ____ varies directly as the square of ____.

77. Following the example, write each direct square variation as an equation or formula.

   y varies directly as the square of x, and the constant of variation is 75. $y = 75x^2$

   a) R varies directly as the square of T, and the variation constant is 1.4.

   b) h varies directly as the square of p and the constant of variation is k.

   P. I

78. Write each of these as a formula.

   a) The distance (s) that an object falls from rest varies directly with the square of the time (t) it falls. The variation constant is 16.

   b) The area (A) of a circle varies directly with the square of the radius (r). The constant of variation is π.

   a) $R = 1.4T^2$
   b) $h = kp^2$

79. The graph of any direct square variation is a parabola or half a parabola. Two examples are shown below.

   a) $s = 16t^2$
   b) $A = \pi r^2$

   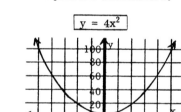
   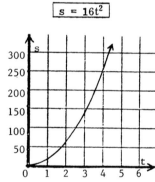

   Which of the following are direct square variations? _____

   a) $b = 14.5v^2$ b) $h = 16t$ c) $PV = 150$ d) $P = 200V^2$

80. Each equation below is also a direct square variation. To show that fact, we can write the fractional constant explicitly. That is:

   $y = \dfrac{x^2}{4}$ can be written $y = \dfrac{1}{4}x^2$

   $y = \dfrac{5x^2}{3}$ can be written $y = \dfrac{5}{3}x^2$

   Both (a) and (d)

Continued on following page.

80. Continued

Write each formula below so that the constant of variation is explicit.

a) $R = \dfrac{T^2}{10}$ _____ b) $M = \dfrac{23v^2}{11}$ _____

a) $R = \dfrac{1}{10}T^2$

b) $m = \dfrac{23}{11}v^2$

81. In direct square variation, the change in the dependent variable is the square of any change in the independent variable. Some specific changes are shown in the table below.

Independent Variable	Dependent Variable
Doubled	Increased four times, from 2^2
Tripled	Increased nine times, from 3^2
Cut in half	Cut to one-fourth, from $\left(\dfrac{1}{2}\right)^2$

As an example, some pairs of values for $y = 2x^2$ are given at the right.

If $\underline{x}$ is doubled from 1 to 2, $\underline{y}$ increases four times from 2 to 8.

If $\underline{x}$ is tripled from 2 to 6, $\underline{y}$ increases nine times from 8 to 72.

If $\underline{x}$ is cut in half from 4 to 2, $\underline{y}$ is cut to one-fourth from ____ to ____.

x	y
1	2
2	8
3	18
4	32
5	50
6	72

82. If we are given one specific pair of values in a direct square variation, we can find $\underline{k}$ by substitution. For example, if $\underline{y}$ varies directly as the square of $\underline{x}$ and $y = 12$ when $x = 2$, we get:

$$y = kx^2$$
$$12 = k(2)^2$$
$$12 = k(4)$$
$$k = 3$$

Since $k = 3$, the direct square variation is $y = 3x^2$. We can use that equation to find $\underline{y}$ for other values of $\underline{x}$.

a) Find $\underline{y}$ when $x = 4$.

   $y = 3x^2$

b) Find $\underline{y}$ when $x = 10$.

   $y = 3x^2$

32 to 8

a) $y = 48$

b) $y = 300$

83. Using the two steps from the last frame, we solved the problem below.

If a varies directly as the square of m and a = 100 when m = 5, find a when m = 8.

$a = km^2$ $a = 4m^2$

$100 = k(5)^2$ $a = 4(8)^2$

$100 = k(25)$ $a = 4(64)$

$k = 4$ $a = 256$

Using the same steps, solve this one.

If R varies directly as the square of V and R = 500 when V = 10, find R when V = 4.

R = 80, from:
R = 5V²

---

84. Using the same steps, solve these.

a) The kinetic energy (E) of a moving object varies directly as the square of its velocity (v). If E = 400 vigs when v = 10 centimeters per second, find E when v = 8 centimeters per second.

b) The power (P) in an electric circuit varies directly as the square of the current (I). If P = 20 watts when I = 2 amperes, find P when I = 3 amperes.

a) E = 256 vigs, from: E = 4v²   b) P = 45 watts, from: P = 5I²

248 Functions, Non-Linear Graphs, Variation

5-14 INVERSE SQUARE VARIATION

Any variation of the form $y = \dfrac{k}{x^2}$ is called "inverse square" variation. We will discuss "inverse square" variation in this section.

85. Any equation or formula of the form $\boxed{y = \dfrac{k}{x^2}}$ is called "inverse square variation". Some examples are:

$$y = \dfrac{10}{x^2} \qquad F = \dfrac{55}{d^2} \qquad P = \dfrac{200}{t^2}$$

In $y = \dfrac{k}{x^2}$, k is also called the "constant of variation" or the "variation constant". The following language is used to state an inverse square variation.

For $y = \dfrac{10}{x^2}$, we say: y varies inversely as the square of x.

For $F = \dfrac{55}{d^2}$, we say: ___ varies inversely as the square of ___.

86. Following the example, write each inverse square variation as an equation or formula.

y varies inversely as the square of x, and the variation constant is 96. $y = \dfrac{96}{x^2}$

a) D varies inversely as the square of R, and the variation constant is 16.5. _____

b) a varies inversely as the square of m, and the constant of variation is k. _____

F . . . . d

87. Using k as the variation constant, write each of these as a formula.

a) The gravitational force (F) between two objects varies inversely as the square of the distance (d) between the two objects. _____

b) The intensity (I) of light varies inversely as the square of the distance (d) from the light source. _____

a) $D = \dfrac{16.5}{R^2}$

b) $a = \dfrac{k}{m^2}$

88. In inverse square variation, the change in the dependent variable is the reciprocal of the square of any change in the independent variable. Some specific changes are shown in the table below.

Independent Variable	Dependent Variable
Doubled	Cut to one-fourth, from $\left(\dfrac{1}{2}\right)^2$
Tripled	Cut to one-ninth, from $\left(\dfrac{1}{3}\right)^2$
Cut in half	Increased four times, from 2^2

a) $F = \dfrac{k}{d^2}$

b) $I = \dfrac{k}{d^2}$

Continued on following page.

88. Continued

As an example, some pairs of values for
$y = \frac{144}{x^2}$ are given at the right.

x	y
1	144
2	36
3	16
4	9

If $\underline{x}$ is doubled from 1 to 2, $\underline{y}$ is cut to one-fourth from 144 to 36.

If $\underline{x}$ is tripled from 1 to 3, $\underline{y}$ is cut to one-ninth from 144 to 16.

If $\underline{x}$ is cut in half from 4 to 2, $\underline{y}$ increases four times from _____ to _____.

---

89. If we are given one specific pair of values in an inverse square variation, we can find $\underline{k}$ by substitution. For example, if $\underline{y}$ varies inversely as the square of $\underline{x}$ and $y = 12$ when $x = 2$, we get:

$$y = \frac{k}{x^2}$$

$$12 = \frac{k}{2^2}$$

$$12 = \frac{k}{4}$$

$$k = 48$$

Since $k = 48$, the inverse square variation is $y = \frac{48}{x^2}$. We can use that equation to find $\underline{y}$ for other values of $\underline{x}$.

    a) Find $\underline{y}$ when $x = 4$.   b) Find $\underline{y}$ when $x = 8$.

9 to 36

---

90. Using the two steps from the last frame, we solved the problem below.

If H varies inversely as the square of P and $H = 4$ when $P = 5$, find H when $P = 2$.

$$H = \frac{k}{P^2} \qquad\qquad H = \frac{100}{P^2}$$

$$4 = \frac{k}{5^2} \qquad\qquad H = \frac{100}{2^2}$$

$$4 = \frac{k}{25} \qquad\qquad H = \frac{100}{4}$$

$$k = 100 \qquad\qquad H = 25$$

a) $y = 3$

b) $y = \frac{3}{4}$ or 0.75

Continued on following page.

250 Functions, Non-Linear Graphs, Variation

90. Continued

   Using the same steps, solve this one.

   If m varies inversely as the square of s and m = 4 when s = 6, find m when s = 4.

   $m = 9$, from: $m = \dfrac{144}{s^2}$

91. Using the same two steps, solve these.

   a) The intensity (I) of a radio signal varies inversely as the square of its distance (d) from the transmitting source. If I = 5 microvolts when d = 10 kilometers, find I when d = 5 kilometers.

   b) The gravitational force (F) between two objects varies inversely as the square of the distance (d) between the two objects. If F = 40 dynes when d = 10 centimeters, find F when d = 4 centimeters.

   a) I = 20 microvolts, from:
   $I = \dfrac{500}{d^2}$

   b) F = 250 dynes, from:
   $F = \dfrac{4,000}{d^2}$

5-15 VARIATION WITH MORE THAN ONE INDEPENDENT VARIABLE

In this section, we will discuss variations that contain more than one independent variable.

92. There are two independent variables in each equation below.

   $y = xz$ $y = \dfrac{z}{x}$

   If we set z equal to the constant 5, we get one of the basic types of variation. That is:

   $y = xz$ becomes $y = 5x$ which is a direct variation.

   $y = \dfrac{z}{x}$ becomes $y = \dfrac{5}{x}$ which is an _____ variation.

   inverse

Functions, Non-Linear Graphs, Variation   251

93. There are two independent variables in each formula below.

$$E = \frac{1}{2}mv^2 \qquad R = \frac{P}{I^2}$$

If we set one of the variables equal to a constant, we get one of the basic types of variation. That is:

If $m = 100$, $E = \frac{1}{2}mv^2$ becomes $E = 50v^2$ which is a direct square variation.

If $P = 48$, $R = \frac{P}{I^2}$ becomes $R = \frac{48}{I^2}$ which is is an _____ variation.

---

94. If we set z equal to 10 in $y = \frac{x}{z}$, we get a direct variation. That is: | inverse square

$$y = \frac{x}{z} \quad \text{becomes} \quad y = \frac{x}{10} \quad \text{or} \quad y = \frac{1}{10}(x)$$

If we set z equal to 4 in $y = \frac{x^2}{z}$, we get a direct square variation. That is:

$$y = \frac{x^2}{z} \quad \text{becomes} \quad \underline{\qquad} \quad \text{or} \quad \underline{\qquad}$$

---

95. There are more than two independent variables in each equation below. | $y = \frac{x^2}{4}$ or $y = \frac{1}{4}(x^2)$

$$y = dxz \qquad y = \frac{1}{2}dx^2z$$

If we set d equal to 10 and z equal to 5, we get one of the basic types of variations. That is:

$y = dxz$ becomes $y = 50x$ which is a direct variation.

$y = \frac{1}{2}dx^2z$ becomes $y = 25x^2$ which is a _____ _____ variation.

---

96. There are more than two independent variables in each formula below. | direct square

$$F_1 = \frac{d_2F_2}{d_1} \qquad P = \frac{pL}{d^2}$$

If we set all but one of the variables equal to a constant, we get one of the basic types of variation. That is:

If $d_2 = 5$ and $F_2 = 50$, $F_1 = \frac{d_2F_2}{d_1}$ becomes $F_1 = \frac{250}{d_1}$ which is an inverse variation.

If $p = 2$ and $L = 100$, $P = \frac{pL}{d^2}$ becomes $P = \frac{200}{d^2}$ which is an _____ variation.

---

inverse square

252 Functions, Non-Linear Graphs, Variation

97. Each formula below is in proportion form.

$$\frac{V_1}{V_2} = \frac{T_1}{T_2} \qquad \frac{I_1}{I_2} = \frac{(d_2)^2}{(d_1)^2}$$

If we solve for one variable and then set all variables but two equal to constants, we get a basic type of variation. That is:

If we solve the left formula for V_1 and set $T_1 = 8$ and $T_2 = 16$, we get a direct variation since:

$$V_1 = \frac{T_1 V_2}{T_2} \quad \text{and} \quad V_1 = \frac{8V_2}{16} \quad \text{or} \quad \frac{1}{2}V_2$$

If we solve the right formula for I_2 and set $I_1 = 100$ and $d_1 = 2$, we get an inverse square variation since:

$$I_2 = \frac{I_1 (d_1)^2}{(d_2)^2} \quad \text{and} \quad I_2 = \frac{100(2^2)}{(d_2)^2} \quad \text{or} \quad \underline{\qquad}$$

$\dfrac{400}{(d_2)^2}$

5-16 JOINT VARIATION

Any variation of the form $y = kxz$ is called "joint" variation. We will discuss "joint" variation in this section.

98. Any equation or formula of the form $\boxed{y = kxz}$ is called "joint variation". Some examples are

$$y = 3xz \qquad V = 10LW \qquad A = \frac{1}{2}bh$$

The following language is used for a joint variation.

For $y = 3xz$, we say: $\underline{y}$ varies jointly as $\underline{x}$ and $\underline{z}$.

For $V = 10LW$, we say: V varies jointly as ___ and ___.

99. When a formula is a joint variation, the units for the variables are frequently chosen so that $k = 1$. For example:

$$E = IR \qquad A = LW \qquad F = ma$$

The same language is used for variations of that type. That is:

For $E = IR$, we say: E varies jointly as I and R.

For $F = ma$, we say: F varies jointly as ___ and ___.

L and W

$\underline{m}$ and $\underline{a}$

100. In a joint variation, the multiplication can contain more than two independent variables. For example:

$$y = 5bxz \qquad V = LWH$$

In $y = 5bxz$, we say: y varies jointly as b, x, and z.

In $V = LWH$, we say: V varies jointly as ___, ___, and ___.

---

101. Following the example, write each joint variation as an equation or formula. | L, W, and H

y varies jointly as x and z, and k = 50. $y = 50xz$

a) H varies jointly as G and P, and k = 2.5.   _____

b) M varies jointly as R, S, and T, and k = 1.   _____

---

102. To solve the problem below, we set up and used a joint variation. Solve the other problem. | a) $H = 2.5GP$
| b) $M = RST$

b varies jointly as c and d, and k = 4. Find b when c = 2 and d = 10.

$$b = 4cd$$
$$b = 4(2)(10)$$
$$b = 80$$

V varies jointly with L, W, and H, and k = 1. Find V when L = 5, W = 2, and H = 9.

---

103. In a joint variation, if any independent variable increases, the dependent variable increases; if any independent variable decreases, the dependent variable decreases. | $V = 90$, from: $V = LWH$

a) In $y = 6xz$, if x increases, y _____ (increases/decreases).

b) In $V = LWH$, if H decreases, V _____ (increases/decreases).

---

a) y increases

b) V decreases

254 Functions, Non-Linear Graphs, Variation

5-17 COMBINED VARIATION

Many variations involving two or more independent variables are "combined variation". We will discuss "combined" variation in this section.

104. Some "combined" variations and the language used to describe them are given below. In this frame and in the remainder of this section, k will equal "1" unless otherwise noted.

$P = I^2R$ P varies jointly as R and the square of I.

$F = \dfrac{mv^2}{r}$ F varies jointly as m and the square of v and inversely as r.

$F = \dfrac{m_1 m_2}{rd^2}$ F varies jointly as m_1 and m_2 and inversely as r and the square of d.

Following the examples, write each of these as a formula.

a) P_1 varies jointly as P_2 and V_2 and inversely as V_1. _____

b) P varies jointly as p and L and inversely as the square of d. _____

c) H varies directly as B and inversely as G and the square of T. _____

Answers:
a) $P_1 = \dfrac{P_2 V_2}{V_1}$
b) $P = \dfrac{pL}{d^2}$
c) $H = \dfrac{B}{GT^2}$

105. Combined variations can involve the language of a "sum" or "difference". Some examples are given.

$D = P(Q + R)$ D varies jointly as P and the sum of Q and R.

$Q = C(T_1 - T_2)$ Q varies jointly as C and the difference of T_1 and T_2.

$H = \dfrac{AKT(t_2 - t_1)}{L}$ H varies jointly as A, K, T, and the difference of t_2 and t_1 and inversely as L.

Following the examples, write each of these as a formula.

a) M varies jointly as P and the difference of L and X. _____

b) A varies jointly as h and the sum of b_1 and b_2, and $k = \dfrac{1}{2}$ _____

c) α varies directly as the difference of L_2 and L_1 and inversely as L_1 and the difference of t_2 and t_1. _____

Answers:
a) $M = P(L - X)$
b) $A = \dfrac{1}{2}h(b_1 + b_2)$
c) $\alpha = \dfrac{L_2 - L_1}{L_1(t_2 - t_1)}$

106. To solve the problem below, we set up and used a combined variation. Solve the other problem.

I varies directly as E and inversely as R.
Find I when E = 80 and R = 40.

$$I = \frac{E}{R}$$

$$I = \frac{80}{40} = 2$$

W varies jointly as K and N and inversely as R². Find W when K = 6, N = 100, and R = 5.

107. If a combined variation contains a fraction, the dependent variable has the same change as an independent variable in the numerator and the opposite change from an independent variable in the denominator. For example:

In $I = \frac{E}{R}$: If E increases, I increases.
If E decreases, I decreases.

If R increases, I decreases.
If R decreases, I increases.

$W = 24$, from: $W = \frac{KN}{R^2}$

Answer "increases" or "decreases" for these.

a) In $T = \frac{Q}{cm}$, if Q decreases, what happens to T? _____

b) In $M = \frac{KP}{S^2}$, if S increases, what happens to M? _____

c) In $F = \frac{m_1 m_2}{rd^2}$, if <u>r</u> decreases, what happens to F? _____

108. The dependent variable also changes with changes in a factor which is a sum or difference. For example:

In $v = a(t_2 - t_1)$: If $(t_2 - t_1)$ increases, <u>v</u> increases.
If $(t_2 - t_1)$ decreases, <u>v</u> decreases.

a) decreases
b) decreases
c) increases

Answer "increases" or "decreases" for these.

a) In $K = G(H + R)$, if (H + R) increases, what happens to K?
_____

b) In $Q = C(T_1 - T_2)$, if $(T_1 - T_2)$ decreases, what happens to Q?
_____

a) increases
b) decreases

109. Answer "increases" or "decreases" for these.

a) In $t = \dfrac{a(v^2 + h^2)}{cr}$, if $(v^2 + h^2)$ increases, what happens to t? _____

b) In $W = \dfrac{FT(P_1 - P_2)}{S}$, if S increases, what happens to W? _____

c) In $S = \dfrac{H(V_2 - V_1)}{A(B_1 + B_2)}$, if H decreases, what happens to S? _____

d) In $G = \dfrac{r_1 + r_2}{d(p_1 - p_2)}$, if $(p_1 - p_2)$ decreases, what happens to G? _____

---

a) increases   b) decreases   c) decreases   d) increases

SELF-TEST 17 (pages 244-257)

1. If h varies directly as the square of t and $h = 80$ when $t = 4$, find h when $t = 10$.

2. If W varies inversely as the square of G and $W = 36$ when $G = 2$, find W when $G = 6$.

3. Which of the following formulas involve "inverse square" variation?

   a) $R = \dfrac{P}{I^2}$ b) $h = \dfrac{v^2}{g}$ c) $E = \dfrac{KH^2}{8\pi}$ d) $F = \dfrac{Q_1 Q_2}{r^2}$

4. The centripetal force F on an object varies directly as the square of its velocity v. If $F = 8$ newtons when $v = 20$ meters per second, find F when $v = 40$ meters per second.

5. The level of illumination E on a surface varies inversely as the square of the distance d from the luminous source. If $E = 12$ lumens per square meter when $d = 2$ meters, find E when $d = 4$ meters.

6. B varies jointly as N and R, and $k = 10$. If $N = 5$ and $R = 3$, find B.

7. f varies jointly as a and p and inversely as m, and $k = 1$. If $a = 12$, $p = 30$, and $m = 24$, find f.

Write each of the following as a formula. In each case, $k = 1$.

8. P varies directly as h and inversely as the square of v.

9. s varies jointly as a and d and inversely as the difference of t_1 and t_2.

10. W varies directly as the sum of b_1 and b_2 and inversely as p^2.

ANSWERS:
1. $h = 500$
2. $W = 4$
3. (a) and (d)
4. $F = 32$ newtons
5. $E = 3$ lumens per square meter
6. $B = 150$
7. $f = 15$
8. $P = \dfrac{h}{v^2}$
9. $s = \dfrac{ad}{t_1 - t_2}$
10. $W = \dfrac{b_1 + b_2}{p^2}$

258 Functions, Non-Linear Graphs, Variation

SUPPLEMENTARY PROBLEMS - CHAPTER 5

Assignment 15

1. In $y = 5x^2 - 2x$, the <u>dependent</u> variable is _____.

2. In $w = 2t + 3$, the <u>independent</u> variable is _____.

3. Which of the following functions are <u>not</u> defined for $x = 0$?

   a) $y = 8x$ b) $y = \dfrac{8}{x}$ c) $y = \dfrac{1}{x^2}$ d) $y = \dfrac{4}{1 - x}$

4. Which of the following functions are <u>not</u> defined for negative values of x?

   a) $y = \sqrt{3x}$ b) $y = \sqrt{1 - x}$ c) $y = \sqrt{2x - 9}$ d) $y = \dfrac{1}{x - 2}$

5. In graphing the function $h = 40p^2$, is "h" or "p" put on the vertical axis?

6. When graphing a two-variable equation, is the <u>independent</u> variable put on the horizontal axis or on the vertical axis?

Complete: 7. If $y = x^3 + 4$, $f(x) = $ _____   8. If $d = 5t^2 - 2t$, $f(t) = $ _____

Evaluate the following functions.

   9. If $f(x) = 1 - 9x$, $f(-1) = $ _____   10. If $f(v) = v^2 + 4v - 3$, $f(0) = $ _____

   11. If $f(p) = p^2 + 2p$, $f(a - 1) = $ _____   12. If $f(t) = \dfrac{2t}{t - 2}$, $f(2h) = $ _____

   13. If $f(r, w) = \dfrac{2r}{w}$, $f(3, 4) = $ _____   14. If $f(x, y) = \dfrac{x - y}{3x}$, $f(1, 0) = $ _____

15. Which of the following are linear functions?

   a) $y = 3x + 2$ b) $y = x$ c) $y = \dfrac{6}{x}$ d) $y = x^2 + 4$

16. The slope of the linear function $y = 7x - 4$ is _____ and its y-intercept is _____.

17. The slope of the linear function $y = 80x$ is _____ and its y-intercept is _____.

Write the <u>equation</u> of each linear function graphed below.

18. 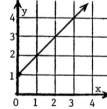   19. 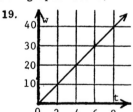   20.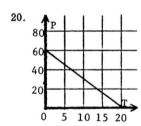

21. Which of the following are quadratic functions?

   a) $y = x^2 - 4$ b) $y = 3x + 2$ c) $y = x^3 + 5x$ d) $y = 7x^2 + x - 1$

Continued on following page.

Functions, Non-Linear Graphs, Variation   259

Assignment 15 (Continued)

22. The parabola $y = \frac{3}{4}x^2 - 3$ is graphed below.

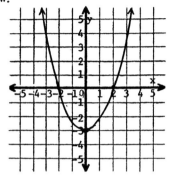

a) Its vertex is at (___ , ___).
b) Its x-intercepts are at (___ , ___) and (___ , ___).
c) Its y-intercept is at (___ , ___).
d) The <u>minimum</u> value of <u>y</u> is _____ .

23. The parabola $y = 4x - x^2$ is graphed below.

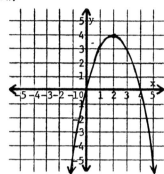

a) Its vertex is at (___ , ___).
b) Its x-intercepts are at (___ , ___) and (___ , ___).
c) Its y-intercept is at (___ , ___).
d) The <u>maximum</u> value of <u>y</u> is _____ .

Assignment 16

Write the equation of each <u>circle</u> graphed below.

1. Center at (0, 0) and radius = 2.

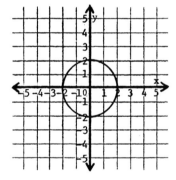

2. Center at (-1, 1) and radius = 4.

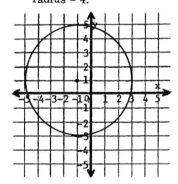

3. Center at (0, 2) and radius = 3.

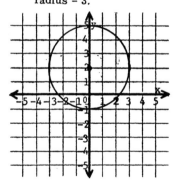

Write the equation of each <u>ellipse</u> graphed below.

4. Passes through (-3, 0), (3, 0), (0, 1) and (0, -1).

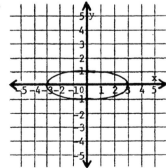

5. Passes through (-2, 0), (2, 0), (0, 4) and (0, -4).

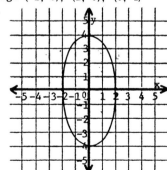

Continued on following page.

Assignment 16 (Continued)

6. Write the equations of the two asymptotes of this hyperbola. $\boxed{\dfrac{x^2}{36} - \dfrac{y^2}{16} = 1}$

7. What two lines are the asymptotes of this hyperbola? $\boxed{xy = 9}$

8. Which of the following equations graph as hyperbolas?

    a) $x^2 - y^2 = 1$ b) $y = \dfrac{5}{x}$ c) $\dfrac{x^2}{25} + \dfrac{y^2}{64} = 1$ d) $y = x^2 - 8x$

9. In $A = \pi r^2$, when r decreases, does A increase or decrease?

10. In $d = \dfrac{2}{v}$, when v increases, does d increase or decrease?

11. If P is <u>directly</u> proportional to V, and if P = 6 when V = 2, find:

    a) The constant of variation k.
    b) The formula relating P and V.

12. If f is <u>inversely</u> proportional to t, and if f = 40 when t = 5, find:

    a) The proportionality constant k.
    b) The formula relating f and t.

13. Which of the following are examples of <u>direct</u> variation?

    a) $h = 4r$ b) $w = \dfrac{1}{2}$ c) $p = 2t + 1$ d) $b = 50a$ e) $d = \dfrac{240}{v}$

14. If G is directly proportional to R, and if G = 40 when R = 8, find G when R = 20.

15. If w is directly proportional to s, and if w = 16 when s = 24, find w when s = 60.

16. The distance d traversed by a car is directly proportional to time t. If d = 150 kilometers when t = 2 hours, find d when t = 5 hours.

17. In an a-c circuit, inductive reactance X_L is directly proportional to frequence f. If X_L = 50 ohms when f = 20 hertz, find X_L when f = 48 hertz.

18. If B is inversely proportional to N, and if B = 6 when N = 4, find B when N = 8.

19. If h is inversely proportional to v, and if h = 40 when v = 15, find h when v = 10.

20. In an a-c circuit, capacitive reactance X_C is inversely proportional to frequency f. If X_C = 80 ohms when f = 400 hertz, find X_C when f = 1,000 hertz.

21. The volume V of a gas is inversely proportional to pressure P. If V = 24 liters when P = 600 grams per square centimeter, find V when P = 240 grams per square centimeter.

Assignment 17

1. The area A of a circle varies directly as the square of its diameter $\underline{d}$. The constant of variation is 0.7854. Write the formula.

2. If $\underline{t}$ varies directly as the square of $\underline{v}$ and $t = 40$ when $v = 2$, find $\underline{t}$ when $v = 5$.

3. If H varies directly as the square of R and $H = 50$ when $R = 5$, find H when $R = 3$.

4. The resistance R of an electric circuit varies inversely as the square of the current I. The constant of variation is 50. Write the formula.

5. If $\underline{d}$ varies inversely as the square of $\underline{h}$ and $d = 8$ when $h = 3$, find $\underline{d}$ when $h = 6$.

6. If W varies inversely as the square of $\underline{p}$ and $W = 10$ when $p = 4$, find W when $p = 2$.

7. The kinetic energy E of a moving object varies directly as the square of its velocity $\underline{v}$. If $E = 32$ joules when $v = 4$ meters per second, find E when $v = 10$ meters per second.

8. The power P in an electric circuit varies directly as the square of the current I. If $P = 100$ watts when $I = 5$ amperes, find P when $I = 3$ amperes.

9. The electrostatic force F between two charged bodies varies inversely as the square of the distance $\underline{d}$ between them. If $F = 50$ newtons when $d = 2$ meters, find F when $d = 5$ meters.

10. The intensity level I of a sound varies inversely as the square of the distance $\underline{d}$ from the source. If $I = 12$ microwatts when $d = 10$ meters, find I when $d = 20$ meters.

11. If T varies jointly as B and H, and $k = 20$, find T when $B = 6$ and $H = 2$.

12. If $\underline{w}$ varies jointly as $\underline{r}$ and $\underline{t}^2$, and $k = 5$, find $\underline{w}$ when $r = 2$ and $t = 7$.

Write each of the following as a formula, using $k = 1$ for each.

13. A varies jointly as $\underline{b}$ and $\underline{h}$ and inversely as the square of $\underline{r}$.

14. $\underline{t}$ varies directly as the square of $\underline{p}$ and inversely as the sum of $\underline{a}$ and $\underline{b}$.

15. P varies jointly as $\underline{m}$ and the difference of $\underline{w}$ and $\underline{r}$ and inversely as the square of $\underline{a}$.

State whether each of the following increases or decreases.

16. In $A = \dfrac{bh}{2}$, if $\underline{b}$ increases, what happens to A?

17. In $N = \dfrac{cp}{s^2}$, if $\underline{s}$ increases, what happens to N?

18. In $F = \dfrac{t}{a(h-v)}$, if $(h-v)$ decreases, what happens to F?

Chapter 6 EXPONENTS, POWERS, AND ROOTS

In this chapter, we will generalize the laws of exponents to powers with negative, fractional, and decimal exponents. As a basis for understanding fractional and decimal exponents, the general meaning of roots is given. Calculator procedures for evaluating powers, roots, and formulas containing powers are shown. Some graphs of exponential functions are discussed.

6-1 POWERS

In this section, we will discuss powers whose exponents are either positive whole numbers, "0", or negative whole numbers.

1. Base-exponent expressions are called "powers" of that base. For example:

    4^3, 4^5, and 4^{12} are called "powers of 4".

    The word names for some powers are given below.

    7^4 is called "7 to the fourth power" or "7 to the fourth".

    x^9 is called "x to the ninth power" or "x to the ninth".

    Write the power corresponding to each word name.

    a) 8 to the fifth power = _____   b) y to the tenth = _____

2. Any power with a positive whole-number exponent stands for a multiplication of identical factors. That is:

    $5^3 = (5)(5)(5)$ (Exponent is "3". Multiply three 5's.)

    $x^4 = (x)(x)(x)(x)$ (Exponent is "4". Multiply four x's.)

    Write each multiplication as a power.

    a) (7)(7) = _____   b) (t)(t)(t)(t)(t)(t) = _____

    a) 8^5 b) y^{10}

3. We converted the power below to an ordinary number by performing the equivalent multiplication.

    $4^3 = (4)(4)(4) = 64$

    Convert each of these to an ordinary number.

    a) $3^4 = (3)(3)(3)(3) = $ _____   b) $2^5 = (2)(2)(2)(2)(2) = $ _____

    a) 7^2 b) t^6

262

Exponents, Powers, and Roots   263

4. Any power whose exponent is "1" equals its base. That is:

$5^1 = 5 \qquad 12^1 = 12 \qquad x^1 = x$

Any power whose exponent is "0" equals "1". That is:

$5^0 = 1 \qquad 12^0 = 1 \qquad x^0 = 1$

Using the definitions, complete these.

a) $7^1 = $ _____ b) $4^0 = $ _____ c) $y^1 = $ _____ d) $m^0 = $ _____

a) 81 b) 32

---

5. Base-exponent expressions with negative exponents are also called "powers" of that base. For example:

4^{-1}, 4^{-5}, and 4^{-10} are called "powers of 4".

The word names for some powers with negative exponents are given below.

8^{-2} is called "8 to the negative-two power" or "8 to the negative-two".

y^{-5} is called "y to the negative-five power" or "y to the negative-five".

Write the power corresponding to each word name.

a) 7 to the negative-nine power = ____ b) t to the negative-one = ___

a) 7
b) 1
c) y
d) 1

---

6. The definition of any power with a negative exponent is given below.

$$a^{-n} = \frac{1}{a^n}$$

Therefore: $9^{-2} = \frac{1}{9^2} \qquad 3^{-5} = \frac{1}{3^5} \qquad y^{-8} = $ _____

a) 7^{-9} b) t^{-1}

---

7. By reversing the definition, we can convert each fraction below to a power with a negative exponent.

$\frac{1}{2^4} = 2^{-4} \qquad \frac{1}{y^9} = y^{-9} \qquad$ a) $\frac{1}{7^5} = $ _____ b) $\frac{1}{x^{10}} = $ _____

$\frac{1}{y^8}$

---

8. Using the definition, we converted each power below to a fraction.

$7^{-2} = \frac{1}{7^2} = \frac{1}{49} \qquad 5^{-3} = \frac{1}{5^3} = \frac{1}{125}$

Convert each of these to a fraction.

a) $6^{-2} = $ _____ b) $4^{-3} = $ _____ c) $2^{-5} = $ _____

a) 7^{-5} b) x^{-10}

---

a) $\frac{1}{36}$ b) $\frac{1}{64}$ c) $\frac{1}{32}$

9. We converted each power below to a fraction.

$$5^{-1} = \frac{1}{5^1} = \frac{1}{5} \qquad x^{-1} = \frac{1}{x^1} = \frac{1}{x}$$

We converted each fraction below to a power.

$$\frac{1}{9} = \frac{1}{9^1} = 9^{-1} \qquad \frac{1}{y} = \frac{1}{y^1} = y^{-1}$$

Convert each power to a fraction and each fraction to a power.

a) $2^{-1} = $ _____ b) $\frac{1}{4} = $ _____ c) $\frac{1}{t} = $ _____ d) $V^{-1} = $ _____

a) $\frac{1}{2}$ b) 4^{-1} c) t^{-1} d) $\frac{1}{V}$

6-2 MULTIPLYING POWERS

In this section, we will discuss the law of exponents for multiplying powers.

10. The following law of exponents is used to multiply powers with the same base.

$$\boxed{a^m \cdot a^n = a^{m+n}}$$

Using the above law, complete these.

a) $5^3 \cdot 5^4 = 5^{3+4} = $ _____ b) $x^2 \cdot x^9 = x^{2+9} = $ _____

11. The same law is used to multiply powers with negative exponents. That is:

$$6^{-3} \cdot 6^5 = 6^{(-3)+5} = 6^2 \qquad y^{-1} \cdot y^{-4} = y^{(-1)+(-4)} = y^{-5}$$

Use the law to complete these.

a) $2^{-4} \cdot 2^{-6} = $ _____ b) $t^{-9} \cdot t^7 = $ _____

a) 5^7 b) x^{11}

12. To perform the multiplications below, we substituted 5^1 for 5 and y^1 for y.

$$5 \cdot 5^2 = 5^1 \cdot 5^2 = 5^3 \qquad y^{-4} \cdot y = y^{-4} \cdot y^1 = y^{-3}$$

Complete these.

a) $x^9 \cdot x = $ _____ b) $4 \cdot 4^{-7} = $ _____ c) $d \cdot d = $ _____

a) 2^{-10} b) t^{-2}

a) x^{10}
b) 4^{-6}
c) d^2

Exponents, Powers, and Roots   265

13. If the product is a base <u>to the first power</u>, the product equals the base. That is:

$$7^5 \cdot 7^{-4} = 7^1 \text{ or } 7$$

If the product is a base <u>to the zero power</u>, the product is "1". That is:

$$x^3 \cdot x^{-3} = x^0 \text{ or } 1$$

Complete these.

a) $5^{-2} \cdot 5^3 = $ _____

b) $y^8 \cdot y^{-7} = $ _____

c) $t^{10} \cdot t^{-10} = $ _____

d) $m^{-1} \cdot m = $ _____

14. The same law is used to multiply more than two powers of the same base. For example:

$$x^4 \cdot x^{-7} \cdot x^5 = x^{4+(-7)+5} = x^2$$

Complete these.

a) $2^{-1} \cdot 2^{-9} \cdot 2^6 = $ _____

b) $y^4 \cdot y \cdot y^{-2} = $ _____

a) 5   c) 1
b) y   d) 1

15. To perform the multiplication below, we multiplied the numerical factors and the power factors.

$$(2y^4)(4y^{-1}) = (2 \cdot 4)(y^4 \cdot y^{-1}) = 8y^3$$

Write each product below.

a) $(4x^{-6})(3x) = $ _____

b) $(7V^{-1})(6V^{-9}) = $ _____

a) 2^{-4} b) y^3

16. To perform the multiplication below, we multiplied the numerical factors and the power factors with the same base.

$$(5a^3b^5)(7ab^{-2}) = (5 \cdot 7)(a^3 \cdot a)(b^5 \cdot b^{-2}) = 35a^4b^3$$

Write each product below.

a) $(m^{-5}p^2)(m^3p^{-9}) = $ _____

b) $(c^4d^{-1})(c^{-4}d^3) = $ _____

c) $(10x^3y^{-5})(9x^{-2}y^{-3}) = $ _____

d) $(2p^{-1}q^{-1})(3p^{-5}q^{-5}) = $ _____

a) $12x^{-5}$ b) $42V^{-10}$

a) $m^{-2}p^{-7}$

b) d^2, from: $c^0d^2 = 1d^2$

c) $90xy^{-8}$, from: $90x^1y^{-8}$

d) $6p^{-6}q^{-6}$

266   Exponents, Powers, and Roots

6-3 DIVIDING POWERS

In this section, we will discuss the law of exponents for dividing powers.

17. The following law of exponents is used to divide powers with the same base.

$$\frac{a^m}{a^n} = a^{m-n}$$

Using the above law, complete these.

a) $\frac{5^7}{5^3} = 5^{7-3} = $ _____

b) $\frac{x^{10}}{x^2} = x^{10-2} = $ _____

18. The same law is used to divide powers with negative exponents. That is:

$$\frac{4^{-3}}{4^5} = 4^{(-3)-5} = 4^{(-3)+(-5)} = 4^{-8}$$

$$\frac{d^{-4}}{d^{-7}} = d^{(-4)-(-7)} = d^{(-4)+7} = d^3$$

Use the law to complete these.

a) $\frac{6^2}{6^4} = $ _____

b) $\frac{9^{-1}}{9^3} = $ _____

c) $\frac{x^4}{x^{-5}} = $ _____

d) $\frac{y^{-10}}{y^{-6}} = $ _____

a) 5^4 b) x^8

a) 6^{-2} c) x^9
b) 9^{-4} d) y^{-4}

19. To perform the divisions below, we substituted 2^1 for 2 and x^1 for x.

$$\frac{2}{2^5} = \frac{2^1}{2^5} = 2^{1-5} = 2^{1+(-5)} = 2^{-4}$$

$$\frac{x^{-4}}{x} = \frac{x^{-4}}{x^1} = x^{(-4)-1} = x^{(-4)+(-1)} = x^{-5}$$

Complete these.

a) $\frac{7^{-1}}{7} = $ _____

b) $\frac{3}{3^6} = $ _____

c) $\frac{y}{y^{-8}} = $ _____

d) $\frac{t^{-3}}{t^{-4}} = $ _____

a) 7^{-2} c) y^9
b) 3^{-5} d) t^1 or t

20. To simplify the fraction below, we multiplied the powers in the numerator and then divided. Simplify the other fraction.

$$\frac{t^4 \cdot t^2}{t^3} = \frac{t^6}{t^3} = t^3 \qquad\qquad \frac{7^{-5} \cdot 7^2}{7^{-1}} = $$ _____

7^{-2}, from: $\frac{7^{-3}}{7^{-1}}$

Exponents, Powers, and Roots   267

21. To simplify the fraction below, we multiplied the powers in the denominator and then divided. Simplify the other fraction.

$$\frac{x^4}{x^2 \cdot x^3} = \frac{x^4}{x^5} = x^{-1} \qquad \frac{2^8}{2^{-5} \cdot 2^7} = \underline{\hspace{2cm}}$$

22. To simplify the fraction below, we performed both multiplications before dividing. Simplify the other fraction.

$$\frac{3^2 \cdot 3^4}{3^5 \cdot 3^{-1}} = \frac{3^6}{3^4} = 3^2 \qquad \frac{x^{-7} \cdot x^4}{x^2 \cdot x^{-8}} = \underline{\hspace{2cm}}$$

2^6, from: $\frac{2^8}{2^2}$

23. We simplified the fraction below by performing two divisions.

$$\frac{b^5 x^{10}}{b^2 x^4} = \left(\frac{b^5}{b^2}\right)\left(\frac{x^{10}}{x^4}\right) = b^3 x^6$$

Simplify each fraction.

a) $\frac{a^{-1} m^2}{a^{-3} m} = \underline{\hspace{2cm}}$ b) $\frac{t^9 y^3}{t^8 y^{-1}} = \underline{\hspace{2cm}}$

x^3, from: $\frac{x^{-3}}{x^{-6}}$

24. We simplified the fraction below by performing two divisions.

$$\frac{10 a^8}{2 a^5} = \left(\frac{10}{2}\right)\left(\frac{a^8}{a^5}\right) = 5a^3$$

Simplify each fraction.

a) $\frac{12 x^2}{4 x^5} = \underline{\hspace{2cm}}$ b) $\frac{8 b^{-1} t^3}{2 b^{-4} t^2} = \underline{\hspace{2cm}}$

a) $a^2 m$ b) ty^4

25. Notice the steps used to simplify the fraction below.

$$\frac{4y^7}{6y^2} = \left(\frac{4}{6}\right)\left(\frac{y^7}{y^2}\right) = \frac{2}{3}(y^5) = \frac{2y^5}{3}$$

Use the same steps to simplify these.

a) $\frac{10x}{8x^3} = \underline{\hspace{2cm}}$ b) $\frac{5m^3}{10m} = \underline{\hspace{2cm}}$

a) $3x^{-3}$ b) $4b^3 t$

26. When a power is divided by itself, the quotient is "1". For example:

$$\frac{7^2}{7^2} = 7^{2-2} = 7^0 = 1$$

$$\frac{y^{-4}}{y^{-4}} = y^{(-4)-(-4)} = y^{(-4)+4} = y^0 = 1$$

Therefore both $\frac{6^{-3}}{6^{-3}}$ and $\frac{t^{10}}{t^{10}}$ equal what number? $\underline{\hspace{1cm}}$

a) $\frac{5x^{-2}}{4}$ b) $\frac{m^2}{2}$

268   Exponents, Powers, and Roots

27. We used the fact that $\frac{x^4}{x^4} = 1$ to simplify the fraction below.

$$\frac{a^2x^4}{b^3x^4} = \left(\frac{a^2}{b^3}\right)\left(\frac{x^4}{x^4}\right) = \left(\frac{a^2}{b^3}\right)(1) = \frac{a^2}{b^3}$$

Simplify each fraction.

a) $\frac{7y^{-2}}{8y^{-2}} = $ _____   b) $\frac{c^5x}{c^5y^2} = $ _____

+1

a) $\frac{7}{8}$ b) $\frac{x}{y^2}$

6-4 CONVERTING DIVISION TO MULTIPLICATION

In this section, we will define the "reciprocal" of a power and then show how any division of powers can be converted to a multiplication of powers.

28. Two powers with the same base are reciprocals if their exponents are opposites. They are reciprocals since their product is +1. That is:

2^5 and 2^{-5} are reciprocals, since $2^5 \cdot 2^{-5} = 2^0 = 1$

x^{-3} and x^3 are reciprocals, since $x^{-3} \cdot x^3 = x^0 = 1$

Using the above fact, write the reciprocal of each of these.

a) 5^6 _____   b) 9^{-2} _____   c) y^4 _____   d) V^{-10} _____

29. Since any number or letter is a base to the first power, its reciprocal is the same base to the negative-one power. That is:

The reciprocal of 6 (or 6^1) is 6^{-1}.

The reciprocal of t (or t^1) is t^{-1}.

Write the reciprocal of each of these as a power.

a) 4 _____   b) 7^{-1} _____   c) x _____   d) y^{-1} _____

a) 5^{-6} c) y^{-4}
b) 9^2 d) V^{10}

30. Instead of using the law of exponents to perform a division, we can convert the division to a multiplication. To do so, we multiply the numerator by the reciprocal of the denominator. For example:

a) $\frac{2^7}{2^4} = (2^7)(\text{the reciprocal of } 2^4) = 2^7 \cdot 2^{-4} = $ _____

b) $\frac{x^2}{x^8} = (x^2)(\text{the reciprocal of } x^8) = x^2 \cdot x^{-8} = $ _____

c) $\frac{t^5}{t^{-3}} = (t^5)(\text{the reciprocal of } t^{-3}) = t^5 \cdot t^3 = $ _____

a) 4^{-1}
b) 7^1 (or 7)
c) x^{-1}
d) y^1 (or y)

Exponents, Powers, and Roots   269

31. Complete these divisions by converting to multiplication first.

a) $\dfrac{x^3}{x^4} = ($ $)($ $) = $ _____   c) $\dfrac{y^{-9}}{y^{-5}} = ($ $)($ $) = $ _____

b) $\dfrac{5^{-2}}{5^7} = ($ $)($ $) = $ _____   d) $\dfrac{3^{-5}}{3^{-2}} = ($ $)($ $) = $ _____

a) 2^3
b) x^{-6}
c) t^8

---

32. We simplified the fraction below by converting to multiplication first.

$\dfrac{x^2 \cdot x^5}{x^3} = (x^2 \cdot x^5)(\text{the reciprocal of } x^3) = x^2 \cdot x^5 \cdot x^{-3} = x^4$

Simplify each of these by converting to multiplication first.

a) $\dfrac{8^{-1} \cdot 8^3}{8^{-2}} = (8^{-1} \cdot 8^3)($ $) = $ _____   b) $\dfrac{y^3 \cdot y^{-7}}{y^2} = (y^3 \cdot y^{-7})($ $) = $ _____

a) $(x^3)(x^{-4}) = x^{-1}$
b) $(5^{-2})(5^{-7}) = 5^{-9}$
c) $(y^{-9})(y^5) = y^{-4}$
d) $(3^{-5})(3^2) = 3^{-3}$

---

33. We can show that a^3b^4 and $a^{-3}b^{-4}$ are reciprocals by showing that their product is $+1$.

$(a^3b^4)(a^{-3}b^{-4}) = (a^3 \cdot a^{-3})(b^4 \cdot b^{-4}) = a^0 \cdot b^0 = 1 \cdot 1 = +1$

You can see that two expressions like those above are reciprocals if each factor in one is the reciprocal of each factor in the other. That is:

a^3b^4 and $a^{-3}b^{-4}$ are reciprocals,

since: a^3 and a^{-3} are reciprocals,
and: b^4 and b^{-4} are reciprocals.

Using the above fact, write the reciprocal for each of these.

a) x^5y^{-2} _____   b) st^{-4} _____   c) $p^{-5}q^{-1}$ _____

a) $(8^{-1} \cdot 8^3)(8^2) = 8^4$
b) $(y^3 \cdot y^{-7})(y^{-2}) = y^{-6}$

---

34. We simplified the fraction below by converting to multiplication first.

$\dfrac{y^6 \cdot y^3}{y^2 \cdot y^4} = y^6 \cdot y^3(\text{the reciprocal of } y^2 \cdot y^4) = y^6 \cdot y^3(y^{-2} \cdot y^{-4}) = y^3$

Simplify each of these by converting to multiplication first.

a) $\dfrac{7^4}{7^{-1} \cdot 7^6} = 7^4($ $) = $ _____   b) $\dfrac{b^4 \cdot b^{-2}}{b^3 \cdot b^{-7}} = b^4 \cdot b^{-2}($ $) = $ _____

a) $x^{-5}y^2$
b) $s^{-1}t^4$
c) p^5q^1 or p^5q

---

a) $7^4(7^1 \cdot 7^{-6}) = 7^{-1}$
b) $b^4 \cdot b^{-2}(b^{-3} \cdot b^7) = b^6$

270   Exponents, Powers, and Roots

35. Though we cannot use the law of exponents to simplify the fraction below, we can convert the division to multiplication. That is:

$$\frac{a^4}{b^2c^{-5}} = a^4(\text{the reciprocal of } b^2c^{-5}) = a^4(b^{-2}c^5) = a^4b^{-2}c^5$$

Convert each division to a multiplication.

a) $\dfrac{cx^5}{y^{-6}} =$ _____   b) $\dfrac{a^8b^{-1}}{c^4d^{-9}} =$ _____

a) cx^5y^6 b) $a^8b^{-1}c^{-4}d^9$

6-5 WRITING EXPRESSIONS WITHOUT NEGATIVE EXPONENTS

In this section, we will show some procedures for writing expressions without negative exponents.

36. In an earlier section we gave the following definition for powers with negative whole-number exponents.

$$y^{-2} = \frac{1}{y^2} \qquad x^{-4} = \frac{1}{x^4} \qquad V^{-7} = \frac{1}{V^7}$$

By substituting for the power with a negative exponent, we wrote each expression below without a negative exponent.

$$x^{-3}y^5 = \left(\frac{1}{x^3}\right)y^5 = \frac{y^5}{x^3} \qquad ab^{-6} = a\left(\frac{1}{b^6}\right) = \frac{a}{b^6}$$

Write each expression below without a negative exponent.

a) $c^{-7}d^2 =$ _____   b) $p^4q^{-2} =$ _____

37. By substituting for both powers with negative exponents we wrote the expression below without negative exponents. Notice how we got "1" as the numerator of the fraction.

$$x^{-2}y^{-3} = \left(\frac{1}{x^2}\right)\left(\frac{1}{y^3}\right) = \frac{1}{x^2y^3}$$

Write each expression below without negative exponents.

a) $c^{-5}d^{-7} =$ _____   b) $P^{-4}Q^{-9} =$ _____

a) $\dfrac{d^2}{c^7}$ b) $\dfrac{p^4}{q^2}$

a) $\dfrac{1}{c^5d^7}$ b) $\dfrac{1}{P^4Q^9}$

38. We substituted for x^{-1} to write the expression below without a negative exponent.

$$x^{-1}y^7 = \left(\frac{1}{x^1}\right)y^7 = \frac{y^7}{x}$$

Write each expression without negative exponents.

a) $p^{10}q^{-1} =$ _____ b) $c^{-1}d^{-2} =$ _____

---

39. To write each expression below without negative exponents, we substituted and then performed a division involving fractions.

$$\frac{x^{-3}}{a^4y^{-2}} = \frac{\frac{1}{x^3}}{a^4\left(\frac{1}{y^2}\right)} = \frac{\frac{1}{x^3}}{\frac{a^4}{y^2}} = \left(\frac{1}{x^3}\right)\left(\frac{y^2}{a^4}\right) = \frac{y^2}{a^4x^3}$$

$$\frac{c^{-5}d}{p^6q^{-9}} = \frac{\left(\frac{1}{c^5}\right)d}{p^6\left(\frac{1}{q^9}\right)} = \frac{\frac{d}{c^5}}{\frac{p^6}{q^9}} = (\quad)(\quad) = $$ _____

a) $\dfrac{p^{10}}{q}$ b) $\dfrac{1}{cd^2}$

---

40. The two problems from the last frame are shown below.

$$\frac{x^{-3}}{a^4y^{-2}} = \frac{y^2}{a^4x^3} \qquad \frac{c^{-5}d}{p^6q^{-9}} = \frac{dq^9}{c^5p^6}$$

You can see this fact from the examples above:

> A power with a negative exponent can be replaced by a power with a positive exponent by moving it from the numerator to the denominator (or from the denominator to the numerator) and changing the exponent to its opposite.

Using the above fact, write each of these without negative exponents.

a) $\dfrac{m^{-5}}{t^{-2}} =$ _____ b) $\dfrac{ax^{-8}}{b^3} =$ _____

$\left(\dfrac{d}{c^5}\right)\left(\dfrac{q^9}{p^6}\right) = \dfrac{dq^9}{c^5p^6}$

---

41. Write each expression without negative exponents.

a) $\dfrac{h^{-1}}{p^2q^{-2}} =$ _____ b) $\dfrac{x^{-3}y^{-1}}{d^{-2}f^{-6}} =$ _____

a) $\dfrac{t^2}{m^5}$ b) $\dfrac{a}{b^3x^8}$

---

a) $\dfrac{q^2}{hp^2}$ b) $\dfrac{d^2f^6}{x^3y}$

272   Exponents, Powers, and Roots

42. When moving powers from the denominator to the numerator, we sometimes get a non-fractional expression. For example:

$$\frac{a^3}{b^{-6}} = a^3b^6 \qquad \frac{c}{x^{-1}y^{-4}} = cxy^4$$

Write each expression without negative exponents.

a) $\dfrac{m}{t^{-1}} =$ _____   b) $\dfrac{p^4q}{a^{-3}d^{-2}} =$ _____

43. When moving powers from the numerator to the denominator, we sometimes get a fraction whose numerator is "1". For example:

$$\frac{x^{-3}}{y^2} = \frac{1x^{-3}}{y^2} = \frac{1}{x^3y^2} \qquad \frac{b^{-1}t^{-2}}{x^4} = \frac{1b^{-1}t^{-2}}{x^4} = \frac{1}{bt^2x^4}$$

Write each expression without negative exponents.

a) $\dfrac{c^{-9}}{t} =$ _____   b) $\dfrac{a^{-4}b^{-5}}{x^2y} =$ _____

a) mt   b) $a^3d^2p^4q$

a) $\dfrac{1}{c^9t}$ b) $\dfrac{1}{a^4b^5x^2y}$

SELF-TEST 18 (pages 262-272)

Convert each power to an ordinary number.

1. 2^4
2. 8^0

Convert each power to a fraction.

3. 3^{-3}
4. 6^{-1}

Find each product.

5. $y \cdot y^2$
6. $R^3 \cdot R^{-5} \cdot R$
7. $(2t^4)(5t^{-4})$
8. $(4a^2bc^{-3})(3ab^{-3}c^4)$

Find each quotient.

9. $\dfrac{w}{w^{-2}}$
10. $\dfrac{t^3 \cdot t^{-1}}{t^{-4} \cdot t^2}$
11. $\dfrac{20P^2}{4P^3}$
12. $\dfrac{6d^2h^{-3}}{3h^{-3}k}$

13. Convert this division to a multiplication. $\dfrac{m^2}{p^{-1}s}$

Write each expression without negative exponents.

14. $\dfrac{d^{-2}}{p^{-1}}$
15. $a^{-3}b^{-2}$
16. $\dfrac{pt^{-3}}{r^{-2}s}$

ANSWERS:
1. 16
2. 1
3. $\dfrac{1}{27}$
4. $\dfrac{1}{6}$
5. y^3
6. R^{-1}
7. 10
8. $12a^3b^{-2}c$
9. w^3
10. t^4
11. $5P^{-1}$
12. $2d^2k^{-1}$
13. m^2ps^{-1}
14. $\dfrac{p}{d^2}$
15. $\dfrac{1}{a^3b^2}$
16. $\dfrac{pr^2}{st^3}$

Exponents, Powers, and Roots   273

6-6 RAISING A POWER TO A POWER

In this section, we will discuss the law of exponents for raising a power to a power.

---

44. In $(5^3)^2$ and $(y^{-4})^3$, we are asked to raise a power to a power. Such expressions also stand for a multiplication of identical factors. That is:

$(5^3)^2 = 5^3 \cdot 5^3$ (There are two 5^3's.)

$(y^{-4})^3 = y^{-4} \cdot y^{-4} \cdot y^{-4}$ (There are three y^{-4}'s.)

Write the expression below as a multiplication of identical factors.

$(2^{-6})^4 = $ _____

---

45. To raise a power to a power, we can convert to a multiplication. That is:

$(3^5)^2 = 3^5 \cdot 3^5 = 3^{5+5} = 3^{10}$

$(y^{-6})^3 = y^{-6} \cdot y^{-6} \cdot y^{-6} = y^{(-6)+(-6)+(-6)} = y^{-18}$

However, it is simpler to raise a power to a power by multiplying the exponents. We get:

$(3^5)^2 = 3^{(5)(2)} = 3^{10}$ $(y^{-6})^3 = y^{(-6)(3)} = y^{-18}$

Therefore we use the following law of exponents to raise a power to a power.

$$\boxed{(a^m)^n = a^{mn}}$$

Using the above law, complete these.

a) $(6^3)^5 = $ _____   b) $(p^{-4})^2 = $ _____   c) $(m^{-2})^8 = $ _____

Answer to frame 44: $2^{-6} \cdot 2^{-6} \cdot 2^{-6} \cdot 2^{-6}$

---

46. Any power raised to the first power equals the original power. That is:

$(2^4)^1 = 2^{(4)(1)} = 2^4$ $(y^{-9})^1 = y^{(-9)(1)} = y^{-9}$

Any power raised to the zero power equals "1". That is:

$(8^3)^0 = 8^{(3)(0)} = 8^0 = 1$ $(m^{-5})^0 = m^{(-5)(0)} = m^0 = 1$

Using the above facts, complete these.

a) $(x^7)^1 = $ ____   b) $(3^{10})^0 = $ ____   c) $(y^{-4})^0 = $ ____   d) $(4^{-3})^1 = $ ____

Answers to frame 45:
a) 6^{15}
b) p^{-8}
c) m^{-16}

---

47. The definition of a power with a negative whole-number exponent also applies when the base is a power. That is:

$(x^4)^{-2} = \dfrac{1}{(x^4)^2}$ $(t^{-1})^{-3} = \dfrac{1}{(t^{-1})^3}$ $(2^{-5})^{-4} = $ _____

Answers to frame 46:
a) x^7 c) 1
b) 1 d) 4^{-3}

274 Exponents, Powers, and Roots

48. Using the definition from the last frame, we raised each power below to a negative power.

$$(y^3)^{-4} = \frac{1}{(y^3)^4} = \frac{1}{y^{12}} = y^{-12}$$

$$(2^5)^{-3} = \frac{1}{(2^5)^3} = \frac{1}{2^{15}} = 2^{-15}$$

The two examples above are shown below. You can see that $(a^m)^n = a^{mn}$ also applies when n is negative. That is:

$$(y^3)^{-4} = y^{(3)(-4)} = y^{-12} \qquad (2^5)^{-3} = 2^{(5)(-3)} = 2^{-15}$$

Using the above fact, multiply exponents to raise each power to a power.

a) $(V^6)^{-5} = $ _____   b) $(4^3)^{-7} = $ _____

$\dfrac{1}{(2^{-5})^4}$

49. Using the same definition, we raised each power below to a negative power.

$$(x^{-5})^{-2} = \frac{1}{(x^{-5})^2} = \frac{1}{x^{-10}} = x^{10}$$

$$(3^{-4})^{-6} = \frac{1}{(3^{-4})^6} = \frac{1}{3^{-24}} = 3^{24}$$

The two examples above are shown below. You can see that $(a^m)^n = a^{mn}$ also applies when both m and n are negative. That is:

$$(x^{-5})^{-2} = x^{(-5)(-2)} = x^{10} \qquad (3^{-4})^{-6} = 3^{(-4)(-6)} = 3^{24}$$

Using the above fact, multiply exponents to raise each power to a power.

a) $(6^{-2})^{-1} = $ _____   b) $(x^{-9})^{-2} = $ _____   c) $(t^{-3})^{-10} = $ _____

a) V^{-30} b) 4^{-21}

a) 6^2 b) x^{18} c) t^{30}

6-7 POWERS OF MONOMIALS

In this section, we will discuss the procedure for raising monomials to a power.

50. To raise the monomial below to a power, we converted to a multiplication.

$$(x^{-4}y^5)^3 = (x^{-4}y^5)(x^{-4}y^5)(x^{-4}y^5) = x^{-12}y^{15}$$

However, it is simpler to multiply each exponent by 3. That is:

$$(x^{-4}y^5)^3 = x^{(-4)(3)}y^{(5)(3)} = x^{-12}y^{15}$$

Using the shorter method, do these.

a) $(c^6 d^{-7})^4 = $ _____   b) $(p^{-1}q^{-2})^5 = $ _____

51. The same method is used to raise a monomial to a negative whole-number power. For example:

$$(x^4y^{-3})^{-2} = x^{(4)(-2)}y^{(-3)(-2)} = x^{-8}y^6$$

Using the same method, do these.

a) $(P^{-2}R^9)^{-3} =$ _____   b) $(c^{-5}t^{-1})^{-1} =$ _____

a) $c^{24}d^{-28}$
b) $p^{-5}q^{-10}$

---

52. Notice how we substituted x^1 for $\underline{x}$ below.

$$(xy^{-3})^4 = (x^1y^{-3})^4 = x^4y^{-12}$$

Following the example, do these.

a) $(p^{-1}q)^8 =$ _____   b) $(ab^4)^{-6} =$ _____

a) P^6R^{-27}
b) c^5t

---

53. We converted 3^3 to an ordinary number below. Complete the other problem.

$$(3x^{-2})^3 = 3^3 \cdot x^{-6} = 27x^{-6} \qquad (2V^2)^4 =$$ _____

a) $p^{-8}q^8$
b) $a^{-6}b^{-24}$

---

54. We did not convert 7^5 to an ordinary number below because the number would be too large. Complete the other problem.

$$(7t^4)^5 = 7^5 \cdot t^{20} \qquad (4y^{-3})^8 =$$ _____

$16V^8$, from: $2^4 \cdot V^8$

---

55. The same procedure is used when the monomial contains more than two factors. For example:

$$(2x^{-1}y^3)^5 = 2^5 \cdot x^{-5}y^{15} = 32x^{-5}y^{15}$$

Following the example, complete these.

a) $(4a^{-2}d)^2 =$ _____   b) $(p^{-4}q^3r)^{-5} =$ _____

$4^8 \cdot y^{-24}$

---

a) $16a^{-4}d^2$
b) $p^{20}q^{-15}r^{-5}$

276 Exponents, Powers, and Roots

6-8 POWERS OF FRACTIONS

In this section, we will discuss the law of exponents for raising a fraction to a power.

56. To raise each fraction below to a power, we converted to a multiplication.

$$\left(\frac{4}{5}\right)^3 = \left(\frac{4}{5}\right)\left(\frac{4}{5}\right)\left(\frac{4}{5}\right) = \frac{4^3}{5^3}$$

$$\left(\frac{a}{b}\right)^4 = \left(\frac{a}{b}\right)\left(\frac{a}{b}\right)\left(\frac{a}{b}\right)\left(\frac{a}{b}\right) = \frac{a^4}{b^4}$$

From the examples, you can see this fact: <u>Any fraction can be raised to a power by raising both terms of the fraction to that power</u>. That is:

$$\left(\frac{4}{5}\right)^3 = \frac{4^3}{5^3} \qquad \left(\frac{a}{b}\right)^4 = \frac{a^4}{b^4}$$

Using the above fact, complete these.

$$\left(\frac{2}{3}\right)^5 = \frac{2^5}{3^5} \qquad a)\ \left(\frac{x}{y}\right)^2 = \underline{\qquad} \qquad b)\ \left(\frac{t}{10}\right)^8 = \underline{\qquad}$$

57. In the last frame, we saw the following law of exponents for raising a fraction to a power.

$$\boxed{\left(\frac{a}{b}\right)^n = \frac{a^n}{b^n}}$$

Using the above law, we raised one fraction to a power below. Raise the other fraction to a power.

$$\left(\frac{1}{2}\right)^5 = \frac{1^5}{2^5} = \frac{1}{32} \qquad \left(\frac{3}{4}\right)^3 = \underline{\qquad}$$

a) $\frac{x^2}{y^2}$ b) $\frac{t^8}{10^8}$

58. Following the example, raise the other fraction to a power.

$$\left(\frac{2}{y}\right)^4 = \frac{2^4}{y^4} = \frac{16}{y^4} \qquad \left(\frac{x}{10}\right)^3 = \underline{\qquad}$$

$\frac{27}{64}$, from: $\frac{3^3}{4^3}$

59. The same law applies when one or both terms are powers. For example:

$$\left(\frac{x^2}{y^3}\right)^4 = \frac{(x^2)^4}{(y^3)^4} = \frac{x^8}{y^{12}}$$

Following the example, complete these.

a) $\left(\frac{3}{t^5}\right)^2 = \underline{\qquad}$ b) $\left(\frac{p^{-2}}{q^4}\right)^5 = \underline{\qquad}$

$\frac{x^3}{1,000}$, from: $\frac{x^3}{10^3}$

Exponents, Powers, and Roots   277

60. The same law also applies when one or both terms are monomials. For example:

$$\left(\frac{x^2y}{b^3}\right)^4 = \frac{(x^2y)^4}{(b^3)^4} = \frac{x^8y^4}{b^{12}}$$

Following the example, complete these.

a) $\left(\frac{b^5}{2d^4}\right)^2 = $ _____

b) $\left(\frac{p^{-1}q^2}{a^3b^{-4}}\right)^6 = $ _____

a) $\frac{9}{t^{10}}$, from: $\frac{3^2}{t^{10}}$

b) $\frac{p^{-10}}{q^{20}}$

---

61. The same law is used to raise a fraction to a negative power. For example:

$$\left(\frac{a^2}{b^{-1}}\right)^{-3} = \frac{(a^2)^{-3}}{(b^{-1})^{-3}} = \frac{a^{-6}}{b^3}$$

Following the example, complete these.

a) $\left(\frac{x}{y}\right)^{-5} = $ _____

b) $\left(\frac{p^{-2}}{q^5}\right)^{-2} = $ _____

a) $\frac{b^{10}}{4d^8}$

b) $\frac{p^{-6}q^{12}}{a^{18}b^{-24}}$

---

62. Another example of raising a fraction to a negative power is shown below.

$$\left(\frac{a^2b^{-1}}{d^3}\right)^{-4} = \frac{(a^2b^{-1})^{-4}}{(d^3)^{-4}} = \frac{a^{-8}b^4}{d^{-12}}$$

Following the example, complete these.

a) $\left(\frac{xy}{c^2}\right)^{-1} = $ _____

b) $\left(\frac{p^{-2}q^3}{m^4t^{-5}}\right)^{-2} = $ _____

a) $\frac{x^{-5}}{y^{-5}}$ b) $\frac{p^4}{q^{-10}}$

---

63. The definition of a power with a negative whole-number exponent also applies when the base is a fraction. That is:

$$\left(\frac{3}{4}\right)^{-2} = \frac{1}{\left(\frac{3}{4}\right)^2} \qquad \left(\frac{x}{y}\right)^{-4} = \frac{1}{\left(\frac{x}{y}\right)^4}$$

Using the above definition, notice how we can convert each power above to a power with a positive exponent.

$$\left(\frac{3}{4}\right)^{-2} = \frac{1}{\left(\frac{3}{4}\right)^2} = \frac{1}{\frac{3^2}{4^2}} = 1\left(\text{the reciprocal of } \frac{3^2}{4^2}\right) = 1\left(\frac{4^2}{3^2}\right) = \frac{4^2}{3^2} = \left(\frac{4}{3}\right)^2$$

$$\left(\frac{x}{y}\right)^{-4} = \frac{1}{\left(\frac{x}{y}\right)^4} = \frac{1}{\frac{x^4}{y^4}} = 1\left(\text{the reciprocal of } \frac{x^4}{y^4}\right) = 1\left(\frac{y^4}{x^4}\right) = \frac{y^4}{x^4} = $$ _____

a) $\frac{x^{-1}y^{-1}}{c^{-2}}$

b) $\frac{p^4q^{-6}}{m^{-8}t^{10}}$

$\left(\frac{y}{x}\right)^4$

278   Exponents, Powers, and Roots

64. Here are the two conversions from the last frame.

$$\left(\frac{3}{4}\right)^{-2} = \left(\frac{4}{3}\right)^{2} \qquad \left(\frac{x}{y}\right)^{-4} = \left(\frac{y}{x}\right)^{4}$$

You can see this fact: <u>Any fraction raised to a negative power can be converted to a fraction raised to a positive power by</u>:

1) interchanging the numerator and denominator, and
2) replacing the exponent with its opposite.

Using the above fact, convert each power below to a power with a positive exponent.

a) $\left(\frac{x}{5}\right)^{-3} =$ _____   b) $\left(\frac{E}{R}\right)^{-7} =$ _____

---

65. Convert each power to a power with a positive exponent.

a) $\left(\frac{3y}{4}\right)^{-2} =$ _____   b) $\left(\frac{m}{pq}\right)^{-1} =$ _____

---

a) $\left(\frac{5}{x}\right)^{3}$ b) $\left(\frac{R}{E}\right)^{7}$

---

a) $\left(\frac{4}{3y}\right)^{2}$ b) $\left(\frac{pq}{m}\right)^{1}$ or $\frac{pq}{m}$

6-9 ROOTS

In this section, we will define "roots" and show how they can be found on a calculator.

66. In an earlier chapter, we discussed the square roots of numbers.

To find the <u>square</u> root (or <u>second</u> root) of a number, we must find one of <u>two</u> identical factors whose product is that number.

Since $(7)(7) = 49$, the <u>square</u> root of 49 is 7.

Numbers have other roots besides the "square" root. For example:

To find the <u>cube</u> root (or <u>third</u> root) of a number, we must find one of <u>three</u> identical factors whose product is that number.

Since $(5)(5)(5) = 125$, the <u>cube</u> root of 125 is _____.

Exponents, Powers, and Roots   279

67. Roots beyond the third root do not have special names like "square" root or "cube" root. They are simply called <u>fourth</u> root, <u>fifth</u> root, <u>sixth</u> root, and so on.

To find the <u>fourth</u> root of a number, we must find one of <u>four</u> identical factors whose product is that number.

Since (3)(3)(3)(3) = 81 , the <u>fourth</u> root of 81 is 3 .

To find the <u>sixth</u> root of a number, we must find one of <u>six</u> identical factors whose product is that number.

Since (4)(4)(4)(4)(4)(4) = 4,096 , the <u>sixth</u> root of 4,096 is _____ .

68. We have seen that a square root radical is used for square roots. That is:

$\sqrt{64}$ means: find the <u>square</u> root of 64 .

The radical $\sqrt[3]{}$ is used for cube roots. The small "3" in a cube root is called the <u>index</u>. That is:

$\sqrt[3]{27}$ means: find the <u>cube</u> root of 27 .

A radical with an index is also used for roots beyond cube roots. That is:

$\sqrt[4]{81}$ means: find the <u>fourth</u> root of 81 .

$\sqrt[5]{32}$ means: find the <u>fifth</u> root of 32 .

$\sqrt[7]{1,089}$ means: find the _____ root of 1,089 .

Answers:
- 4
- seventh

69. Though the index of a square root is actually "2", the "2" is not usually written. That is:

Instead of $\sqrt[2]{824}$, we simply write $\sqrt{824}$.

Write each of these in radical form.

a) the ninth root of 45 _____   b) the square root of 33 _____

70. Since (4)(4)(4) = 64 , $\sqrt[3]{64}$ = 4

Since (6)(6)(6)(6) = 1,296 , $\sqrt[4]{1,296}$ = 6

Since (8)(8)(8)(8)(8) = 32,768 , $\sqrt[5]{32,768}$ = _____

Answers:
a) $\sqrt[9]{45}$
b) $\sqrt{33}$ (not $\sqrt[2]{33}$)

71. In each multiplication below, the identical factors are "1" and the product is "1". Therefore, any root of "1" is "1". That is:

Since (1)(1) = 1 , $\sqrt{1}$ = 1

Since (1)(1)(1) = 1 , $\sqrt[3]{1}$ = 1

Since (1)(1)(1)(1)(1) = 1 , $\sqrt[5]{1}$ = _____

Answer: 8

280 Exponents, Powers, and Roots

72. In each multiplication below, the identical factors are "0" and the product is "0". Therefore, any root of "0" is "0". That is:

Since $(0)(0)(0) = 0$, $\sqrt[3]{0} = 0$

Since $(0)(0)(0)(0) = 0$, $\sqrt[4]{0} = 0$

Since $(0)(0)(0)(0)(0)(0) = 0$, $\sqrt[6]{0} = $ _____

1

73. The root of a number can be a decimal number. For example:

Since $(2.5)(2.5) = 6.25$, $\sqrt{6.25} = 2.5$

Since $(1.2)(1.2)(1.2) = 1.728$, $\sqrt[3]{1.728} = 1.2$

Since $(.3)(.3)(.3)(.3)(.3) = .00243$, $\sqrt[5]{.00243} = $ _____

0

74. To find the square root of a number on a calculator, we use either the "square root" key $\boxed{\sqrt{x}}$ or $\boxed{\text{INV}}\boxed{x^2}$. As examples, let's do $\sqrt{81} = 9$ and $\sqrt{17.64} = 4.2$.

If your calculator has a "square root" key $\boxed{\sqrt{x}}$, the steps are:

Enter	Press	Display
81	$\boxed{\sqrt{x}}$	9.
17.64	$\boxed{\sqrt{x}}$	4.2

If your calculator does not have a $\boxed{\sqrt{x}}$ key, use $\boxed{\text{INV}}\boxed{x^2}$. The steps are:

Enter	Press	Display
81	$\boxed{\text{INV}}\boxed{x^2}$	9.
17.64	$\boxed{\text{INV}}\boxed{x^2}$	4.2

.3

75. Use a calculator for these.

a) $\sqrt{348,100} = $ _____

b) $\sqrt{.000729} = $ _____

76. When square roots (or any roots) are non-ending decimal numbers, we usually round to a definite place.

a) Rounding to tenths, $\sqrt{549} = $ _____

b) Rounding to thousandths, $\sqrt{.917} = $ _____

a) 590   b) .027

a) 23.4
b) .958

77. To find roots beyond square roots on a calculator, we use either the "root" key $\boxed{\sqrt[x]{y}}$ or $\boxed{\text{INV}}\ \boxed{y^x}$. As an example, let's do $\sqrt[4]{625} = 5$.

If your calculator has a "root" key $\boxed{\sqrt[x]{y}}$, the steps are:

Enter	Press	Display
625	$\boxed{\sqrt[x]{y}}$	625.
4	$\boxed{=}$	5.

If your calculator does not have a $\boxed{\sqrt[x]{y}}$ key, use $\boxed{\text{INV}}\ \boxed{y^x}$. The steps are:

Enter	Press	Display
625	$\boxed{\text{INV}}\ \boxed{y^x}$	625.
4	$\boxed{=}$	5.

Using either $\boxed{\sqrt[x]{y}}$ or $\boxed{\text{INV}}\ \boxed{y^x}$ and the proper index, do these.

a) $\sqrt[3]{1{,}331} =$ _____   b) $\sqrt[8]{6{,}561} =$ _____

---

78. Do these on a calculator. Round to tenths.

a) $\sqrt[4]{256{,}740} =$ _____   b) $\sqrt[7]{12{,}250{,}000} =$ _____

a) 11	b) 3

---

79. Do these on a calculator. Round to hundredths.

a) $\sqrt[5]{15.89} =$ _____   b) $\sqrt[9]{0.00466} =$ _____

a) 22.5	b) 10.3

a) 1.74	b) 0.55

6-10 ROOTS OF POWERS

In this section, we will show how the definition of roots can be extended to include the roots of powers.

80. The definition of roots can be extended to include the roots of powers. For example:

The <u>square</u> root of 2^6 is one of <u>two</u> identical factors whose product is 2^6.

Since $(2^3)(2^3) = 2^6$, the square root of 2^6 is 2^3.

The <u>cube</u> root of y^{12} is one of <u>three</u> identical factors whose product is y^{12}.

Since $(y^4)(y^4)(y^4) = y^{12}$, the <u>cube</u> root of y^{12} is y^4.

The <u>fourth</u> root of t^{20} is one of <u>four</u> identical factors whose product is t^{20}.

Since $(t^5)(t^5)(t^5)(t^5) = t^{20}$, the <u>fourth</u> root of t^{20} is _____.

t^5

81. The roots of powers can also be written in radical notation. That is:

$\sqrt{3^4}$ means the square root of 3^4.

a) $\sqrt[3]{y^6}$ means the _____ root of y^6.

b) $\sqrt[5]{m^{20}}$ means the _____ root of m^{20}.

---

82. Complete these.

Since $(x^5)(x^5) = x^{10}$, $\sqrt{x^{10}} = x^5$

a) Since $(t^2)(t^2)(t^2) = t^6$, $\sqrt[3]{t^6} = $ _____

b) Since $(2^6)(2^6)(2^6)(2^6)(2^6) = 2^{30}$, $\sqrt[5]{2^{30}} = $ _____

a) cube
b) fifth

---

83. The concept of roots also applies to powers with negative exponents. That is:

Since $(x^{-4})(x^{-4}) = x^{-8}$, $\sqrt{x^{-8}} = x^{-4}$

a) Since $(t^{-2})(t^{-2})(t^{-2}) = t^{-6}$, $\sqrt[3]{t^{-6}} = $ _____

b) Since $(V^{-3})(V^{-3})(V^{-3})(V^{-3}) = V^{-12}$, $\sqrt[4]{V^{-12}} = $ _____

a) t^2
b) 2^6

---

84. To find the square root of a power, we can divide its exponent by 2. That is:

$\sqrt{x^6} = x^{\frac{6}{2}}$ or x^3, since $(x^3)(x^3) = x^6$

$\sqrt{y^{-8}} = y^{\frac{-8}{2}}$ or y^{-4}, since $(y^{-4})(y^{-4}) = y^{-8}$

Using the above fact, complete these.

a) $\sqrt{5^4} = $ _____   b) $\sqrt{7^{-10}} = $ _____   c) $\sqrt{d^{20}} = $ _____

a) t^{-2}
b) V^{-3}

---

85. To find the cube root of a power, we can divide its exponent by 3. That is:

$\sqrt[3]{5^{12}} = 5^{\frac{12}{3}}$ or 5^4, since $(5^4)(5^4)(5^4) = 5^{12}$

$\sqrt[3]{x^{-9}} = x^{\frac{-9}{3}}$ or x^{-3}, since $(x^{-3})(x^{-3})(x^{-3}) = x^{-9}$

Using the above fact, complete these.

a) $\sqrt[3]{t^6} = $ _____   b) $\sqrt[3]{10^{-12}} = $ _____   c) $\sqrt[3]{y^{60}} = $ _____

a) 5^2
b) 7^{-5}
c) d^{10}

---

a) t^2
b) 10^{-4}
c) y^{20}

Exponents, Powers, and Roots   283

86. To find the **fourth** **root** of a power, we can divide its exponent by 4.
That is:

$\sqrt[4]{3^8} = 3^{\frac{8}{4}}$ or 3^2, since $(3^2)(3^2)(3^2)(3^2) = 3^8$

$\sqrt[4]{x^{-20}} = x^{\frac{-20}{4}}$ or x^{-5}, since $(x^{-5})(x^{-5})(x^{-5})(x^{-5}) = x^{-20}$

Using the above fact, complete these.

a) $\sqrt[4]{5^{12}} =$ _____   b) $\sqrt[4]{V^{-16}} =$ _____   c) $\sqrt[4]{y^{36}} =$ _____

---

87. The general law of radicals for finding the roots of powers is given below.

$$\sqrt[n]{a^m} = a^{\frac{m}{n}}$$

That is, to find any root of a power, we divide the exponent of the power by the desired root. (That is, we divide by the index.)

Using the above law, complete these.

a) $\sqrt[5]{x^{15}} = x^{\frac{15}{5}} =$ _____   b) $\sqrt[8]{m^{-40}} = m^{\frac{-40}{8}} =$ _____

a) 5^3
b) V^{-4}
c) y^9

---

88. Use the general law to find each root.

a) $\sqrt[4]{t^{-20}} =$ ___   b) $\sqrt[5]{x^{10}} =$ ___   c) $\sqrt[7]{y^{-21}} =$ ___   d) $\sqrt[10]{V^{80}} =$ ___

a) x^3 b) m^{-5}

---

89. The root of a power can be a power whose exponent is "1". That is:

Since $(y)(y) = y^2$, $\sqrt{y^2} = y^1$ or y

Since $(x)(x)(x)(x) = x^4$, $\sqrt[4]{x^4} = x^1$ or x

The above examples are consistent with the law of radicals for powers. That is:

$\sqrt{y^2} = y^{\frac{2}{2}} = y^1$ or y $\sqrt[4]{x^4} = x^{\frac{4}{4}} = x^1$ or x

Complete these.

a) $\sqrt[3]{7^3} =$ _____   b) $\sqrt[5]{t^5} =$ _____   c) $\sqrt[12]{P^{12}} =$ _____

a) t^{-5}
b) x^2
c) y^{-3}
d) V^8

---

a) 7^1 or 7
b) t^1 or t
c) P^1 or P

284 Exponents, Powers, and Roots

90. The root of a power can be a power whose exponent is -1. That is:

Since $(x^{-1})(x^{-1}) = x^{-2}$, $\sqrt{x^{-2}} = x^{-1}$

Since $(y^{-1})(y^{-1})(y^{-1})(y^{-1}) = y^{-4}$, $\sqrt[4]{y^{-4}} = y^{-1}$

The above examples are consistent with the law of radicals for powers. That is:

$\sqrt{x^{-2}} = x^{\frac{-2}{2}} = x^{-1}$ $\sqrt[4]{y^{-4}} = y^{\frac{-4}{4}} = y^{-1}$

Complete these.

a) $\sqrt[3]{5^{-3}} = $ _____   b) $\sqrt[7]{t^{-7}} = $ _____   c) $\sqrt[10]{V^{-10}} = $ _____

91. The same law of radicals applies to the roots of monomials. That is:

$\sqrt{(xy)^6} = (xy)^{\frac{6}{2}} = (xy)^3$ $\sqrt[5]{(2x)^{-20}} = (2x)^{\frac{-20}{5}} = $ _____

a) 5^{-1}
b) t^{-1}
c) V^{-1}

92. The same law also applies to the roots of fractions. That is:

$\sqrt{\left(\frac{y}{4}\right)^8} = \left(\frac{y}{4}\right)^{\frac{8}{2}} = \left(\frac{y}{4}\right)^4$ $\sqrt[4]{\left(\frac{a}{b}\right)^{-12}} = \left(\frac{a}{b}\right)^{\frac{-12}{4}} = $ _____

$(2x)^{-4}$

$\left(\frac{a}{b}\right)^{-3}$

SELF-TEST 19 (pages 273-284)

Raise each power to the power shown.

1. $(x^2)^{-5}$ 2. $(w^6)^0$

Raise each monomial to the power shown.

3. $(3p^4t)^3$ 4. $(a^2b^{-1}c^{-3})^{-2}$

Raise each fraction to the power shown.

5. $\left(\frac{w^2}{r}\right)^4$ 6. $\left(\frac{dh^{-2}}{p^{-1}v^4}\right)^{-3}$

Convert each power to a power with a positive exponent.

7. $\left(\frac{x}{y}\right)^{-1}$ 8. $\left(\frac{2R}{P}\right)^{-7}$

Using a calculator, find these roots.

9. $\sqrt{136,900}$ 10. $\sqrt[3]{6,859}$ 11. Round to hundredths. $\sqrt[5]{8.13}$ 12. Round to thousandths. $\sqrt[8]{0.000539}$

Find these roots of powers.

13. $\sqrt{10^{16}}$ 14. $\sqrt[6]{A^{-6}}$ 15. $\sqrt[3]{(5x)^{-9}}$ 16. $\sqrt[4]{\left(\frac{b}{d}\right)^8}$

Exponents, Powers, and Roots   285

ANSWERS:	1. x^{-10}	5. $\dfrac{w^8}{r^4}$	8. $\left(\dfrac{P}{2R}\right)^7$	13. 10^8
	2. 1	6. $\dfrac{d^{-3}h^6}{p^3v^{-12}}$	9. 370	14. A^{-1}
	3. $27p^{12}t^3$		10. 19	15. $(5x)^{-3}$
	4. $a^{-4}b^2c^6$	7. $\left(\dfrac{y}{x}\right)^1$ or $\dfrac{y}{x}$	11. 1.52	16. $\left(\dfrac{b}{d}\right)^2$
			12. 0.390	

6-11 POWERS WITH FRACTIONAL EXPONENTS

In this section, we will discuss powers with fractional exponents.

93. The law of radicals for finding the roots of powers is given below.

$$\boxed{\sqrt[n]{a^m} = a^{\frac{m}{n}}}$$

When converting a radical to a power, we can get a fractional exponent. For example:

$$\sqrt{3^7} = 3^{\frac{7}{2}} \qquad \sqrt[3]{x^5} = x^{\frac{5}{3}} \qquad \sqrt[7]{t^4} = t^{\frac{4}{7}}$$

Convert each radical to a power.

a) $\sqrt{5^3} = $ _____   b) $\sqrt[4]{x^3} = $ _____   c) $\sqrt[9]{y^{10}} = $ _____

94. If the exponent of the power is not explicitly shown, its exponent is "1". That is:

$$\sqrt{5} = \sqrt{5^1} = 5^{\frac{1}{2}} \qquad \sqrt[4]{x} = \sqrt[4]{x^1} = x^{\frac{1}{4}}$$

Convert each radical to a power.

a) $\sqrt{y} = $ _____   b) $\sqrt[3]{t} = $ _____   c) $\sqrt[8]{v} = $ _____

a) $5^{\frac{3}{2}}$

b) $x^{\frac{3}{4}}$

c) $y^{\frac{10}{9}}$

95. We reversed the law of radicals for the roots of powers below.

$$\boxed{a^{\frac{m}{n}} = \sqrt[n]{a^m}}$$

Using the reversed form, we can convert any power with a fractional exponent to a radical. Some examples are shown. Notice that the denominator of the fractional exponent is the index of the radical.

$$7^{\frac{3}{2}} = \sqrt{7^3} \qquad x^{\frac{2}{3}} = \sqrt[3]{x^2} \qquad y^{\frac{4}{9}} = \sqrt[9]{y^4}$$

Convert each power to a radical.

a) $t^{\frac{5}{2}} = $ _____   b) $5^{\frac{3}{4}} = $ _____   c) $m^{\frac{4}{5}} = $ _____

a) $y^{\frac{1}{2}}$

b) $t^{\frac{1}{3}}$

c) $v^{\frac{1}{8}}$

286   Exponents, Powers, and Roots

96. In each conversion below, the numerator of the fractional exponent is "1". $$5^{\frac{1}{2}} = \sqrt{5^1} \text{ or } \sqrt{5} \qquad t^{\frac{1}{6}} = \sqrt[6]{t^1} \text{ or } \sqrt[6]{t}$$ Convert each power to a radical. a) $x^{\frac{1}{2}} = $ _____ b) $7^{\frac{1}{3}} = $ _____ c) $y^{\frac{1}{5}} = $ _____	a) $\sqrt{t^5}$ b) $\sqrt[4]{5^3}$ c) $\sqrt[5]{m^4}$
97. Convert each power to a radical and each radical to a power. a) $5^{\frac{7}{2}} = $ _____ b) $\sqrt[6]{2^5} = $ _____ c) $x^{\frac{1}{4}} = $ _____ d) $\sqrt[3]{m} = $ _____	a) $\sqrt{x}$ b) $\sqrt[3]{7}$ c) $\sqrt[5]{y}$
98. By converting to a radical and then finding the root mentally, we converted each power below to an ordinary number. $$9^{\frac{1}{2}} = \sqrt{9} = 3 \qquad 8^{\frac{2}{3}} = \sqrt[3]{8^2} = \sqrt[3]{64} = 4$$ However, we usually need a calculator to find the roots in such conversions. Use $\boxed{\sqrt[x]{y}}$ or $\boxed{\text{INV}}\boxed{y^x}$ to complete these. Round to hundredths. a) $10^{\frac{1}{3}} = \sqrt[3]{10} = $ _____ b) $2^{\frac{5}{4}} = \sqrt[4]{2^5} = \sqrt[4]{32} = $ _____	a) $\sqrt{5^7}$ c) $\sqrt[4]{x}$ b) $2^{\frac{5}{6}}$ d) $m^{\frac{1}{3}}$
99. Each radical below contains a negative exponent. Therefore, when converting the radical to a power, we get a power with a negative fractional exponent. $$\sqrt{5^{-1}} = 5^{-\frac{1}{2}} \qquad \sqrt[3]{7^{-2}} = 7^{-\frac{2}{3}} \qquad \sqrt[4]{y^{-7}} = y^{-\frac{7}{4}}$$ Convert each radical to a power. a) $\sqrt{x^{-1}} = $ _____ b) $\sqrt[5]{10^{-2}} = $ _____ c) $\sqrt[3]{t^{-4}} = $ _____	a) 2.15 b) 2.38
100. If we convert a power with a negative fractional exponent to a radical, we get a negative exponent in the radical. For example: $$2^{-\frac{1}{3}} = \sqrt[3]{2^{-1}} \qquad x^{-\frac{3}{2}} = \sqrt{x^{-3}} \qquad y^{-\frac{7}{8}} = \sqrt[8]{y^{-7}}$$ Convert each power to a radical. a) $7^{-\frac{1}{4}} = $ _____ b) $m^{-\frac{5}{2}} = $ _____ c) $V^{-\frac{3}{7}} = $ _____	a) $x^{-\frac{1}{2}}$ b) $10^{-\frac{2}{5}}$ c) $t^{-\frac{4}{3}}$
101. Convert each power to a radical and each radical to a power. a) $6^{-\frac{1}{2}} = $ _____ b) $\sqrt{t^{-7}} = $ _____ c) $x^{-\frac{9}{5}} = $ _____ d) $\sqrt[5]{2^{-3}} = $ _____	a) $\sqrt[4]{7^{-1}}$ b) $\sqrt{m^{-5}}$ c) $\sqrt[7]{V^{-3}}$

102. The law of exponents for multiplication is shown below.

$$a^m \cdot a^n = a^{m+n}$$

The law also applies to powers with fractional exponents. That is:

$$x^{\frac{1}{2}} \cdot x^{\frac{1}{4}} = x^{\frac{1}{2}+\frac{1}{4}} = x^{\frac{2}{4}+\frac{1}{4}} = x^{\frac{3}{4}}$$

Use the law of exponents for these.

a) $y^{\frac{2}{3}} \cdot y^{\frac{1}{2}} =$ _____ b) $m^{\frac{4}{5}} \cdot m^{\frac{6}{5}} =$ _____

a) $\sqrt{6^{-1}}$

b) $t^{-\frac{7}{2}}$

c) $\sqrt[5]{x^{-9}}$

d) $2^{-\frac{3}{5}}$

---

103. The law of exponents for division is shown below.

$$\frac{a^m}{a^n} = a^{m-n}$$

The law also applies to powers with fractional exponents. That is:

$$\frac{x^{\frac{3}{4}}}{x^{\frac{1}{2}}} = x^{\frac{3}{4}-\frac{1}{2}} = x^{\frac{3}{4}-\frac{2}{4}} = x^{\frac{1}{4}}$$

Use the law of exponents for these.

a) $\dfrac{y^{\frac{5}{3}}}{y^{\frac{1}{3}}} =$ _____ b) $\dfrac{m^{\frac{7}{6}}}{m^{\frac{2}{3}}} =$ _____

a) $y^{\frac{7}{6}}$, from: $y^{\frac{2}{3}+\frac{1}{2}}$

b) m^2, from: $m^{\frac{4}{5}+\frac{6}{5}}$

---

104. The law of exponents for raising a power to a power is shown below.

$$(a^m)^n = a^{mn}$$

The law also applies when fractional exponents are involved. For example:

$$\left(x^{\frac{1}{2}}\right)^3 = x^{\left(\frac{1}{2}\right)(3)} = x^{\frac{3}{2}} \qquad \left(y^4\right)^{\frac{2}{3}} = y^{(4)\left(\frac{2}{3}\right)} = y^{\frac{8}{3}}$$

Use the law of exponents for these.

a) $\left(m^{\frac{3}{5}}\right)^2 =$ _____ b) $\left(x^5\right)^{\frac{1}{6}} =$ _____ c) $\left(d^{\frac{1}{2}}\right)^{\frac{4}{3}} =$ _____

a) $y^{\frac{4}{3}}$, from: $y^{\frac{5}{3}-\frac{1}{3}}$

b) $m^{\frac{1}{2}}$, from: $m^{\frac{7}{6}-\frac{2}{3}}$

---

a) $m^{\frac{6}{5}}$, from: $m^{\left(\frac{3}{5}\right)(2)}$

b) $x^{\frac{5}{6}}$, from: $x^{(5)\left(\frac{1}{6}\right)}$

c) $d^{\frac{2}{3}}$, from: $d^{\left(\frac{1}{2}\right)\left(\frac{4}{3}\right)}$

105. The law of exponents for raising a fraction to a power is shown below.

$$\left(\frac{a}{b}\right)^n = \frac{a^n}{b^n}$$

The law also applies to fractional powers. That is:

$$\left(\frac{y}{4}\right)^{\frac{1}{3}} = \frac{y^{\frac{1}{3}}}{4^{\frac{1}{3}}} \qquad \left(\frac{x}{t}\right)^{\frac{4}{5}} = \frac{x^{\frac{4}{5}}}{t^{\frac{4}{5}}} \qquad \left(\frac{m}{d}\right)^{\frac{3}{2}} = \underline{}$$

$\dfrac{m^{\frac{3}{2}}}{d^{\frac{3}{2}}}$

6-12 POWERS WITH DECIMAL EXPONENTS

In this section, we will discuss powers with decimal exponents.

106. Any power with a fractional exponent can be converted to a power with a decimal exponent. For example:

$$\sqrt{9} = 9^{\frac{1}{2}} = 9^{0.5} \qquad \sqrt[4]{x^5} = x^{\frac{5}{4}} = x^{1.25}$$

Following the examples, convert each radical to a power with a fractional exponent and then to a power with a decimal exponent.

a) $\sqrt[4]{5} = \underline{} = \underline{}$ b) $\sqrt[5]{y^6} = \underline{} = \underline{}$

107. When converting a fractional exponent to a decimal exponent, we frequently get a non-ending decimal number. In such cases, we round the decimal number. For example, we rounded each exponent below to hundredths.

$$y^{\frac{1}{3}} = y^{0.33} \qquad x^{\frac{7}{6}} = x^{1.17}$$

Convert each radical to a power with a fractional exponent and then to a power with a decimal exponent. <u>Round to hundredths</u>.

a) $\sqrt[3]{m^2} = \underline{} = \underline{}$ b) $\sqrt[7]{d^9} = \underline{} = \underline{}$

a) $5^{\frac{1}{4}} = 5^{0.25}$
b) $y^{\frac{6}{5}} = y^{1.2}$

108. By reversing the procedure in the last two frames, we can convert powers with decimal exponents to radicals. For example:

$$4^{0.5} = 4^{\frac{1}{2}} = \sqrt{4} \qquad x^{1.25} = x^{\frac{5}{4}} = \sqrt[4]{x^5}$$

Using the same steps, convert each power to a radical.

a) $7^{0.25} = \underline{} = \underline{}$ b) $y^{2.5} = \underline{} = \underline{}$

a) $m^{\frac{2}{3}} = m^{0.67}$
b) $d^{\frac{9}{7}} = d^{1.29}$

109. By converting to a radical and then finding the root mentally, we converted each power below to an ordinary number.

$$9^{0.5} = 9^{\frac{1}{2}} = \sqrt{9} = 3 \qquad 16^{0.25} = 16^{\frac{1}{4}} = \sqrt[4]{16} = 2$$

However, we ordinarily need a calculator to find the roots in such conversions. Use a calculator for these.

a) $5^{1.5} = 5^{\frac{3}{2}} = \sqrt{5^3} = \sqrt{125} = $ _____ (Round to tenths)

b) $3^{0.8} = 3^{\frac{4}{5}} = \sqrt[5]{3^4} = \sqrt[5]{81} = $ _____ (Round to hundredths)

a) $7^{\frac{1}{4}} = \sqrt[4]{7}$

b) $y^{\frac{5}{2}} = \sqrt{y^5}$

---

110. In each conversion below, we got a power with a negative decimal exponent.

$$\sqrt{x^{-1}} = x^{-\frac{1}{2}} = x^{-0.5} \qquad \sqrt[4]{y^{-5}} = y^{-\frac{5}{4}} = y^{-1.25}$$

Using the same steps, convert each radical to a power.

a) $\sqrt[5]{m^{-2}} = $ _____ = _____ 	b) $\sqrt{R^{-3}} = $ _____ = _____

a) 11.2

b) 2.41

---

111. In each conversion below, we got a negative exponent in the radical.

$$t^{-0.25} = t^{-\frac{1}{4}} = \sqrt[4]{t^{-1}} \qquad p^{-1.2} = p^{-\frac{6}{5}} = \sqrt[5]{p^{-6}}$$

Using the same steps, convert each power to a radical.

a) $x^{-0.2} = $ _____ = _____ 	b) $y^{-2.5} = $ _____ = _____

a) $m^{-\frac{2}{5}} = m^{-0.4}$

b) $R^{-\frac{3}{2}} = R^{-1.5}$

---

112. The law of exponents for multiplication applies to powers with decimal exponents. That is:

$$x^{0.5} \cdot x^{1.2} = x^{0.5+1.2} = x^{1.7}$$

Using the same law, complete these.

a) $y^{1.5} \cdot y^{0.4} = $ _____ 	b) $m^{-1.3} \cdot m^{2.5} = $ _____

a) $x^{-\frac{1}{5}} = \sqrt[5]{x^{-1}}$

b) $y^{-\frac{5}{2}} = \sqrt{y^{-5}}$

---

113. The law of exponents for division applies to powers with decimal exponents. That is:

$$\frac{x^{0.67}}{x^{0.25}} = x^{0.67-0.25} = x^{0.42}$$

Using the same law, complete these.

a) $\dfrac{y^{2.5}}{y^{-1.4}} = $ _____ 	b) $\dfrac{t^{0.75}}{t^{0.25}} = $ _____

a) $y^{1.9}$ 	b) $m^{1.2}$

---

a) $y^{3.9}$

b) $t^{0.50}$ or $t^{0.5}$

290 Exponents, Powers, and Roots

114. The law of exponents for raising a power to a power applies to powers with decimal exponents. That is:

$$\left(x^{1.2}\right)^{0.3} = x^{(1.2)(0.3)} = x^{0.36}$$

Using the same law, complete these.

a) $\left(y^{0.5}\right)^5 =$ _____   b) $\left(t^{2.4}\right)^{-0.2} =$ _____

115. The law of exponents for raising a fraction to a power also applies to decimal powers. That is:

$$\left(\frac{x}{2}\right)^{0.25} = \frac{x^{0.25}}{2^{0.25}} \qquad \left(\frac{p}{q}\right)^{-1.2} = \underline{\qquad}$$

a) $y^{2.5}$

b) $t^{-0.48}$

$\dfrac{p^{-1.2}}{q^{-1.2}}$

6-13 FINDING POWERS WITH A CALCULATOR

In this section, we will show the procedure for finding powers with a calculator.

116. To square a number on a calculator, we use the $\boxed{x^2}$ key. Following the steps below, do $7^2 = 49$ and $2.5^2 = 6.25$.

Enter	Press	Display
7	$\boxed{x^2}$	49.
2.5	$\boxed{x^2}$	6.25

117. Use a calculator for these:

a) $256^2 =$ _____   b) $0.52^2 =$ _____

118. To find other powers on a calculator, we use the "power" key $\boxed{y^x}$.

The calculator steps for $12^4 = 20{,}736$ are:

Enter	Press	Display
12	$\boxed{y^x}$	12.
4	$\boxed{=}$	20736.

Use $\boxed{y^x}$ to convert each power below to an ordinary number.

a) $15^5 =$ _____   b) $3^9 =$ _____

a) 65,536

b) 0.2704

119. The $\boxed{y^x}$ key can also be used to convert powers with decimal bases to ordinary numbers.

   The calculator steps for $(2.2)^4 = 23.4256$ are:

Enter	Press	Display
2.2	$\boxed{y^x}$	2.2
4	$\boxed{=}$	23.4256

   Use $\boxed{y^x}$ to convert each power to an ordinary number.

   a) $(1.8)^5 = $ _____   b) $(0.9)^7 = $ _____

   a) 759,375
   b) 19,683

120. We use the $\boxed{y^x}$ key to convert a power with a decimal exponent to an ordinary number. For example:

   To do $36^{0.5} = 6$, follow these steps.

Enter	Press	Display
36	$\boxed{y^x}$	36.
0.5	$\boxed{=}$	6.

   Use a calculator for these.

   a) $4^{1.5} = $ _____   b) $81^{0.25} = $ _____

   a) 18.8957
   b) 0.478297

121. Do these on a calculator. Round to hundredths.

   a) $15^{0.4} = $ _____   b) $4.8^{1.3} = $ _____

   a) 8   b) 3

122. Do these on a calculator. Round to the nearest whole number.

   a) $75^{2.2} = $ _____   b) $19.4^{1.8} = $ _____

   a) 2.95   b) 7.68

   a) 13,339   b) 208

292 Exponents, Powers, and Roots

123. Using the definition of a power with a negative exponent, we converted each power below to an ordinary number.

$$2^{-1} = \frac{1}{2^1} = \frac{1}{2} = 0.5$$

$$5^{-2} = \frac{1}{5^2} = \frac{1}{25} = 0.04$$

We can use the $\boxed{y^x}$ key and the $\boxed{+/-}$ key to perform the same conversions on a calculator. The steps are:

Enter	Press	Display
2	$\boxed{y^x}$	2.
1	$\boxed{+/-}\boxed{=}$	0.5

Enter	Press	Display
5	$\boxed{y^x}$	5.
2	$\boxed{+/-}\boxed{=}$	0.04

Use a calculator to convert these to ordinary numbers.

a) $4^{-1} =$ _____ b) $20^{-2} =$ _____ c) $5^{-3} =$ _____

---

124. Use a calculator for these. Round to millionths.

a) $12^{-4} =$ _____ b) $9^{-5} =$ _____

a) 0.25
b) 0.0025
c) 0.008

---

125. The same calculator procedure is used for powers with negative decimal exponents. Do these. Round to four decimal places.

a) $85^{-0.8} =$ _____ b) $17.5^{-1.2} =$ _____

a) 0.000048
b) 0.000017

---

126. To evaluate $(1 + 0.05)^{12}$, we perform the addition within the parentheses first. The calculator steps are shown. Notice that we pressed $\boxed{=}$ to complete the addition before pressing $\boxed{y^x}$. Rounded to hundredths, the answer is 1.80.

Enter	Press	Display
1	$\boxed{+}$	1.
0.05	$\boxed{=}\boxed{y^x}$	1.05
12	$\boxed{=}$	1.79586

Do these on a calculator. Round to hundredths.

a) $(1 + 0.06)^{20} =$ _____ b) $(27 + 3.5)^{0.1} =$ _____

a) 0.0286
b) 0.0322

---

a) 3.21 b) 1.41

Exponents, Powers, and Roots   293

127. To evaluate $\left(\dfrac{342}{295}\right)^{1.67}$, we perform the division within the parentheses first. The calculator steps are shown. Notice that we pressed $\boxed{=}$ to complete the division before pressing $\boxed{y^x}$. Rounded to hundredths, the answer is 1.28.

Enter	Press	Display
342	$\boxed{\div}$	342.
295	$\boxed{=}\ \boxed{y^x}$	1.15932
1.67	$\boxed{=}$	1.28003

Use a calculator for these. Round to hundredths.

a) $\left(\dfrac{450}{299}\right)^{1.9}$ = _____   b) $\left(\dfrac{228}{1.22}\right)^{0.2}$ = _____

a) 2.17   b) 2.85

6-14 EVALUATING FORMULAS CONTAINING POWERS

In this section, we will use a calculator for evaluations with formulas containing powers.

128. We used $\boxed{y^x}$ for the evaluation below. The steps are shown.

In $\boxed{y = x^a}$, find "y" when $x = 5$ and $a = 4$.

Enter	Press	Display
5	$\boxed{y^x}$	5.
4	$\boxed{=}$	625. (y = 625)

Use a calculator for these. Round to tenths.

a) In $\boxed{M = T^a}$, when $T = 18$ and $a = 1.3$, $M =$ _____

b) In $\boxed{P = V^k}$, when $V = 15.9$ and $k = 1.64$, $P =$ _____

129. Do these. Round to hundredths.

a) In $\boxed{t = m^{0.4}}$, when $m = 27.1$, $t =$ _____

b) In $\boxed{H = 1.14^d}$, when $d = 3.75$, $H =$ _____

a) 42.8
b) 93.4

a) 3.74
b) 1.63

294 Exponents, Powers, and Roots

130. In the evaluation below, we divide first. Notice that we press $\boxed{=}$ to complete the division before using $\boxed{y^x}$. We rounded the answer to hundredths.

In $\boxed{D = \left(\dfrac{a}{b}\right)^{1.8}}$, find D when a = 72.9 and b = 58.6.

Enter	Press	Display	
72.9	$\boxed{\div}$	72.9	
58.6	$\boxed{=}\boxed{y^x}$	1.24403	
1.8	$\boxed{=}$	1.48147	(D = 1.48)

Use the same steps for this one. Round to thousandths.

In $\boxed{F = \left(\dfrac{S}{T}\right)^{0.1}}$, when S = 13.9 and T = 47.6, F = _____

131. In the evaluation below, we add first. Notice that we press $\boxed{=}$ to complete the addition before using $\boxed{y^x}$. We rounded the answer to the nearest whole number.

0.884

In $\boxed{V = (x + 1)^n}$, find V when x = 7.8 and n = 2.5.

Enter	Press	Display	
7.8	$\boxed{+}$	7.8	
1	$\boxed{=}\boxed{y^x}$	8.8	
2.5	$\boxed{=}$	229.724	(V = 230)

Use the same steps for this one. Round to hundredths.

In $\boxed{F = (c + d)^{0.2}}$, when c = 11.9 and d = 48.3, F = _____

132. In the formula below, Q^a is a factor in a multiplication. The steps for the evaluation are shown. We rounded the answer to tenths. (Note: On some calculators, you have to evaluate Q^a first and then multiply by P.)

2.27

In $\boxed{R = PQ^a}$, find R when P = 1.59, Q = 2.24, and a = 3.18.

Enter	Press	Display	
1.59	$\boxed{\times}$	1.59	
2.24	$\boxed{y^x}$	2.24	
3.18	$\boxed{=}$	20.6626	(R = 20.7)

Use the same steps for these. Round to ten-thousandths in (a) and to the nearest whole number in (b).

a) In $\boxed{I = KP^{1.5}}$, when K = 0.001 and P = 19, I = _____

b) In $\boxed{b = TV^{0.4}}$, when T = 400 and V = 100, b = _____

Exponents, Powers, and Roots   295

133. Two methods for the evaluation below are shown. We rounded the answer to the nearest whole number.

In $\boxed{M = T(a + b)^{0.1}}$, find M when T = 25, a = 37, and b = 49.

1) If a calculator has parentheses symbols, we can use the parentheses to evaluate (a + b) before pressing $\boxed{y^x}$.

Enter	Press	Display
25	$\boxed{x}$ $\boxed{(}$	25.
37	$\boxed{+}$	37.
49	$\boxed{)}$ $\boxed{y^x}$	86.
0.1	$\boxed{=}$	39.0292 (M = 39)

2) If a calculator does not have parentheses symbols, we must evaluate $(a + b)^{0.1}$ first and then multiply by the value of T.

Enter	Press	Display
37	$\boxed{+}$	37.
49	$\boxed{=}$ $\boxed{y^x}$	86.
0.1	$\boxed{x}$	1.56117
25	$\boxed{=}$	39.0292 (M = 39)

Use a calculator for these. Round to the nearest whole number.

a) In $\boxed{A = P(1 + i)^n}$, when P = 1,000, i = 0.085, and n = 10,

A = _____

b) In $\boxed{D = K(t + 1)^n}$, when K = 250, t = 60, and n = 0.5,

D = _____

a) 0.0828

b) 2,524

a) 2,261   b) 1,953

6-15 GRAPHS OF EXPONENTIAL FUNCTIONS

In this section, we will discuss the graphs of exponential functions.

134. Any equation or formula of the form $y = a^x$ or $y = a^{-x}$ is called an "exponential" function. That is:

$y = 2^x$ and $y = 2^{-x}$ are exponential functions.

Which of the following are exponential functions? _____

a) $y = 7^x$ b) $xy = 10$ c) $t = 5^{-d}$ d) $y = 14x^2$

Both (a) and (c)

135. We found some pairs of values for $y = 2^x$ below.

If $x = 1$, $y = 2^x = 2^1 = 2$

If $x = -2$, $y = 2^x = 2^{-2} = \frac{1}{2^2} = \frac{1}{4} = 0.25$

Using the same steps, complete these.

a) If $x = -1$, $y = 2^x = 2^{-1} = $ _____

b) If $x = 3$, $y = 2^x = 2^3 = $ _____

---

136. We found some pairs of values for $y = 2^{-x}$ below. Notice that the "−" in front of x means that we use the opposite of the value substituted for x.

If $x = 1$, $y = 2^{-x} = 2^{-1} = \frac{1}{2} = 0.5$

If $x = -2$, $y = 2^{-x} = 2^{-(-2)} = 2^2 = 4$

Using the same steps, complete these.

a) If $x = 2$, $y = 2^{-x} = 2^{-2} = $ _____

b) If $x = -1$, $y = 2^{-x} = 2^{-(-1)} = 2^1 = $ _____

a) 0.5, from: $\frac{1}{2^1}$

b) 8

---

137. When $x = 0$, $y = 1$ for both $y = 2^x$ and $y = 2^{-x}$. That is:

If $x = 0$, $y = 2^x = 2^0 = 1$

If $x = 0$, $y = 2^{-x} = 2^{-0} = 2^0 = $ _____

a) 0.25

b) 2

1

138. We prepared tables for $y = 2^x$ and $y = 2^{-x}$ below.

$y = 2^x$			$y = 2^{-x}$	
x	y		x	y
3	8		3	0.125
2	4		2	0.25
1	2		1	0.5
0	1		0	1
-1	0.5		-1	2
-2	0.25		-2	4
-3	0.125		-3	8

Using the tables above, we graphed both functions below.

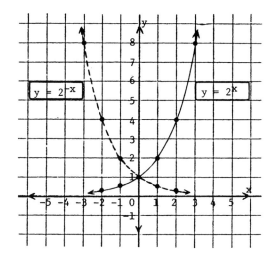

Write the coordinates of the y-intercept of each graph. _____

$(0, 1)$

SELF-TEST 20 (pages 285-298)

Convert each power to a radical.

1. $x^{\frac{5}{3}}$
2. $10^{-\frac{2}{7}}$

Convert each radical to a power with a fractional exponent.

3. $\sqrt[6]{P}$
4. $\sqrt{a^{-3}}$

Simplify. Write each answer as a single power with a fractional exponent.

5. $N^{\frac{2}{5}} \cdot N^{\frac{1}{2}}$
6. $\dfrac{t^{\frac{3}{2}}}{t^{\frac{1}{4}}}$
7. $(R^2)^{\frac{1}{4}}$

8. Convert $\sqrt[6]{y^{-2}}$ to a power with a decimal exponent. Round to hundredths.

9. Convert $B^{0.6}$ to a radical.

Simplify. Write each answer as a single power with a decimal exponent.

10. $\dfrac{h^{2.4}}{h^{1.5}}$

11. $(T^{3.5})^{0.4}$

Use a calculator to convert these powers to ordinary numbers.

12. Round to tenths.
    1.28^{10}

13. Round to the nearest whole number.
    $31.6^{1.5}$

14. Round to hundredths.
    $(0.417)^{-2.3}$

15. In $\boxed{F = CP^t}$, find F when C = 0.184, P = 21.4, and t = 1.75. Round to tenths.

16. In $\boxed{W = \left(\dfrac{d}{r}\right)^a}$, find W when d = 12.9, r = 47.2, and a = 0.85. Round to thousandths.

ANSWERS:

1. $\sqrt[3]{x^5}$
2. $\sqrt[7]{10^{-2}}$
3. $P^{\frac{1}{6}}$
4. $a^{-\frac{3}{2}}$
5. $N^{\frac{9}{10}}$
6. $t^{\frac{5}{4}}$
7. $R^{\frac{1}{2}}$
8. $y^{-0.33}$
9. $\sqrt[5]{B^3}$
10. $h^{0.9}$
11. $T^{1.4}$
12. 11.8
13. 178
14. 7.48
15. F = 39.2
16. W = 0.332

Exponents, Powers, and Roots 299

SUPPLEMENTARY PROBLEMS - CHAPTER 6

Assignment 18

Convert each power to an ordinary number.

1. 2^4 2. 9^1 3. 4^3 4. 25^0 5. 3^5

Convert each power to a fraction.

6. 2^{-3} 7. 5^{-2} 8. 10^{-1} 9. x^{-4} 10. w^{-1}

Convert each fraction to a power with a negative exponent.

11. $\frac{1}{y^3}$ 12. $\frac{1}{R}$ 13. $\frac{1}{5}$ 14. $\frac{1}{4^2}$

Using the law of exponents for multiplying powers, find each product.

15. $a^5 \cdot a^{-3}$ 16. $h \cdot h^3$ 17. $y^2 \cdot y^{-4} \cdot y$ 18. $(8x^3)(5x^{-2})$

19. $(d^2t)(dt^{-3})$ 20. $(p^{-6}w^4)(p^3w^{-4})$ 21. $(7x^{-2}y)(9xy)$ 22. $(8a^2b^{-1})(3ab^{-3})$

Using the law of exponents for dividing powers, find each quotient.

23. $\frac{x^3}{x^5}$ 24. $\frac{P}{P^{-1}}$ 25. $\frac{r^4}{r^2 \cdot r^3}$ 26. $\frac{20a^2b}{4ab^{-3}}$

Simplify each fraction.

27. $\frac{15y^7}{3y^4}$ 28. $\frac{6m}{8m^3}$ 29. $\frac{3d^2p^{-3}}{6d^2p^{-4}}$ 30. $\frac{12h^4k^{-1}}{18hk^{-1}}$

Write the reciprocal of each of the following.

31. t^3 32. 10^{-8} 33. $a^{-2}b^5$ 34. xy^2

Convert each division to a multiplication.

35. $\frac{s^2}{w^3}$ 36. $\frac{k}{p^2t}$ 37. $\frac{ax^4}{y^{-2}}$ 38. $\frac{2d}{hm^{-1}}$

Write each expression without negative exponents.

39. xy^{-2} 40. $\frac{R^{-1}}{T^{-1}}$ 41. $\frac{m^{-3}}{n^3}$ 42. $\frac{a^5b^{-2}}{c^{-4}d^3}$

300 Exponents, Powers, and Roots

Assignment 19

Raise each power to the power shown.

1. $(w^5)^4$
2. $(x^{-3})^0$
3. $(y^2)^{-3}$
4. $(P^{-1})^{-2}$
5. $(a^3)^1$

Raise each monomial to the power shown.

6. $(x^2y)^4$
7. $(a^{-1}b^2)^3$
8. $(wv^{-1})^{-2}$
9. $(c^{-2}d^5)^0$
10. $(5rw^{-1})^3$
11. $(2m^{-2}n^3)^5$
12. $(d^{-3}h^2p)^2$
13. $(s^{-1}v^4w^{-2})^{-3}$

Raise each fraction to the power shown.

14. $\left(\dfrac{1}{2}\right)^4$
15. $\left(\dfrac{x}{4}\right)^3$
16. $\left(\dfrac{a^{-2}}{b^3}\right)^4$
17. $\left(\dfrac{p^2 r^{-1}}{d^{-3}h}\right)^{-2}$

Convert each power to a power with a positive exponent.

18. $\left(\dfrac{w}{k}\right)^{-2}$
19. $\left(\dfrac{m}{2}\right)^{-1}$
20. $\left(\dfrac{ab}{c}\right)^{-3}$
21. $\left(\dfrac{s}{2t}\right)^{-5}$

Write each of the following in radical form.

22. The cube root of 50.
23. The fifth root of "x".
24. The square root of A.

Use a calculator to find the following roots.

25. $\sqrt[3]{2{,}744}$
26. $\sqrt[7]{16{,}384}$
27. $\sqrt{0.001764}$
28. $\sqrt[5]{79.62624}$

29. Round to hundredths.

$\sqrt[4]{38.2}$

30. Round to thousandths.

$\sqrt[10]{0.872}$

31. Round to tenths.

$\sqrt[3]{72{,}500}$

Find the following roots of powers.

32. $\sqrt{P^{10}}$
33. $\sqrt[3]{a^{-12}}$
34. $\sqrt[4]{t^{-8}}$
35. $\sqrt[6]{b^{30}}$
36. $\sqrt[5]{x^5}$
37. $\sqrt{m^{-2}}$
38. $\sqrt[3]{(3y)^{-15}}$
39. $\sqrt[4]{\left(\dfrac{y}{p}\right)^8}$

Assignment 20

Convert each power to a radical.

1. $8^{\frac{3}{4}}$ 2. $x^{\frac{1}{3}}$ 3. $a^{-\frac{3}{2}}$ 4. $10^{-\frac{2}{5}}$ 5. $p^{\frac{1}{2}}$

Convert each radical to a power with a fractional exponent.

6. $\sqrt[3]{R^2}$ 7. $\sqrt[8]{b}$ 8. $\sqrt{7}$ 9. $\sqrt[4]{k^{-5}}$ 10. $\sqrt{y^{-1}}$

Using the laws of exponents, write each answer as a single power with a fractional exponent.

11. $t^{\frac{1}{2}} \cdot t^{\frac{3}{4}}$ 12. $\dfrac{A^{\frac{5}{6}}}{A^{\frac{2}{3}}}$ 13. $\left(y^{\frac{1}{2}}\right)^{\frac{4}{3}}$ 14. $\left(p^{\frac{3}{8}}\right)^4$

Convert each radical to a power with a decimal exponent.

15. $\sqrt{x^3}$ 16. $\sqrt[4]{N}$ 17. $\sqrt[6]{r^{-3}}$ 18. $\sqrt[5]{p^{-8}}$

Convert each power to a radical.

19. $p^{0.5}$ 20. $y^{1.4}$ 21. $b^{-0.75}$ 22. $V^{-2.5}$

Using the laws of exponents, write each answer as a single power with a decimal exponent.

23. $y^{2.7} \cdot y^{0.9}$ 24. $\dfrac{w^{0.48}}{w^{-0.12}}$ 25. $(x^{0.15})^4$ 26. $(M^{2.5})^{-0.2}$

Use a calculator to convert each power to an ordinary number.

27. 2^{18} 28. 0.6^5 29. 20^{-3} 30. 4^{-2}

31. Round to the nearest whole number.

   $10^{2.38}$

32. Round to hundredths.

   $17.4^{0.52}$

33. Round to thousandths.

   $3.8^{-1.5}$

Use a calculator to evaluate each formula.

34. In $\boxed{P = b^t}$, find P when b = 50 and t = 1.1 . Round to tenths.

35. In $\boxed{Q = \left(\dfrac{w}{r}\right)^a}$, find Q when w = 540 , r = 130 , and a = 0.4 . Round to hundredths.

36. In $\boxed{F = CT^h}$, find F when C = 0.65 , T = 4.9 , and h = 2.5 . Round to tenths.

37. In $\boxed{A = P(1 + i)^n}$, find A when P = 10 , i = 0.14 , and n = 24 . Round to the nearest whole number.

38. In $\boxed{w = k(p + r)^{0.2}}$, find "w" when k = 25 , p = 47 , and r = 18 . Round to tenths.

Chapter 7 COMMON LOGARITHMS

In this chapter, we will discuss common (base 10) logarithms, logarithmic notation, and the laws of logarithms. Evaluations and rearrangements are performed with "log" formulas. Logarithmic scales, semi-log graphs, and log-log graphs are discussed. The log principle for equations is introduced and used to solve exponential equations and to do evaluations with exponential formulas.

7-1 POWERS OF TEN

In this section, we will discuss powers of ten with positive and negative whole-number exponents.

1. Any power of ten with a positive whole-number exponent is a short way of writing a multiplication of 10's. The <u>exponent</u> tells us how many 10's to use. For example:

    $10^2 = 10 \cdot 10$ (The exponent tells us to multiply <u>two</u> 10's.)

    $10^5 = 10 \cdot 10 \cdot 10 \cdot 10 \cdot 10$ (The exponent tells us to multiply <u>five</u> 10's.)

    Powers of ten with positive whole-number exponents equal ordinary numbers like 10 , 100 , 1,000 , and so on. For example:

    $10^1 = 10$

    $10^2 = 10 \cdot 10 = 100$

    $10^3 = 10 \cdot 10 \cdot 10 =$ _____

2. By examining the table below, you can see this fact: <u>The number of 0's in the ordinary number equals the exponent of the power of ten</u>.

    $10^6 = 1,000,000$
    $10^5 = 100,000$
    $10^4 = 10,000$
    $10^3 = 1,000$
    $10^2 = 100$
    $10^1 = 10$

    Using the above fact, convert these to an ordinary number.

    a) $10^7 =$ _____   b) $10^8 =$ _____

1,000

a) 10,000,000
b) 100,000,000

3. Using the number of 0's to determine the exponent, convert these to powers of ten.

   a) 100 = _____   b) 10,000,000 = _____

   a) 10^2 b) 10^7

4. By examining the pattern at the right, you can see that the following definition makes sense for 10^0.

   $10^3 = 1,000$
   $10^2 = 100$
   $10^1 = 10$
   $10^0 = 1$

   $\boxed{10^0 = 1}$

   Convert each power of ten to an ordinary number and each ordinary number to a power of ten.

   a) $10^0 =$ _____   d) $10^5 =$ _____

   b) $10,000 =$ _____   e) $10,000,000,000 =$ _____

   c) $10^1 =$ _____   f) $10^9 =$ _____

   a) 1
   b) 10^4
   c) 10
   d) 100,000
   e) 10^{10}
   f) 1,000,000,000

5. Using the definition of powers with negative whole-number exponents, we get:

   $10^{-1} = \dfrac{1}{10^1}$ $10^{-3} = \dfrac{1}{10^3}$ $10^{-5} = \dfrac{1}{10^5}$

   Because of the definition, powers of ten with negative whole-number exponents equal ordinary numbers like .1 , .01 , .001 , and so on. For example:

   $10^{-1} = \dfrac{1}{10^1} = \dfrac{1}{10} = .1$

   $10^{-2} = \dfrac{1}{10^2} = \dfrac{1}{100} = .01$

   $10^{-4} = \dfrac{1}{10^4} = \dfrac{1}{10,000} =$ _____

   .0001

6. By examining the table below, you can see this fact: The number of decimal places in each ordinary number equals the absolute value of the exponent.

   $10^{-1} = .1$
   $10^{-2} = .01$
   $10^{-3} = .001$
   $10^{-4} = .0001$
   $10^{-5} = .00001$
   $10^{-6} = .000001$

   Using the fact above, convert these to ordinary numbers.

   a) $10^{-7} =$ _____   b) $10^{-8} =$ _____

7. Using the number of decimal places to determine the absolute value of the exponent, convert these to powers of ten.

   a) .001 = _____   b) .000000001 = _____

   a) .0000001
   b) .00000001

8. Convert each power of ten to an ordinary number and each ordinary number to a power of ten.

   a) 10^{-1} = _____   d) .00001 = _____

   b) .01 = _____   e) 10^{-10} = _____

   c) 10^{-4} = _____   f) .00000001 = _____

   a) 10^{-3} b) 10^{-9}

   a) .1   c) .0001   e) .0000000001
   b) 10^{-2} d) 10^{-5} f) 10^{-8}

7-2 POWERS OF TEN WITH DECIMAL EXPONENTS

Any power of ten with a decimal exponent can be converted to an ordinary number, and any ordinary number can be converted to a power of ten. We will discuss the conversion methods in this section.

9. Some powers of ten can be converted to an ordinary number by converting to a radical. Use your calculator to complete these. Round to hundredths.

   a) $10^{0.5} = 10^{\frac{1}{2}} = \sqrt{10} = $ _____

   b) $10^{0.75} = 10^{\frac{3}{4}} = \sqrt[4]{10^3} = \sqrt[4]{1,000} = $ _____

   c) $10^{1.5} = 10^{\frac{3}{2}} = \sqrt{10^3} = \sqrt{1,000} = $ _____

10. However, we usually use the $\boxed{y^x}$ key to convert powers of ten with decimal exponents to ordinary numbers. Do these.

    a) $10^{1.7892} = $ _____   (Round to tenths.)

    b) $10^{-3.5168} = $ _____   (Round to millionths.)

    a) 3.16
    b) 5.62
    c) 31.62

11. $10^0 = 1$ and $10^1 = 10$. Therefore any power of ten with an exponent between 0 and 1 equals an ordinary number between 1 and 10. To show that fact, do these. Round to hundredths.

    a) $10^{0.35} = $ _____   b) $10^{0.7829} = $ _____

    a) 61.5
    b) 0.000304

    a) 2.24   b) 6.07

12. $10^1 = 10$ and $10^2 = 100$. Therefore, any power of ten with an exponent between 1 and 2 equals an ordinary number between 10 and 100. To show that fact, do these. Round to tenths.

    a) $10^{1.2} = $ _____   b) $10^{1.7091} = $ _____

---

13. a) Since $10^{2.68}$ lies between 10^2 and 10^3, $10^{2.68}$ equals a number between 100 and 1,000.

    To the nearest whole number, $10^{2.68} = $ _____

    b) Since $10^{5.3799}$ lies between 10^5 and 10^6, $10^{5.3799}$ equals a number between 100,000 and 1,000,000.

    To the nearest thousand, $10^{5.3799} = $ _____

a) 15.8   b) 51.2

---

14. Any power of ten with a <u>negative</u> decimal exponent equals an ordinary number between 0 and 1. To show that fact, do these.

    a) $10^{-1.3} = $ _____ (Round to four decimal places.)

    b) $10^{-3.85} = $ _____ (Round to six decimal places.)

    c) $10^{-5.2075} = $ _____ (Round to seven decimal places.)

a) 479

b) 240,000

---

15. To convert an ordinary number to a power of ten, we use the [log] key to find the exponent. We did so below for 475 and 0.0936.

    Note: [log] is an abbreviation for "logarithm" which means exponent.

Enter	Press	Display
475	[log]	2.6766936
0.0936	[log]	-1.0287242

    Rounding each exponent to four decimal places, we get:

    $475 = 10^{2.6767}$ $0.0936 = $ _____

a) 0.0501

b) 0.000141

c) 0.0000062

---

16. $100 = 10^2$ and $1,000 = 10^3$. Therefore, the exponent of the power-of-ten form of any number between 100 and 1,000 lies between 2 and 3. To show that fact, convert these to powers of ten. Round each exponent to four decimal places.

    a) $225 = $ _____   b) $879 = $ _____

$10^{-1.0287}$

---

a) $10^{2.3522}$ b) $10^{2.9440}$

Common Logarithms   305

306   Common Logarithms

17. $10{,}000 = 10^4$ and $100{,}000 = 10^5$. Therefore, the exponent of the power-of-ten form of any number between 10,000 and 100,000 lies between 4 and 5. To show that fact, convert these to powers of ten. Round each exponent to four decimal places.

   a) $35{,}700 = $ _____   b) $69{,}699 = $ _____

18. a) Since 5.49 lies between 1 and 10, the exponent of its power-of-ten form lies between 0 and 1.

   Rounding the exponent to four decimal places, $5.49 = $ _____

   b) Since 7,825 lies between 1,000 and 10,000, the exponent of its power-of-ten form lies between 3 and 4.

   Rounding the exponent to four decimal places, $7{,}825 = $ _____

a) $10^{4.5527}$ b) $10^{4.8432}$

19. The power-of-ten form of any number between 0 and 1 has a negative exponent. To show that fact, convert these to powers to ten. Round the exponents to four decimal places.

   a) $0.679 = $ _____   b) $0.000175 = $ _____

a) $10^{0.7396}$
b) $10^{3.8935}$

20. Use the $\boxed{\text{log}}$ key to try to find the exponent of the power-of-ten form of the negative numbers below.

   $-275 = $ _____ $-9.63 = $ _____

   Since negative numbers cannot be converted to power-of-ten form, finding that form for a negative number is an IMPOSSIBLE operation. A calculator shows that fact with a flashing display, by printing out "Error", or in some other way.

a) $10^{-0.1681}$
b) $10^{-3.7570}$

7-3   GRAPHS OF BASE-10 EXPONENTIAL FUNCTIONS

In this section, we will discuss the graphs of two base-10 exponential functions, $y = 10^x$ and $y = 10^{-x}$.

21. For $\boxed{y = 10^x}$, we can mentally find the values of y corresponding to whole-number values of x. For example:

   If $x = 2$, $y = 10^2 = 100$ If $x = -3$, $y = 10^{-3} = .001$

   For the same function, find y for these values of x.

   a) If $x = 0$, $y = $ _____   c) If $x = -1$, $y = $ _____

   b) If $x = 3$, $y = $ _____   d) If $x = -2$, $y = $ _____

22. For $\boxed{y = 10^x}$, we use the $\boxed{y^x}$ key to find y when x is a decimal number. Complete these.

   a) If $x = 1.25$, $y =$ _____ (Round to tenths.)

   b) If $x = 0.4$, $y =$ _____ (Round to hundredths.)

   c) If $x = -2.5$, $y =$ _____ (Round to five decimal places.)

   a) y = 1
   b) y = 1,000
   c) y = .1
   d) y = .01

23. We found some pairs of values for $\boxed{y = 10^{-x}}$ below. Notice that the "–" in front of x means that <u>we use the opposite of the value substituted for x</u>.

   If $x = 1$, $y = 10^{-1} = .01$
   If $x = -2$, $y = 10^2 = 100$

   Using the same steps, complete these.

   a) If $x = 0$, $y = 10^{-0} = 10^0 =$ _____

   b) If $x = 2$, $y = 10^{-2} =$ _____

   c) If $x = -1$, $y = 10^1 =$ _____

   a) 17.8
   b) 2.51
   c) 0.00316

24. For $\boxed{y = 10^{-x}}$, use the $\boxed{y^x}$ key to complete these.

   a) If $x = 1.2$, $y = 10^{-1.2} =$ _____ (Round to thousandths.)

   b) If $x = -0.4$, $y = 10^{0.4} =$ _____ (Round to hundredths.)

   a) 1
   b) .01
   c) 10

25. Some tables for $y = 10^x$ and $y = 10^{-x}$ are shown below. To aid in graphing, we rounded all y-values to tenths.

   a) .063
   b) 2.51

   $\boxed{y = 10^x}$

x	y
-1.0	0.1
-0.5	0.3
-0.3	0.5
0.0	1.0
0.2	1.6
0.5	3.2
0.7	5.0
0.9	7.9
1.0	10.0

   $\boxed{y = 10^{-x}}$

x	y
-1.0	10.0
-0.9	7.9
-0.7	5.0
-0.5	3.2
-0.2	1.6
0.0	1.0
0.3	0.5
0.5	0.3
1.0	0.1

Continued on following page.

25. Continued

Using these values, we graphed each function below.

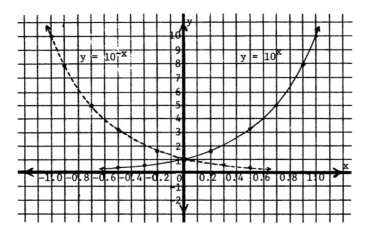

Notice these points about the graphs:

1. The graph of $y = 10^x$ is called an <u>ascending</u> exponential curve.
   The graph of $y = 10^{-x}$ is called a <u>descending</u> exponential curve.

2. Since powers of ten <u>always</u> stand for positive numbers, the graphs appear only in Quadrants 1 and 2.

3. The y-intercept for both graphs is (0,1) because $y = 10^0$ or 1 when $x = 0$.

7-4 COMMON LOGARITHMS

We will define common logarithms and discuss logarithmic notation in this section. The discussion will be limited to whole-number and "0" logarithms.

26. The exponent of the power-of-ten form of a number is called the <u>common logarithm</u> (or simply <u>logarithm</u>) of the number.

   Since $1 = 10^0$, the logarithm of 1 is 0.

   Since $10 = 10^1$, the logarithm of 10 is 1.

   Since $100 = 10^2$, the logarithm of 100 is _____.

27. The logarithm of a number is <u>only the exponent</u> of the power of ten. It does not include the base "10". That is:

   Since $10,000 = 10^4$: the logarithm of 10,000 is 4.
   (the logarithm <u>is not</u> 10^4)

   $100,000 = 10^5$ Is 5 or 10^5 the logarithm of 100,000? _____

2

28. By examining the table below, you can see this fact: <u>The logarithm equals the number of 0's in the ordinary number</u>.

NUMBER	LOGARITHM
1,000,000	6
100,000	5
10,000	4
1,000	3
100	2
10	1
1	0

Using the above fact, complete these.

a) The logarithm of 10,000,000 is _____.

b) The logarithm of 100,000,000 is _____.

5

---

29. The logarithm of any number between 0 and 1 is a negative number.

Since $.1 = 10^{-1}$, the logarithm of .1 is -1.

Since $.01 = 10^{-2}$, the logarithm of .01 is -2.

Since $.001 = 10^{-3}$, the logarithm of .001 is _____.

a) 7

b) 8

---

30. By examining the table below, you can see this fact: <u>The absolute value of the logarithm equals the number of decimal places in the ordinary number</u>.

NUMBER	LOGARITHM
.1	−1
.01	−2
.001	−3
.0001	−4
.00001	−5
.000001	−6

Using the above fact, complete these.

a) The logarithm of .0000001 is _____.

b) The logarithm of .00000001 is _____.

−3

---

31. The phrase "<u>the logarithm of</u>" is usually abbreviated to "<u>log</u>". For example:

log 1,000 means: the logarithm of 1,000

Using the fact above, complete these.

a) log 100,000 = _____   c) log .01 = _____

b) log 1 = _____   d) log .0001 = _____

a) −7

b) −8

Common Logarithms 309

310   Common Logarithms

32. Following the example, complete these.

   Since log 100 = 2 , 100 = 10^2

   a) Since log 1,000 = 3 , 1,000 = $10^{\boxed{}}$

   b) Since log .001 = -3 , .001 = $10^{\boxed{}}$

   a) 5   c) -2
   b) 0   d) -4

33. The logarithm of a number is 4. The number is 10^4 or 10,000.

   a) The logarithm of a number is 1. The number is _____ or _____.

   b) The logarithm of a number is -1. The number is _____ or _____.

   c) The logarithm of a number is -4. The number is _____ or _____.

   a) 10^3
   b) 10^{-3}

34. Any basic power-of-ten equation can be written as a "log" equation. For example:

   $100 = 10^2$ can be written: log 100 = 2
   $10^{-3} = .001$ can be written: -3 = log .001

   Write each equation as a "log" equation.

   a) $.1 = 10^{-1}$ b) $1 = 10^0$ c) $10^4 = 10,000$

   a) 10^1 or 10
   b) 10^{-1} or .1
   c) 10^{-4} or .0001

35. Any basic "log" equation can be written as a power-of-ten equation. For example:

   log 1,000 = 3   can be written: $1,000 = 10^3$
   -2 = log .01   can be written: $10^{-2} = .01$

   Write each equation as a power-of-ten equation.

   a) log 100,000 = 5   b) 1 = log 10   c) log .0001 = -4

   a) log .1 = -1
   b) log 1 = 0
   c) 4 = log 10,000

   a) $100,000 = 10^5$
   b) $10^1 = 10$
   c) $.0001 = 10^{-4}$

Common Logarithms 311

7-5 DECIMAL LOGARITHMS

In this section, we will discuss decimal logarithms and show that the same notation is used with them.

36. To find the <u>logarithm</u> of a number on a calculator, enter the number and press the [log] key. Do these. Round to four decimal places. a) The logarithm of 761,000 is _____ . b) The logarithm of 0.0829 is _____ .	
37. log 1 = 0 and log 10 = 1. Therefore the logarithm of any number between 1 and 10 is a number between 0 and 1. To show that fact, do these. Round to four decimal places. a) log 6 = _____ b) log 2.76 = _____	a) 5.8814 b) -1.0814
38. log 10 = 1 and log 100 = 2. Therefore the logarithm of any number between 10 and 100 is a number between 1 and 2. To show that fact, do these. Round to four decimal places. a) log 27 = _____ b) log 78.3 = _____	a) 0.7782 b) 0.4409
39. a) 375 lies between 100 and 1,000. Therefore its logarithm lies between 2 and 3 since log 100 = 2 and log 1,000 = 3. Rounding to four decimal places, log 375 = _____ . b) 836,000 lies between 100,000 and 1,000,000. Therefore its logarithm lies between 5 and 6 since log 100,000 = 5 and log 1,000,000 = 6. Rounding to four decimal places, log 836,000 = _____	a) 1.4314 b) 1.8938
40. The logarithm of any number between 0 and 1 is negative. To show that fact, do these. Round to four decimal places. a) log 0.25 = _____ b) log 0.0829 = _____	a) 2.5740 b) 5.9222
41. A logarithm is <u>the exponent of the power-of-ten form</u> of a number. Therefore: Since log 27.9 = 1.4456, 27.9 = $10^{1.4456}$ Since log 0.55 = -0.2596, 0.55 = $10^{-0.2596}$ Therefore, when given the logarithm of a number, we can write its power-of-ten form. For example: If log N = 2.4099, N = $10^{2.4099}$ If log N = -1.6387, N = _____	a) -0.6021 b) -1.0814

42. If the logarithm of a number is 2.7959 :

   a) In power-of-ten form, the number is _____ .

   b) Rounded to the nearest whole number, the number is _____ .

43. If the logarithm of a number is -0.7622 :

   a) In power-of-ten form, the number is _____ .

   b) Rounded to thousandths, the number is _____ .

44. a) If the logarithm of a number is 0.8549 , the number is _____ .
       (Round to hundredths.)

   b) If the logarithm of a number is -2.0633 , the number is _____ .
       (Round to five decimal places.)

45. To find N in log N = 1.9238 , we must find the number whose logarithm is 1.9238 . That process is called "finding the antilogarithm" or "finding the inverse logarithm". To find the inverse logarithm, we can use [INV] [log] on a calculator. Use [INV] [log] to complete these.

   a) log N = 1.9238   Rounded to tenths, N = _____

   b) log N = -0.56   Rounded to thousandths, N = _____

46. When converting from a logarithm to an ordinary number, the information in the following table is helpful.

LOGARITHM	NUMBER
NEGATIVE	BETWEEN 0 AND 1
0 TO 1	BETWEEN 1 AND 10
1 TO 2	BETWEEN 10 AND 100
2 TO 3	BETWEEN 100 AND 1,000
3 TO 4	BETWEEN 1,000 AND 10,000
4 TO 5	BETWEEN 10,000 AND 100,000
5 TO 6	BETWEEN 100,000 AND 1,000,000

   Using the facts in the table, complete these.

   If log N = 2.6033 , N is a number between 100 and 1,000.

   a) If log N = 0.9655 , N is a number between _____ and _____ .

   b) If log N = 5.2471 , N is a number between _____ and _____ .

   c) If log N = -1.9127 , N is a number between _____ and _____ .

$10^{-1.6387}$

a) $10^{2.7959}$

b) 625

a) $10^{-0.7622}$

b) 0.173

a) 7.16 , from: $10^{0.8549}$

b) 0.00864 , from: $10^{-2.0633}$

a) 83.9

b) 0.275

Common Logarithms   313

47. Negative numbers and "0" do not have logarithms. Therefore, finding the logarithm of a negative number or "0" is an IMPOSSIBLE operation. A calculator shows that fact with a flashing display, by printing out "Error", or in some other way. Try these on a calculator.

$$\log(-155) \qquad \log 0$$

Note: When you are using a calculator for a "log" problem and the calculator display shows an IMPOSSIBLE operation, you know that you have made a mistake.

a) 1 and 10
b) 100,000 and 1,000,000
c) 0 and 1

48. Any basic power-of-ten equation can be written as a "log" equation. For example:

$$56.2 = 10^{1.7497} \quad \text{can be written:} \quad \log 56.2 = 1.7497$$

$$10^{-0.6925} = 0.203 \quad \text{can be written:} \quad -0.6925 = \log 0.203$$

Write each equation as a "log" equation.

a) $0.0097 = 10^{-2.0132}$

b) $10^{5.8136} = 651,000$

49. Any basic "log" equation can be converted to a basic power-of-ten equation. For example:

$$\log 299 = 2.4757 \quad \text{can be written:} \quad 299 = 10^{2.4757}$$

$$-0.1871 = \log 0.65 \quad \text{can be written:} \quad 10^{-0.1871} = 0.65$$

Write each equation as a power-of-ten equation.

a) $\log 0.017 = -1.7696$

b) $4.5846 = \log 38,420$

a) $\log 0.0097 = -2.0132$
b) $5.8136 = \log 651,000$

a) $0.017 = 10^{-1.7696}$ b) $10^{4.5846} = 38,420$

7-6   GRAPH OF THE LOGARITHMIC FUNCTION

In this section, we will discuss the graph of $y = \log x$.

50. The equation $\boxed{y = \log x}$ is called the logarithmic function. Let's find some pairs of values for that function.

If $x = 10$, $y = \log 10 = 1$

If $x = .1$, $y = \log .1 = -1$

a) If $x = 100$, $y = \log 100 = $ _____

b) If $x = .01$, $y = \log .01 = $ _____

314   Common Logarithms

51. Some pairs of values for y = log x are given in the table at the left below. Using these pairs of values, we graphed the function at the right.

a) 2
b) -2

y = log x	
x	y
0	-----
0.1	-1.0
0.2	-0.70
0.3	-0.52
0.5	-0.30
0.8	-0.10
1.0	0
1.5	0.18
2.0	0.30
2.5	0.40
3.0	0.48
4.0	0.60
5.0	0.70
6.0	0.78
8.0	0.90
10.0	1.0

Notice these points about the graph:

1. It is called a logarithmic graph.

2. Since there are no logarithms for "0" and negative numbers, x is always positive. Therefore the graph appears only in Quadrants 1 and 4.

3. There is no y-intercept. The x-intercept is (1, 0) because y = 0 when x = 1, since log 1 = 0.

SELF-TEST 21 (pages 302-315)

1. Convert 10^{-5} to an ordinary number.

2. Convert 10,000 to a power of ten.

3. Convert $10^{1.7319}$ to an ordinary number. Round to tenths.

4. Convert 0.00163 to a power of ten. Round to four decimal places.

Continued on following page.

Common Logarithms 315

SELF-TEST 21 (Continued)

5. For the function $y = 10^x$, find y when $x = -0.86$. Round to thousandths.

6. If the logarithm of a number is 6, write the number in power-of-ten form.

Find each logarithm. Round to four decimal places.

7. log 9.24

8. log 14,530

9. log 0.0674

10. If the logarithm of a number is -2.1738, write the number in power-of-ten form.

11. Find P. Round to thousands.
    log P = 5.4910

12. Find G. Round to four decimal places.
    log G = -1.2793

13. In which of the following is N a number lying between 1 and 10?

    a) log N = 2   b) log N = 0.73   c) log N = -0.46   d) log N = 1.67   e) log N = 0.10

14. Write this power-of-ten equation as a "log" equation.
    $10^{2.5038} = 319$

15. Write this "log" equation as a power-of-ten equation.
    log 0.704 = -0.1524

16. For the function $y = \log x$, find y when $x = 1$.

ANSWERS:
1. .00001
2. 10^4
3. 53.9
4. $10^{-2.7878}$
5. y = 0.138
6. 10^6
7. 0.9657
8. 4.1623
9. -1.1713
10. $10^{-2.1738}$
11. P = 310,000
12. G = 0.0526
13. (b) and (e)
14. 2.5038 = log 319
15. 0.704 = $10^{-0.1524}$
16. y = 0

7-7 SOLVING POWER-OF-TEN AND "LOG" EQUATIONS

In this section, we will discuss the methods for solving power-of-ten and "log" equations that contain a letter.

52. In each power-of-ten equation below, the letter stands for an ordinary number. Use the y^x key to solve each equation.

    a) Round to tenths.

    $10^{1.6294} = x$

    x = _____

    b) Round to five decimal places.

    $y = 10^{-2.1095}$

    y = _____

316   Common Logarithms

53. In each power-of-ten equation below, the letter stands for an exponent. Use the ⎡log⎤ key to solve each equation. Round each exponent to four decimal places.

    a) $0.016 = 10^x$ b) $10^b = 81{,}900$

    x = _____   b = _____

a) $x = 42.6$
b) $y = 0.00777$

54. Any "log" equation containing a letter can be written in power-of-ten form. For example:

    $\log t = 1.4075$ can be written $t = 10^{1.4075}$

    $x = \log 0.714$ can be written $10^x = 0.714$

Write each equation as a power-of-ten equation.

    a) $-1.5277 = \log d$ b) $\log 803 = a$

a) $x = -1.7959$
b) $b = 4.9133$

55. When a "log" equation contains a letter, we can solve it by writing the equation in power-of-ten form. Let's use that method to solve the equation below.

    $\log V = 2.7427$

    a) Write the equation in power-of-ten form. _____

    b) Therefore, rounded to the nearest whole number, V = _____

a) $10^{-1.5277} = d$
b) $803 = 10^a$

56. Let's solve $\log 0.456 = y$.

    a) Write the equation in power-of-ten form. _____

    b) Therefore, rounded to four decimal places, y = _____

a) $V = 10^{2.7427}$
b) $V = 553$

57. Let's solve $-1.7086 = \log x$.

    a) Write the equation in power-of-ten form. _____

    b) Therefore, rounded to four decimal places, x = _____

a) $0.456 = 10^y$
b) $y = -0.3410$

58. Let's solve $m = \log 3{,}190$.

    a) Write the equation in power-of-ten form. _____

    b) Therefore, rounded to four decimal places, m = _____

a) $10^{-1.7086} = x$
b) $x = 0.0196$

Common Logarithms   317

59. When the letter is on the opposite side of "log" in an equation, it stands for the logarithm (or exponent). For example:

In log 3.87 = d , "d" stands for the logarithm since: $3.87 = 10^d$

In h = log 0.025 , "h" stands for the logarithm since: $10^h = 0.025$

In such cases, we can simply use the [log] key without converting to a power-of-ten equation. Solve these. Round to four decimal places.

a) log 0.051 = R

b) G = log 995,000

R = _____

G = _____

a) $10^m = 3,190$
b) m = 3.5038

---

60. When the letter follows "log" in an equation, it stands for the ordinary number. For example:

In log x = -1.53 , "x" stands for the ordinary number since: $x = 10^{-1.53}$

In 2.25 = log y , "y" stands for the ordinary number since: $10^{2.25} = y$

In such cases, we can convert the "log" equation to a power-of-ten equation and use [y^x] . Solve these.

a) Round to tenths.

log t = 1.7539

t = _____

b) Round to thousandths.

-0.49 = log m

m = _____

a) R = -1.2924 , from:

$0.051 = 10^R$

b) G = 5.9978 , from:

$10^G = 995,000$

---

61. In each equation below, N stands for the ordinary number. Instead of converting to power-of-ten form, we can find N directly on a calculator by using [INV] [log] .

a) log N = 6.37   Rounding to thousands, N = _____ .

b) -0.207 = log N   Rounding to thousandths, N = _____ .

a) t = 56.7 , from:

$t = 10^{1.7539}$

b) m = 0.324 , from:

$10^{-0.49} = m$

a) N = 2,344,000   b) N = 0.621

7-8 EVALUATING "LOG" FORMULAS

In this section, we will show how a calculator can be used to evaluate formulas containing "log" expressions.

62. The formula below contains a "log" expression.

$$D = 10 \log R$$

To show that there are two factors on the right side, we put them in parentheses below.

$$D = (10)(\log R)$$

Continued on following page.

318   Common Logarithms

62. Continued

To find D when R = 752, we multiply 10 times log 752. The steps are shown. Notice that we entered 752 and then pressed $\boxed{\log}$ before pressing $\boxed{=}$. We rounded the answer to tenths.

Enter	Press	Display	
10	$\boxed{x}$	10.	
752	$\boxed{\log}$ $\boxed{=}$	28.762178	(D = 28.8)

Use a calculator for these. Round to tenths in (a) and to the nearest whole number in (b).

a) In $\boxed{D = 10 \log R}$, when R = 0.0386, D = _____

b) In $\boxed{H = w \log T}$, when w = 100 and T = 32.5, H = _____

a) -14.1

b) 151

63. The steps for the evaluation below are shown. Notice again that we entered the value for Q and then pressed $\boxed{\log}$. We rounded the answer to tenths.

In $\boxed{P = A - K \log Q}$, find P when A = 59, K = 19, and Q = 0.215.

Enter	Press	Display	
59	$\boxed{-}$	59.	
19	$\boxed{x}$	19.	
0.215	$\boxed{\log}$ $\boxed{=}$	71.683669	(P = 71.7)

Use a calculator for this one. Round to tenths.

In $\boxed{P = A - K \log Q}$, when A = 94, K = 30, and Q = 3.9,

P = _____

76.3

64. In the formula below, $\log\left(\dfrac{P_2}{P_1}\right)$ is the log of a fraction or division. Two calculator methods are shown. We rounded the answer to hundredths.

In $\boxed{D = 10 \log\left(\dfrac{P_2}{P_1}\right)}$, find D when P_2 = 750 and P_1 = 4,875.

1) If the calculator has parentheses symbols, we can use them to evaluate $\dfrac{P_2}{P_1}$ before pressing $\boxed{\log}$.

Enter	Press	Display	
10	$\boxed{x}$ $\boxed{(}$	10.	
750	$\boxed{\div}$	750.	
4,875	$\boxed{)}$ $\boxed{\log}$ $\boxed{=}$	-8.1291336	(D = -8.13)

Continued on following page.

64. Continued

2) If a calculator does not have parentheses symbols, we must evaluate $\log\left(\frac{P_2}{P_1}\right)$ first and then multiply by 10. Notice that we pressed $\boxed{=}$ to complete the division before pressing $\boxed{\log}$.

Enter	Press	Display	
750	$\boxed{\div}$	750.	
4,875	$\boxed{=}$ $\boxed{\log}$ $\boxed{\times}$	-.81291336	
10	$\boxed{=}$	-8.1291336	(D = -8.13)

Use a calculator for this one. Round to hundredths.

In $\boxed{M = 2.5 \log\left(\frac{I_1}{I}\right)}$, when $I_1 = 79.3$ and $I = 16.4$, M = _____

---

65. In the formula below, log(W + H) is the log of an addition. Two calculator methods are shown. We rounded the answer to hundredths.

In $\boxed{B = K \log(W + H)}$, find B when K = 2.79, W = 18.3, and H = 41.9.

1) If a calculator has parentheses symbols, we can use them to evaluate (W + H) before pressing $\boxed{\log}$.

Enter	Press	Display	
2.79	$\boxed{\times}$ $\boxed{(}$	2.79	
18.3	$\boxed{+}$	18.3	
41.9	$\boxed{)}$ $\boxed{\log}$ $\boxed{=}$	4.9650742	(B = 4.97)

2) If a calculator does not have parentheses symbols, we must evaluate log(W + H) first and then multiply by the value of K. Notice that we pressed $\boxed{=}$ to complete the addition before pressing $\boxed{\log}$.

Enter	Press	Display	
18.3	$\boxed{+}$	18.3	
41.9	$\boxed{=}$ $\boxed{\log}$ $\boxed{\times}$	1.7795965	
2.79	$\boxed{=}$	4.9650742	(B = 4.97)

Use a calculator for this one. Round to hundredths.

In $\boxed{G = 1.75 \log(a + b)}$, when a = 3.25 and b = 9.87,

G = _____

1.71

1.96

320 Common Logarithms

66. In $\boxed{P_H = -\log A_H}$, "$-\log A_H$" means "the <u>opposite</u> of log A_H". Therefore, after finding log A_H, we press $\boxed{+/-}$ to get its opposite. An example is given. We rounded the answer to hundredths.

In $\boxed{P_H = -\log A_H}$, find P_H when $A_H = 0.000095$.

<u>Enter</u>   <u>Press</u>   <u>Display</u>
0.000095 $\boxed{\log}$ $\boxed{+/-}$ 4.0222764   ($P_H = 4.02$)

Using the same steps, do this one. Round to hundredths.

In $\boxed{P_H = -\log A_H}$, when $A_H = 0.000061$, $P_H = $ _____

67. A two-step process is needed to find "R" in the evaluation below. We rounded the answer to tenths.

In $\boxed{\log R = \dfrac{D}{10}}$, find R when $D = 17.5$.

1) First we find "log R" by substituting.

$\log R = \dfrac{D}{10} = \dfrac{17.5}{10} = 1.75$

2) Then we use $\boxed{y^x}$ to find R. The steps are:

<u>Enter</u>   <u>Press</u>   <u>Display</u>
10       $\boxed{y^x}$ 10.
1.75 $\boxed{=}$ 56.234133   (R = 56.2)

Note: We could also use $\boxed{INV}$ $\boxed{\log}$ to find R above.

Use the same steps for this one. Round to the nearest whole number.

In $\boxed{\log R = \dfrac{D}{10}}$, when $D = 28.9$, $R = $ _____

4.21

776, since: log R = 2.89

7-9 EVALUATIONS REQUIRING EQUATION-SOLVING

In this section, we will discuss evaluations with "log" formulas that require solving an equation.

68. In the formula below, we found R when D = 16.5 . We used a calculator for the last step. Using the same steps, find R when D = -16.3 . Round to ten-thousandths.

$\boxed{D = 10 \log R}$ $\boxed{D = 10 \log R}$

16.5 = 10 log R

$\frac{16.5}{10}$ = log R

1.65 = log R

R = 44.7

69. We started two evaluations with $\boxed{H = w \log T}$ below. We used a calculator to find log 25.4 at the left. Use a calculator to complete each solution.

R = 0.0234

a) Find "w" when H = 130 and T = 25.4 . Round to tenths.

H = w log T

130 = w log 25.4

130 = w(1.4048)

w = $\frac{130}{1.4048}$

w = _____

b) Find T when H = 186 and w = 100 . Round to tenths.

H = w log T

186 = 100 log T

log T = $\frac{186}{100}$

log T = _____

T = _____

70. We started two evaluations with $\boxed{P = A - K \log Q}$ below. We used a calculator to find log 12.9 at the right. Use a calculator to complete each evaluation.

a) w = 92.5

b) log T = 1.86
   T = 72.4

a) Find Q when P = 700 , A = 550 , and K = 180 . Round log Q to four decimal places and Q to thousandths.

P = A - K log Q

700 = 550 - 180 log Q

150 = -180 log Q

log Q = $\frac{150}{-180}$

log Q = _____

Q = _____

b) Find A when P = 41.3 , K = 9.74 , and Q = 12.9 . Round to tenths.

P = A - K log Q

41.3 = A - 9.74(log 12.9)

41.3 = A - 9.74(1.1106)

41.3 = A - 10.8

A = _____

322   Common Logarithms

71. In the formula below, we want to find K when W = 13.8, H = 16.4, and B = 4.98. Notice that we added 13.8 and 16.4 and then used a calculator to find log 30.2. Use a calculator to complete the solution. Round to hundredths.

$$\boxed{K \log (W + H) = B}$$

$$K \log(13.8 + 16.4) = 4.98$$
$$K \log 30.2 = 4.98$$
$$K(1.48) = 4.98$$
$$K = \frac{4.98}{1.48}$$
$$K = \underline{\qquad}$$

a) log Q = -0.8333
   Q = 0.147

b) A = 52.1

72. In $\boxed{P_H = -\log A_H}$, log A_H is not solved for because it has a "-" in front of it. Therefore to solve for log A_H at the left below, we took the opposite of both sides. Complete the other evaluation. Round to four decimal places.

Find A_H when $P_H = 2.84$.   Find A_H when $P_H = 1.53$.

$P_H = -\log A_H$ $P_H = -\log A_H$
$2.84 = -\log A_H$
$-2.84 = \log A_H$
$A_H = 0.00145$

K = 3.36

$A_H = 0.0295$

7-10 EVALUATIONS REQUIRING A CONVERSION TO POWER-OF-TEN FORM

In this section, we will discuss evaluations that require converting the log equation to power-of-ten form.

73. In each equation below, the variable is part of the "log" expression.

$$\log\left(\frac{P_2}{200}\right) = 0.75 \qquad \log(W + 12.6) = 1.98$$

To solve for the variable in each case, we must convert the log equation to power-of-ten form. That is:

$\log\left(\frac{P_2}{200}\right) = 0.75$ is converted to: $\frac{P_2}{200} = 10^{0.75}$

$\log(W + 12.6) = 1.98$ is converted to: $W + 12.6 = 10^{1.98}$

Convert each log equation to power-of-ten form.

a) $\log\left(\frac{400}{P_1}\right) = 1.51$ b) $\log(17.8 + H) = -1.45$

Common Logarithms 323

74. To solve each equation below, we converted to power-of-ten form and then replaced the power of ten with an ordinary number. Complete each solution.

a) $\log\left(\dfrac{T}{1,200}\right) = 0.64$

$\dfrac{T}{1,200} = 10^{0.64}$

$\dfrac{T}{1,200} = 4.37$

T = _____

b) $\log(R + 5.88) = 1.71$

$R + 5.88 = 10^{1.71}$

$R + 5.88 = 51.29$

R = _____

a) $\dfrac{400}{P_1} = 10^{1.51}$

b) $17.8 + H = 10^{-1.45}$

---

75. In $\boxed{D = 10 \log\left(\dfrac{P_2}{P_1}\right)}$, let's find P_2 when $D = 27.5$ and $P_1 = 600$.

Substituting the known values and simplifying, we get:

$27.5 = 10 \log\left(\dfrac{P_2}{600}\right)$

$\dfrac{27.5}{10} = \log\left(\dfrac{P_2}{600}\right)$

$\log\left(\dfrac{P_2}{600}\right) = 2.75$

Convert to power-of-ten form and complete the solution. When converting the power of ten to an ordinary number, round to the nearest whole number.

a) T = 5,244

b) R = 45.41

---

76. In $\boxed{K \log(W + H) = B}$, let's find H when $K = 13.4$, $W = 12.2$, and $B = 15.9$.

Substituting the known values and simplifying, we get:

$13.4 \log(12.2 + H) = 15.9$

$\log(12.2 + H) = \dfrac{15.9}{13.4}$

$\log(12.2 + H) = 1.1866$

Convert to power-of-ten form and complete the solution. When converting the power of ten to an ordinary number, round to tenths.

$\dfrac{P_2}{600} = 10^{2.75}$

$\dfrac{P_2}{600} = 562$

$P_2 = 600(562)$

$P_2 = 337,200$

$12.2 + H = 10^{1.1866}$

$12.2 + H = 15.4$

$H = 3.2$

7-11 REARRANGING "LOG" FORMULAS

In this section, we will show how a "log" formula can be rearranged to solve for a variable.

77. We solved for log R below. Solve for $\log\left(\frac{P_2}{P_1}\right)$ in the other formula.

$$D = 10 \log R \qquad D = 20 \log\left(\frac{P_2}{P_1}\right)$$

$$\frac{D}{10} = \log R$$

$$\log R = \frac{D}{10}$$

78. When rearranging a formula, any "log" expression is treated as a single quantity. For example, log(W + H) below should be treated as a single quantity.

$$\boxed{k \log(W + H) = B}$$

a) Solve for log(W + H).   b) Solve for "k".

$\log\left(\frac{P_2}{P_1}\right) = \frac{D}{20}$

79. To solve for log A below, we replaced each side with its opposite. Solve for log t in the other formula.

$$P = -\log A \qquad V = -\log t$$

$$-P = \log A$$

$$\log A = -P$$

a) $\log(W + H) = \frac{B}{k}$

b) $k = \frac{B}{\log(W + H)}$

80. To solve for "k" below, we isolated k log T first. Solve for log F in the other formula.

$$R = V + k \log T \qquad D = C + k \log F$$

$$R - V = k \log T$$

$$k = \frac{R - V}{\log T}$$

$\log t = -V$

$\log F = \frac{D - C}{k}$

Common Logarithms 325

81. To solve for log Q below, we isolated −k log Q and then replaced each side with its opposite. Solve for "k" in the other formula.

$$P = A - k \log Q \qquad S = T - k \log M$$

$$P - A = -k \log Q$$

$$A - P = k \log Q$$

$$\log Q = \frac{A - P}{k}$$

82. To solve for A below, we added k log Q to both sides.

$$P = A - k \log Q$$

$$P + k \log Q = A - k \log Q + k \log Q$$

$$P + k \log Q = A + 0$$

$$A = P + k \log Q$$

Solve for D in the formula below.

$$F = D + k \log B$$

$k = \dfrac{T - S}{\log M}$

83. When evaluating to find the value of a non-solved-for variable in a "log" formula, we can use either of two methods. For example:

In $\boxed{V = T + k \log R}$, find "k" when: $\begin{array}{l} V = 200 \\ T = 150 \\ R = 125 \end{array}$

Method 1: Substituting in the formula as it stands and then solving for "k".

$$200 = 150 + k \log 125$$

$$50 = k(2.0969)$$

$$k = \frac{50}{2.0969} = 23.8$$

Method 2: Rearranging to solve for "k" before substituting.

$$k = \frac{V - T}{\log R} = \frac{200 - 150}{\log 125} = \frac{50}{2.0969} = 23.8$$

Did we get the same value for "k" with both methods? _____

D = F − k log B , from:

D = F + (−k log B)

Yes

SELF-TEST 22 (pages 315-326)

1. Find "t". Round to hundredths.

    $10^t = 1{,}520$

2. Find F. Round to thousandths.

    $-1.48 = \log F$

3. In $\boxed{G = 20 \log\left(\dfrac{A}{K}\right)}$, find G when $A = 628$ and $K = 196$. Round to tenths.

4. In $\boxed{W = -\log T}$, find W when $T = 0.000045$. Round to hundredths.

5. In $\boxed{P = a \log R}$, find "a" when $P = 5.27$ and $R = 21.8$. Round to hundredths.

6. In $\boxed{S = B + C \log V}$, find V when $S = 52$, $B = 19$, and $C = 25$. Round to tenths.

7. In $\boxed{A = \log\left(\dfrac{F}{G}\right)}$, find G when $A = 2.38$ and $F = 71.2$. Round to thousandths.

8. Solve for "log W".

    $\boxed{C = -\log W}$

9. Solve for M.

    $\boxed{H = A - M \log B}$

10. Solve for "log(a + b)".

    $\boxed{t = p + r \log(a + b)}$

ANSWERS:
1. $t = 3.18$
2. $F = 0.033$
3. $G = 10.1$
4. $W = 4.35$
5. $a = 3.94$
6. $V = 20.9$
7. $G = 0.297$
8. $\log W = -C$
9. $M = \dfrac{A - H}{\log B}$
10. $\log(a + b) = \dfrac{t - p}{r}$

7-12 LOGARITHMIC SCALES

In this section, we will discuss logarithmic scales and give practice in reading them.

84. The scale below is called a "logarithmic" scale. It is not a uniform scale. That is, the divisions do not have the same lengths. For example, the distance between 1 and 2 is longer than the distance between 2 and 3 .

On a logarithmic scale, the logarithms of the numbers are plotted rather than the numbers themselves. To show that fact, we can use the table below to generate the above scale.

log 1 = 0	log 6 = 0.778
log 2 = 0.301	log 7 = 0.845
log 3 = 0.477	log 8 = 0.903
log 4 = 0.602	log 9 = 0.954
log 5 = 0.699	log 10 = 1

Since the logarithms of the ordinary numbers from 1 to 10 range from 0.00 to 1.00 , we begin by plotting those logarithms on a scale from 0 to 1 . We get:

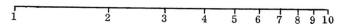

Then we can redraw the scale with the ordinary numbers from 1 to 10 plotted where their logarithms lie. We get:

The new scale is identical to the original scale above. It contains whole numbers plotted where their logarithms lie on a scale from 0 to 1 .

85. We usually need more values than 1 to 10 on a log scale. To get more values, we use a series of compressed log scales called "cycles". The dividing points for cycles are numbers whose logarithms are whole numbers. For example:

The scale below is a two-cycle log scale. The dividing points are 1 , 10 , and 100 whose logarithms are 0, 1, and 2 respectively.

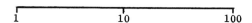

Continued on following page.

328   Common Logarithms

85. Continued

The scale below is a three-cycle log scale. The dividing points are 1, 10, 100, and 1,000 whose logarithms are 0, 1, 2, and 3 respectively.

```
├────────┬────────┬────────┐
1       10      100    1,000
```

How many cycles are there in the scale below? _____

```
├────────┬────────┬────────┬────────┐
1       10      100    1,000   10,000
```

86. A log scale does not have to begin at "1". It can begin at any number whose logarithm is a positive or negative whole number. Two examples are shown below. ``` ├────────┬────────┬────────┐ 10 100 1,000 10,000 ``` ``` ├────┬────┬────┬────┬────┐ 0.01 0.1 1 10 100 1,000 ``` a) The top scale begins at 10 whose log is 1. It is a _____-cycle log scale. b) The bottom scale begins at 0.01 whose log is -2. It is a _____-cycle log scale.	four
87. The log scale below is a two-cycle scale. On the cycle from 1 to 10, each whole number is calibrated. On the cycle from 10 to 100, each 10's number is calibrated. On the scale, point A is 3. What numbers are at these points? a) Point B _____ b) Point C _____ c) Point D _____	a) three b) five
88. On the three-cycle scale below, only the even numbers from 1 to 10 are calibrated. Point A is 20. What numbers are at these points? a) Point B _____ b) Point C _____ c) Point D _____	a) 7 b) 40 c) 80

Common Logarithms   329

89. On the four-cycle scale below, 2 is the only number calibrated between 1 and 10.

[scale: 1, 2, 10, 100, 1,000, 10,000 with points A, B, C]

What numbers are calibrated at these points?

a) Point A _____   b) Point B _____   c) Point C _____

a) 60
b) 400
c) 800

---

90. The three-cycle scale below begins at 10. Only 20, 40, 60, and 80 are calibrated between 10 and 100.

[scale: 10, 20, 60, 100, 1,000, 10,000 with points A, B, C, D]

Point A is 200. What numbers are at these points?

a) Point B _____   b) Point C _____   c) Point D _____

a) 20
b) 200
c) 2,000

---

91. The four-cycle scale below begins at 0.01. There is only one subdivision in each cycle.

[scale: 0.01, 0.02, 0.1, 1, 10, 100 with points A, B, C]

What numbers are at these points?

a) Point A _____   b) Point B _____   c) Point C _____

a) 600
b) 4,000
c) 8,000

---

92. On a logarithmic scale, numbers are plotted according to the size of their logarithms. Therefore, 1.5 and 2.5 (on the next page) are <u>not plotted halfway</u> between the whole numbers since:

$$\log 1 = 0.00 \qquad \log 2 = 0.30$$
$$\log 1.5 = 0.18 \qquad \log 2.5 = 0.40$$
$$\log 2 = 0.30 \qquad \log 3 = 0.48$$

Therefore: $\log 1.5$ is plotted $\dfrac{0.18}{0.30}$ or $\dfrac{3}{5}$ of the distance from 1 to 2.

$\log 2.5$ is plotted $\dfrac{0.10}{0.18}$ or $\dfrac{5}{9}$ of the distance from 2 to 3.

Continued on following page.

a) 0.2
b) 2
c) 20

330 Common Logarithms

92. Continued

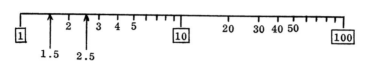

On the scale above, plot and label 15 and 25 .

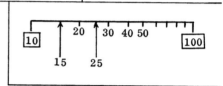

7-13 SEMI-LOG AND LOG-LOG GRAPHS

Sometimes a logarithmic scale is used on one or both axes of a graph. Such graphs are called "semi-log" and "log-log" graphs. We will discuss graphs of that type in this section.

93. A graph with a log scale on one axis is called a "semi-log" graph. A graph with log scales on both axes is called a "log-log" graph.

The graph at the right is a semi-log graph with a three-cycle log scale on the horizontal axis.

Write the coordinates of the following points:

A: _____

B: _____

C: _____

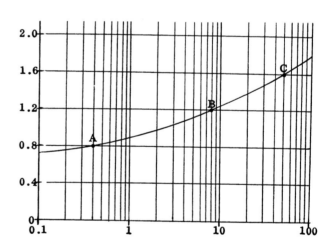

A: (0.4 , 0.8)

B: (8 , 1.2)

C: (50 , 1.6)

94. The graph at the right is a <u>semi-log</u> graph with a three-cycle log scale on the vertical axis.

Write the coordinates of the following points.

A: _____

B: _____

C: _____

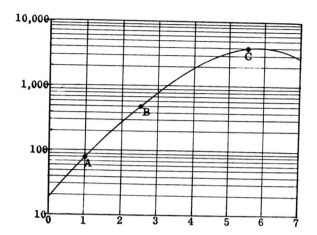

---

95. The graph at the right is a <u>log-log</u> graph with a two-cycle log scale on each axis.

Write the coordinates of the following points:

A: _____

B: _____

C: _____

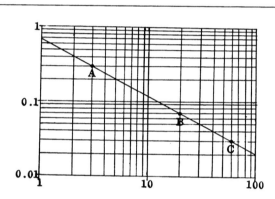

A: (1, 80)

B: (2.5, 500)

C: (5.5, 4,000)

---

96. The graph at the right is a <u>log-log</u> graph with a two-cycle scale on the horizontal axis and a one-cycle scale on the vertical axis.

Write the coordinates of these points:

A: _____

B: _____

C: _____

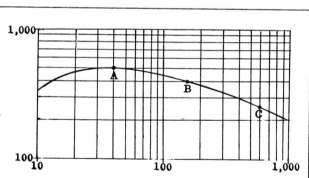

A: (3, 0.3)

B: (20, 0.07)

C: (60, 0.03)

A: (40, 500)   B: (150, 400)   C: (600, 250)

332 Common Logarithms

7-14 LAWS OF LOGARITHMS

In this section, we will discuss the laws of common logarithms for multiplication, division, and powers.

97. The law of logarithms for multiplication is shown below. It says this: <u>The logarithm of a multiplication equals the sum of the logarithms of the factors</u>.

$$\boxed{\log(ab) = \log a + \log b}$$

Two numerical examples of the above law are shown below.

$$\log(2 \times 3) = \log 2 + \log 3 \qquad \log(8 \times 12) = \log 8 + \log 12$$
$$\log 6 = 0.3010 + 0.4771 \qquad \log 96 = 0.9031 + 1.0792$$
$$0.7781 = 0.7781 \qquad 1.9823 = 1.9823$$

98. We used the law of logarithms for multiplication below.

$$\log 3x = \log 3 + \log x$$

Using the same law, complete these.

a) $\log 5t = $ _____ + _____

b) $\log PQ = $ _____ + _____

99. The law of logarithms for division is shown below. It says this: <u>The logarithm of a fraction (or division) equals the difference between the logarithm of the numerator and the logarithm of the denominator</u>.

$$\boxed{\log\left(\frac{a}{b}\right) = \log a - \log b}$$

a) $\log 5 + \log t$

b) $\log P + \log Q$

Two numerical examples of the above law are shown below:

$$\log\left(\frac{6}{2}\right) = \log 6 - \log 2 \qquad \log\left(\frac{96}{12}\right) = \log 96 - \log 12$$
$$\log 3 = 0.7781 - 0.3010 \qquad \log 8 = 1.9823 - 1.0792$$
$$0.4771 = 0.4771 \qquad 0.9031 = 0.9031$$

100. We used the law of logarithms for division below.

$$\log\left(\frac{x}{15}\right) = \log x - \log 15$$

Using the same law, complete these.

a) $\log\left(\frac{7}{y}\right) = $ _____ − _____

b) $\log\left(\frac{V}{D}\right) = $ _____ − _____

a) $\log 7 - \log y$ b) $\log V - \log D$

101. The law of logarithms for powers is shown below. It says this: The logarithm of a power equals the exponent times the logarithm of the base.

$$\log(b^a) = a \log b$$

Two numerical examples of the above law are shown below.

$\log(3^2) = 2 \log 3$ $\log(64)^{0.5} = 0.5 \log 64$
$\log 9 = 2(0.4771)$ $\log 8 = 0.5(1.8062)$
$0.9542 = 0.9542$ $0.9031 = 0.9031$

---

102. We used the law of logarithms for powers below.

$$\log(x^8) = 8 \log x$$

Using the same law, complete these.

a) $\log(25^{1.2}) = $ _____ b) $\log(V^5) = $ _____

---

103. Using the laws of logarithms, complete these.

a) $\log(xy) = $ _____

b) $\log\left(\dfrac{m}{n}\right) = $ _____

c) $\log(p^q) = $ _____

a) 1.2 log 25
b) 5 log V

---

104. Using the laws of logarithms, complete these.

a) $\log(R^d) = $ _____

b) $\log\left(\dfrac{T}{S}\right) = $ _____

c) $\log(CD) = $ _____

a) log x + log y
b) log m - log n
c) q log p

---

105. There is no law of logarithms for addition. Therefore, $\log(a+b)$ cannot be written in an equivalent form. That is:

$$\log(a+b) \neq \log a + \log b$$

Which statement is true? _____

a) $\log MN = \log M + \log N$ b) $\log(M+N) = \log M + \log N$

a) d log R
b) log T - log S
c) log C + log D

(a)

334   Common Logarithms

106. There is no law of logarithms for subtraction. Therefore, $\log(a - b)$ cannot be written in an equivalent form. That is:

$$\log(a-b) \neq \log a - \log b$$

Which statement is true? _____

a) $\log(C - D) = \log C - \log D$ b) $\log\left(\dfrac{C}{D}\right) = \log C - \log D$

107. If possible, use a law of logarithms to write each expression in an equivalent form.

a) $\log\left(\dfrac{m}{t}\right) =$ _____

b) $\log(R - S) =$ _____

c) $\log(V + T) =$ _____

d) $\log(ay) =$ _____

(b)

a) $\log m - \log t$ c) Not possible
b) Not possible   d) $\log a + \log y$

7-15 USING THE LAWS OF LOGARITHMS TO REARRANGE "LOG" FORMULAS

When a variable is part of a logarithmic expression, we sometimes need the laws of logarithms to solve for the variable or its "log". We will discuss rearrangements of that type in this section.

108. To solve for $\log C$ below, we used the law of logarithms for multiplication. Solve for $\log S$ in the other formula.

$$\log(CD) = F \qquad\qquad \log(RS) = T$$

$$\log C + \log D = F$$

$$\log C = F - \log D$$

109. To solve for "y" below, we used the law of logarithms for powers. Solve for $\log P$ in the other formula.

$$V = \log(x^y) \qquad\qquad \log(P^a) = 3t$$

$$V = y \log x$$

$$y = \dfrac{V}{\log x}$$

$\log S = T - \log R$

$\log P = \dfrac{3t}{a}$

110. To solve for $\log P_2$ below, we used the law of logarithms for division. Notice that we isolated $\log\left(\frac{P_2}{P_1}\right)$ before applying the law. Solve for $\log I_1$ in the other formula.

$$D = 10 \log\left(\frac{P_2}{P_1}\right) \qquad\qquad M = 2.5 \log\left(\frac{I_1}{I}\right)$$

$$\frac{D}{10} = \log\left(\frac{P_2}{P_1}\right)$$

$$\frac{D}{10} = \log P_2 - \log P_1$$

$$\log P_2 = \frac{D}{10} + \log P_1$$

111. We also used the law of logarithms for division to solve for $\log P_1$ below. Notice how we replaced each side with its opposite to get rid of the "−" in front of $\log P_1$. Solve for $\log I$ in the other formula.

$$D = 10 \log\left(\frac{P_2}{P_1}\right) \qquad\qquad M = 2.5 \log\left(\frac{I_1}{I}\right)$$

$$\frac{D}{10} = \log\left(\frac{P_2}{P_1}\right)$$

$$\frac{D}{10} = \log P_2 - \log P_1$$

$$\frac{D}{10} - \log P_2 = -\log P_1$$

$$\log P_1 = \log P_2 - \frac{D}{10}$$

$\log I_1 = \dfrac{M}{2.5} + \log I$

112. When a law of logarithms must be used to solve for a variable in a "log" formula, the log expression should be isolated first. Use that method for these.

a) Solve for $\log p$.

$$20 \log(pq) = t$$

b) Solve for $\log R$.

$$k \log\left(\frac{R}{S}\right) = V$$

$\log I = \log I_1 - \dfrac{M}{2.5}$

a) $\log p = \dfrac{t}{20} - \log q$ \qquad b) $\log R = \dfrac{V}{k} + \log S$

336   Common Logarithms

113. Isolate the log expression first when solving these.

    a) Solve for "b".   b) Solve for log S.

$$R = k \log (T^b) \qquad\qquad a \log\left(\frac{Q}{S}\right) = D$$

114. We solved for the log expression in the formula below.

$$k \log(P + Q) = 40$$
$$\log(P + Q) = \frac{40}{k}$$

Can we use a law of logarithms to simplify $\log(P + Q)$ so that we can solve for log P ? _____

a) $b = \dfrac{R}{k \log T}$

b) $\log S = \log Q - \dfrac{D}{a}$,

from:

$-\log S = \dfrac{D}{a} - \log Q$

115. Two methods are shown for the evaluation below.

In $\boxed{D = 10 \log\left(\dfrac{E_2}{E_1}\right)}$, find E_2 when $D = 1.35$ and $E_1 = 14.6$.

1. Substituting in the formula as it stands and then solving.

$$1.35 = 10 \log\left(\frac{E_2}{14.6}\right)$$
$$0.135 = \log\left(\frac{E_2}{14.6}\right)$$
$$10^{0.135} = \frac{E_2}{14.6}$$
$$1.365 = \frac{E_2}{14.6}$$
$$E_2 = 1.365(14.6) = 19.9$$

2. Rearranging to solve for $\log E_2$ before substituting and solving.

$$\log E_2 = \frac{D}{10} + \log E_1$$
$$\log E_2 = \frac{1.35}{10} + \log 14.6$$
$$\log E_2 = 0.135 + 1.1644$$
$$\log E_2 = 1.2994$$
$$E_2 = 19.9$$

Did we get the same value for E_2 in both methods? _____

No. There is no law of logarithms for addition.

Yes

116. We used the law of logarithms for powers to write the formula below in a simpler form. Write the other formula in a simpler form.

$T = 10 \log(R^2)$ $\qquad$ $V = 10 \log\left(\dfrac{R_2}{R_1}\right)^2$

$T = 10(2 \log R)$

$T = 20 \log R$

$V = 20 \log\left(\dfrac{R_2}{R_1}\right)$

SELF-TEST 23 (pages 327-338)

1. Is the graph at the right a "log-log" or a "semi-log" graph? _____

2. How many cycles are on the horizontal axis? _____

Write the coordinates of these points:

3. A: _____
4. B: _____
5. C: _____

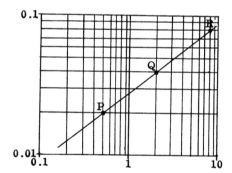

6. Is the graph at the right a "log-log" or a "semi-log" graph? _____

How many cycles are on each axis?

7. Horizontal axis _____   8. Vertical axis _____

Write the coordinates of these points:

9. P: _____
10. Q: _____
11. R: _____

Using the laws of logarithms, rearrange these formulas.

12. Solve for "log t".

$\boxed{\log rt = d}$

13. Solve for "log K".

$\boxed{W = 10 \log\left(\dfrac{K}{P}\right)}$

14. Solve for "a".

$\boxed{F = R \log V^a}$

Continued on following page.

338 Common Logarithms

SELF-TEST 23 (Continued)

15. In $\boxed{B = \log G^t}$, find "t" when $B = 1.75$ and $G = 850$. Round to thousandths.

16. In $\boxed{N = 20 \log\left(\dfrac{P_2}{P_1}\right)}$, find P_1 when $N = 48$ and $P_2 = 850$. Round to hundredths.

ANSWERS:
1. semi-log graph
2. four cycles
3. (2, 30)
4. (50, 65)
5. (7,000, 50)
6. log-log graph
7. two cycles
8. one cycle
9. (0.5, 0.02)
10. (2, 0.04)
11. (8, 0.08)
12. $\log t = d - \log r$
13. $\log K = \dfrac{W}{10} + \log P$
14. $a = \dfrac{F}{R \log V}$
15. $t = 0.597$
16. $P_1 = 3.38$

7-16 EXPONENTIAL EQUATIONS

In this section, we will show how the "log principle for equations" can be used to solve exponential equations.

117. Each equation below contains a power. Equations of that type are called "exponential" equations.

$x = 14^3 \qquad 7 = y^2 \qquad 12 = 5^b$

To solve equations like $x = 14^3$, we saw that we can use the $\boxed{y^x}$ key. Use that key to do these.

a) $x = 14^3$
x = _____

b) $y = 2^{10}$
y = _____

c) $t = 5^6$
t = _____

118. Using the $\boxed{y^x}$ key, solve these. Round to the indicated place.

a) Round to tenths.

$x = 2.4^5$

x = _____

b) Round to the nearest whole number.

$T = 12^{1.9}$

T = _____

a) $x = 2,744$
b) $y = 1,024$
c) $t = 15,625$

a) $x = 79.6$
b) $T = 112$

119. In $7 = y^2$ and $12 = 5^b$, the variable is either the base or exponent of a power. To solve equations of that type, we use the "log principle for equations". That principle says this:

> If two positive quantities are equal, their logarithms are equal.

Some examples of the "log principle for equations" are shown below.

If $7 = y^2$,   If $12 = 5^b$,

then $\log 7 = \log y^2$ then $\log 12 = \log 5^b$

The law of logarithms for a power can be applied to the expression $\log y^2$ and $\log 5^b$ in the equations above. We get:

$\log 7 = 2 \log y$ $\log 12 = $ _____

---

120. Using the log principle for equations, we solved for "log y" and "log x" below. We used a calculator to find log 7 and log 41.

$7 = y^2$ $41 = x^3$

$\log 7 = \log y^2$ $\log 41 = \log x^3$

$\log 7 = 2 \log y$ $\log 41 = 3 \log x$

$\log y = \dfrac{\log 7}{2}$ $\log x = \dfrac{\log 41}{3}$

$\log y = \dfrac{0.8451}{2}$ $\log x = \dfrac{1.6128}{3}$

$\log y = 0.4226$ $\log x = 0.5376$

Now use either the $\boxed{y^x}$ key or $\boxed{\text{INV}}$ $\boxed{\log}$ to solve for "y" and "x". Round to hundredths.

a) y = _____   b) x = _____

$\log 12 = b \log 5$

---

121. Using the steps from the last frame, we solved one equation below. Solve the other equation. Round to hundredths.

$14.5 = R^{1.5}$ $60.7 = t^{2.6}$

$\log 14.5 = \log R^{1.5}$

$\log 14.5 = 1.5 \log R$

$\log R = \dfrac{\log 14.5}{1.5}$

$\log R = \dfrac{1.1614}{1.5}$

$\log R = 0.7742$

$R = 5.95$

a) y = 2.65
b) x = 3.45

t = 4.85

340 Common Logarithms

122. When using the log principle for equations, we take the log of <u>both</u> <u>sides</u>. A common error is shown below. The error is taking the log of only that side that can be rearranged by a law of logarithms.

$$16.9 = m^{0.75}$$

ERROR ⟶ $\begin{cases} 16.9 = \log m^{0.75} \\ 16.9 = 0.75 \log m \end{cases}$

Applying the log principle correctly to the above equation, we get:

$$16.9 = m^{0.75}$$

___

123. Using the log principle for equations, we solved one equation below. We used a calculator to find both log 12 and log 5. Solve the other equation. Round to hundredths.

$$12 = 5^b \qquad\qquad 49 = 3^x$$
$$\log 12 = \log 5^b$$
$$\log 12 = b \log 5$$
$$b = \frac{\log 12}{\log 5}$$
$$b = \frac{1.0792}{0.6990}$$
$$b = 1.54$$

$\log 16.9 = \log m^{0.75}$

$\log 16.9 = 0.75 \log m$

___

124. In the last frame, we got these expressions:

$$\frac{\log 12}{\log 5} \qquad\qquad \frac{\log 49}{\log 3}$$

Don't confuse the above expressions with those below:

$$\log\left(\frac{12}{5}\right) \qquad\qquad \log\left(\frac{49}{3}\right)$$

The law of logarithms for division applies only to the logarithm of a division, not to a division of two logarithms. That is:

$$\log\left(\frac{12}{5}\right) = \log 12 - \log 5 \qquad \frac{\log 12}{\log 5} \neq \log 12 - \log 5$$

Which statement is true? _____

a) $\dfrac{\log 49}{\log 3} = \log 49 - \log 3$ \qquad b) $\log\left(\dfrac{49}{3}\right) = \log 49 - \log 3$

$x = 3.54$, from:

$$x = \frac{1.6902}{0.4771}$$

(b)

125. We solved one equation below. Use the same steps to solve the other equation. Round to thousandths.

$$69.7 = 1.4^y \qquad\qquad 456 = 559^t$$

$$\log 69.7 = \log 1.4^y$$

$$\log 69.7 = y \log 1.4$$

$$y = \frac{\log 69.7}{\log 1.4}$$

$$y = \frac{1.8432}{0.1461}$$

$$y = 12.6$$

$t = 0.968$, from: $t = \dfrac{2.6590}{2.7474}$

7-17 COMBINED USE OF THE LAWS OF LOGARITHMS

In this section, we will show how more than one law of logarithms can be used to write some logarithmic expressions in equivalent forms. We will show how that combined use of the laws is sometimes needed when applying the log principle for equations.

126. We used the laws of logarithms for multiplication and a power to write the expression below in an equivalent form.

$$\log(cd^a) = \log c + \log d^a$$
$$= \log c + a \log d$$

Using the same steps, write this expression in an equivalent form.

$$\log TV^{0.4} = \underline{}$$

127. We used the laws of logarithms for division and a power to write the expression below in an equivalent form.

$\log T + 0.4 \log V$

$$\log\left(\frac{x}{y}\right)^3 = 3\log\left(\frac{x}{y}\right)$$
$$= 3(\log x - \log y)$$

Using the same steps, write this expression in an equivalent form.

$$\log\left(\frac{t}{2}\right)^{1.9} = \underline{}$$

$1.9(\log t - \log 2)$

342 Common Logarithms

128. We used the laws of logarithms for multiplication and a power to write the expression below in an equivalent form.

$$\log P(1+i)^n = \log P + \log(1+i)^n$$
$$= \log P + n \log(1+i)$$

Using the same steps, write this expression in an equivalent form.

$\log 7(x+y)^{19} = $ _____

---

129. Using as many laws as possible, write each of these in an equivalent form.

a) $\log 1.2R^{1.5} = $ _____

b) $\log\left(\dfrac{357}{298}\right)^{1.67} = $ _____

c) $\log 20(1+i)^{15} = $ _____

$\log 7 + 19 \log(x+y)$

---

130. After applying the log principle for equations, sometimes two laws of logarithms can be applied to one side. An example is shown.

$$\dfrac{T_1}{T_2} = \left(\dfrac{V_1}{V_2}\right)^{1.56}$$

$$\log\left(\dfrac{T_1}{T_2}\right) = \log\left(\dfrac{V_1}{V_2}\right)^{1.56}$$

$$\log T_1 - \log T_2 = 1.56 \log\left(\dfrac{V_1}{V_2}\right)$$

$$\log T_1 - \log T_2 = 1.56(\log V_1 - \log V_2)$$

Apply the log principle and then as many laws as possible to each side of these equations.

a) $M = \left(\dfrac{c}{d}\right)^{1.5}$ b) $\left(\dfrac{48}{P_1}\right)^{0.6} = \left(\dfrac{T_2}{65}\right)^{1.4}$

a) $\log 1.2 + 1.5 \log R$

b) $1.67(\log 357 - \log 298)$

c) $\log 20 + 15 \log(1+i)$

---

131. We applied the log principle and then as many laws as possible to the equation below. Do the same to the other equation.

$R = BV^{2.5}$ $6.8T^{0.4} = k$

$\log R = \log BV^{2.5}$

$\log R = \log B + \log V^{2.5}$

$\log R = \log B + 2.5 \log V$

a) $\log M$
$= 1.5(\log c - \log d)$

b) $0.6(\log 48 - \log P_1) = 1.4(\log T_2 - \log 65)$

Common Logarithms   343

132. We applied the log principle and then as many laws as possible to the equation below. Do the same to the other equation.

$$A = P(1 + i)^{19}$$

$$\log A = \log P(1 + i)^{19}$$

$$\log A = \log P + \log(1 + i)^{19}$$

$$\log A = \log P + 19 \log(1 + i)$$

$$15 = 12(1 + i)^{21}$$

$\log 6.8 + 0.4 \log T = \log k$

$\log 15 = \log 12 + 21 \log(1 + i)$

7-18 EVALUATING EXPONENTIAL FORMULAS

We discussed some evaluations with exponential formulas in the last chapter. In this section, we will briefly review that type of evaluation and then discuss evaluations which require using the log principle for equations.

133. In Section 6-14 of the last chapter, we showed the calculator procedures for evaluations like those in this frame and the next. If necessary, review those procedures to do the evaluations.

Use a calculator for these. Round to hundredths.

a) Find M when T = 24.5 and a = 0.4 .

$$M = T^a$$

b) Find D when a = 69.3 and b = 56.1 .

$$D = \left(\frac{a}{b}\right)^{1.8}$$

134. Use a calculator for these.

a) Find I when k = 0.001 and P = 18 . Round to thousandths.

$$I = kP^{1.5}$$

b) Find A when P = \$1,000 , i = 0.12 , and n = 15 . Round to the nearest cent.

$$A = P(1 + i)^n$$

a) M = 3.59

b) D = 1.46

a) I = 0.076

b) A = \$5,473.57

344   Common Logarithms

135. Following the example, do the other evaluation.

Find P when R = 20.7, Q = 2.24, and a = 3.18. Round to hundredths.

$R = PQ^a$

$20.7 = P(2.24)^{3.18}$

$20.7 = P(12.995)$

$P = \dfrac{20.7}{12.995}$

$P = 1.59$

Find T when V = 200 and b = 2,750. Round to the nearest whole number.

$TV^{0.4} = b$

---

136. In the evaluation below, we used a calculator to simplify $\left(\dfrac{2,775}{1,850}\right)^{1.36}$. Do the other evaluation.

Find R_1 when $R_2 = 35.5$, $T_1 = 2,775$, and $T_2 = 1,850$. Round to tenths.

$\dfrac{R_1}{R_2} = \left(\dfrac{T_1}{T_2}\right)^{1.36}$

$\dfrac{R_1}{35.5} = \left(\dfrac{2,775}{1,850}\right)^{1.36}$

$\dfrac{R_1}{35.5} = 1.736$

$R_1 = (35.5)(1.736)$

$R_1 = 61.6$

Find T_2 when $T_1 = 1,925$, $V_1 = 8,250$, and $V_2 = 6,750$. Round to the nearest whole number.

$\dfrac{T_1}{T_2} = \left(\dfrac{V_1}{V_2}\right)^{1.25}$

T = 330, from:

$T = \dfrac{2,750}{8.326}$

---

137. In the evaluation below, we used a calculator to simplify $(1 + 0.12)^{10}$. Do the other evaluation.

Find P when A = $3,750, i = 0.12, and n = 10. Round to the nearest dollar.

$A = P(1 + i)^n$

$\$3,750 = P(1 + 0.12)^{10}$

$\$3,750 = P(3.106)$

$P = \dfrac{\$3,750}{3.106}$

$P = \$1,207$

Find P when A = $10,790, i = 0.15, and n = 5. Round to the nearest dollar.

$A = P(1 + i)^n$

$T_2 = 1,498$, since:

$\dfrac{1,925}{T_2} = 1.285$

$T_2 = \dfrac{1,925}{1.285}$

P = $5,365, from:

$P = \dfrac{\$10,790}{2.011}$

138. In the remaining frames in this section, we will discuss evaluations that require using the log principle for equations. Following the example, do the other evaluation.

Find T when S = 35.7 and k = 2.1. Round to hundredths.

$S = T^k$

$35.7 = T^{2.1}$

$\log 35.7 = \log T^{2.1}$

$1.5527 = 2.1 \log T$

$\log T = \dfrac{1.5527}{2.1}$

$\log T = 0.7394$

$T = 5.49$

Find M when R = 90.6 and a = 1.7. Round to tenths.

$R = M^a$

---

139. Following the example, do the other evaluation.

Find "v" when f = 67.5 and t = 21.2. Round to hundredths.

$f = t^v$

$67.5 = 21.2^v$

$\log 67.5 = \log 21.2^v$

$\log 67.5 = v \log 21.2$

$v = \dfrac{\log 67.5}{\log 21.2}$

$v = \dfrac{1.8293}{1.3263}$

$v = 1.38$

Find "a" when R = 97.9 and S = 17.3. Round to hundredths.

$R = S^a$

M = 14.2, from:

$\log M = \dfrac{1.9571}{1.7}$

---

a = 1.61, from:

$a = \dfrac{\log 97.9}{\log 17.3}$

140. Following the example, do the other evaluation.

Find Q when R = 25.1, P = 1.75, and a = 3.2. Round to hundredths.

$R = PQ^a$

$25.1 = 1.75 Q^{3.2}$

$\log 25.1 = \log 1.75 + 3.2 \log Q$

$1.3997 = 0.2430 + 3.2 \log Q$

$3.2 \log Q = 1.3997 - 0.2430$

$3.2 \log Q = 1.1567$

$\log Q = \dfrac{1.1567}{3.2}$

$\log Q = 0.3615$

$Q = 2.30$

Find V when T = 600 and b = 3,050. Round to tenths.

$TV^{0.4} = b$

V = 58.3, since:

$2.7782 + 0.4 \log V = 3.4843$

$0.4 \log V = 0.7061$

$\log V = \dfrac{0.7061}{0.4}$

$\log V = 1.7653$

---

141. In the evaluation below, we simplified before applying the log principle. Do the other evaluation.

Find "i" when A = 11,523, P = 1,000, and n = 20. Round to hundredths.

$A = P(1 + i)^n$

$11,523 = 1,000(1 + i)^{20}$

$\dfrac{11,523}{1,000} = (1 + i)^{20}$

$11.523 = (1 + i)^{20}$

$\log 11.523 = \log(1 + i)^{20}$

$1.0616 = 20 \log(1 + i)$

$\dfrac{1.0616}{20} = \log(1 + i)$

$\log(1 + i) = 0.05308$

$1 + i = 1.1300$

$i = 0.1300$ or 0.13

Find "n" when A = 29,500, P = 10,000, and i = 0.11. Round to tenths.

$A = P(1 + i)^n$

n = 10.4, from:

$n = \dfrac{\log 2.95}{\log 1.11}$

142. In the evaluation below, we used a calculator to find $\left(\dfrac{295}{233}\right)^{1.33}$ before applying the log principle. Complete the evaluation.

Find P_2 when $P_1 = 24.6$, $T_2 = 295$, and $T_1 = 233$.

$$\left(\dfrac{P_2}{P_1}\right)^{0.67} = \left(\dfrac{T_2}{T_1}\right)^{1.33}$$

$$\left(\dfrac{P_2}{24.6}\right)^{0.67} = \left(\dfrac{295}{233}\right)^{1.33}$$

$$\left(\dfrac{P_2}{24.6}\right)^{0.67} = 1.3686$$

$$\log\left(\dfrac{P_2}{24.6}\right)^{0.67} = \log 1.3686$$

$$0.67(\log P_2 - \log 24.6) = \log 1.3686$$

$$\log P_2 - 1.3909 = \dfrac{0.1363}{0.67}$$

$$\log P_2 - 1.3909 = 0.2034$$

$$\log P_2 = \underline{\hspace{2cm}}$$

$$P_2 = \underline{\hspace{2cm}} \text{ (Round to tenths.)}$$

$\log P_2 = 1.5943$

$P_2 = 39.3$

SELF-TEST 24 (pages 338-348)

1. Find "y". Round to tenths.

   $245 = y^{1.7}$

2. Find "t". Round to hundredths.

   $3.26^t = 87.5$

Using as many of the laws of logarithms as possible, write each expression in an equivalent form.

3. $\log\left(\dfrac{w}{t}\right)^5$

4. $\log k(r+1)^n$

5. In $\boxed{H = CR^s}$, find C when $H = 7{,}190$, $R = 2.18$, and $s = 4.9$. Round to the nearest whole number.

6. In $\boxed{\dfrac{F_1}{F_2} = \left(\dfrac{A_1}{A_2}\right)^{2.15}}$, find F_2 when $F_1 = 9.73$, $A_1 = 81.3$, and $A_2 = 52.7$. Round to hundredths.

7. In $\boxed{W = BG^t}$, find G when $W = 2.81$, $B = 0.035$, and $t = 1.28$. Round to tenths.

8. In $\boxed{A = P(1 + i)^n}$, find "i" when $A = 2{,}000$, $P = 400$, and $n = 10$. Round to thousandths.

ANSWERS:
1. $y = 25.4$
2. $t = 3.78$
3. $5(\log w - \log t)$
4. $\log k + n \log(r+1)$
5. $C = 158$
6. $F_2 = 3.83$
7. $G = 30.8$
8. $i = 0.175$

Common Logarithms 349

SUPPLEMENTARY PROBLEMS - CHAPTER 7

Assignment 21

Convert each power of ten to an ordinary number.

1. 10^3 2. 10^{-5} 3. 10^1 4. 10^0

Convert each number to a power of ten.

5. 100 6. 1,000,000 7. 0.001 8. 0.1

Convert each power of ten to an ordinary number. Round as directed.

9. Round to hundreds.   10. Round to hundred-thousandths.   11. Round to hundredths.
    $10^{4.8}$ $10^{-2.1836}$ $10^{0.0937}$

Convert each number to a power of ten. Round each exponent to four decimal places.

12. 58.1 13. 62,400,000 14. 0.000083 15. 0.296

For the exponential function $\boxed{y = 10^{-x}}$, find "y" for the following values of "x".

16. If $x = 0$, $y = ?$ 17. If $x = -3$, $y = ?$ 18. If $x = 2$, $y = ?$
19. If $x = 3.82$, $y = ?$ (Round to millionths.)   20. If $x = -1.417$, $y = ?$ (Round to tenths.)

Find these logarithms. Where necessary, round to four decimal places.

21. log 10,000,000   22. log 16.5   23. log 0.0001   24. log 0.9172

25. If the logarithm of a number is 4, write the number in power-of-ten form.
26. If the logarithm of a number is -0.5172, write the number in power-of-ten form.
27. If log T = 2.8073, find T. (Round to the nearest whole number.)
28. If log D = 0.3516, find D. (Round to hundredths.)
29. If log R = -1.4252, find R. (Round to ten-thousandths.)
30. If log A = -0.9419, find A. (Round to thousandths.)
31. If the logarithm of a number lies between 2 and 3, the number lies between:

   a) 0 and 1   b) 1 and 10   c) 10 and 100   d) 100 and 1,000   e) 1,000 and 10,000

Write each power-of-ten equation as a "log" equation.

32. $10^2 = 100$ 33. $0.1 = 10^{-1}$ 34. $59.2 = 10^{1.7723}$ 35. $10^{-2.68} = 0.00209$

Write each "log" equation as a power-of-ten equation.

36. log 1 = 0   37. -3 = log 0.001   38. 4.13 = log 13,490   39. log 0.875 = -0.058

For the logarithmic function $\boxed{y = \log x}$, find "y" for the following values of "x".

40. If $x = 100{,}000$, $y = ?$ 41. If $x = 0.01$, $y = ?$ 42. If $x = -100$, $y = ?$

350 Common Logarithms

Assignment 22

Solve each equation. Round each answer to four decimal places.

1. $10^x = 80.7$ 2. $0.0516 = 10^t$ 3. $\log 72{,}900 = A$ 4. $r = \log 0.000146$

Solve each equation. Round as directed.

5. $\log P = 3.35$ Round to tens.
6. $-0.7162 = \log V$ Round to thousandths.
7. $w = 10^{1.90}$ Round to tenths.
8. $10^{-2.512} = d$ Round to hundred-thousandths.

Evaluate these formulas for the solved-for variable. Round as directed.

9. In $\boxed{S = K \log A}$, find S when $K = 18$ and $A = 65$. Round to tenths.
10. In $\boxed{P = B + C \log W}$, find P when $B = 1.93$, $C = 7.08$, and $W = 2.68$.
    Round to hundredths.
11. In $\boxed{v = -\log t}$, find "v" when $t = 0.00545$. Round to hundredths.
12. In $\boxed{H = k \log\left(\dfrac{F}{G}\right)}$, find H when $k = 20$, $F = 880$, and $G = 1.75$.
    Round to the nearest whole number.

Evaluate these formulas for a non-solved-for variable. Round as directed.

13. In $\boxed{d = 10 \log M}$, find M when $d = 41.9$. Round to hundreds.
14. In $\boxed{R = -\log V}$, find V when $R = 3.52$. Round to millionths.
15. In $\boxed{s = h - k \log N}$, find "k" when $s = 180$, $h = 510$, and $N = 130$.
    Round to the nearest whole number.
16. In $\boxed{W = R + P \log A}$, find A when $W = 41.3$, $R = 17.8$, and $P = 12.6$. Round to tenths.
17. In $\boxed{t = a \log s}$, find "a" when $t = 2.57$ and $s = 6.39$. Round to hundredths.

In evaluating these "log" formulas, a conversion to power-of-ten form can be used. Round as directed.

18. In $\boxed{D = 20 \log\left(\dfrac{E_2}{E_1}\right)}$, find E_2 when $D = 70$ and $E_1 = 0.0129$. Round to tenths.
19. In $\boxed{N = 10 \log\left(\dfrac{P_2}{P_1}\right)}$, find P_1 when $N = 28.7$ and $P_2 = 150$. Round to thousandths.
20. In $\boxed{w = r \log(a + 1)}$, find "a" when $w = 41.6$ and $r = 72.9$. Round to hundredths.
21. In $\boxed{T = K \log(P - R)}$, find R when $T = 525$, $K = 195$, and $P = 850$.
    Round to the nearest whole number.

Rearrange these formulas and solve for the designated quantity.

22. Solve for "log P". 23. Solve for "log A". 24. Solve for "m".
    $\boxed{a = k \log P}$ $\boxed{H = -\log A}$ $\boxed{s = m \log w}$

25. Solve for "c". 26. Solve for H. 27. Solve for "log t".
    $\boxed{t = p - c \log F}$ $\boxed{V = B \log R - H}$ $\boxed{a = r - d \log t}$

Assignment 23

For Problems 1 to 10, refer to the graph.

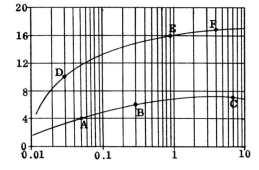

1. Is the horizontal axis a uniform scale or a logarithmic scale?

2. Is the vertical axis a uniform scale or a logarithmic scale?

3. How many cycles are there on the horizontal axis?

4. Is the graph a "log-log" or a "semi-log" graph?

Write the coordinates of the following points.

5. Point A   6. Point B   7. Point C   8. Point D   9. Point E   10. Point F

Using the laws of logarithms, write each expression in an equivalent form.

11. $\log\left(\dfrac{G}{N}\right)$ 12. $\log V^p$ 13. $\log 3R$ 14. $\log(B + 1)$

Using the laws of logarithms, rearrange each formula.

15. Solve for "log t".

$\boxed{v = \log ht}$

16. Solve for "log G".

$\boxed{P = \log\left(\dfrac{A}{G}\right)}$

17. Solve for "d".

$\boxed{N = \log R^d}$

18. Solve for "log M".

$\boxed{K \log\left(\dfrac{M}{W}\right) = S}$

19. Solve for "r".

$\boxed{A \log T^r = f}$

20. Solve for "log P".

$\boxed{b \log 2P = w}$

Evaluate each formula. Rearrange the formula first and then plug in the known values. Round as directed.

21. In $\boxed{F = \log AR}$, find R when F = 3.26 and A = 61.8. Round to tenths.

22. In $\boxed{w = k \log G^a}$, find "a" when w = 50.3, k = 21.6, and G = 3.51. Round to hundredths.

23. In $\boxed{t = 20 \log Q}$, find Q when t = 47.9. Round to the nearest whole number.

24. In $\boxed{D = 10 \log\left(\dfrac{P_2}{P_1}\right)}$, find P_1 when D = 32 and P_2 = 188. Round to thousandths.

352 Common Logarithms

Assignment 24

Solve each equation. Round as directed.

1. Round to tenths.
$y^{2.5} = 790$

2. Round to hundredths.
$83.4 = x^{5.17}$

3. Round to the nearest whole number.
$V^{0.63} = 21.6$

4. Round to hundredths.
$2^t = 500$

5. Round to tenths
$15.3 = 1.14^w$

6. Round to thousandths.
$378^s = 136$

Using as many of the laws of logarithms as possible, write each expression in an equivalent form.

7. $\log(xy^3)$

8. $\log\left(\dfrac{A}{B}\right)^4$

9. $\log R(d+1)^t$

10. $\log\left(\dfrac{w^2}{p}\right)$

Apply the "log principle for equations" and then as many laws of logarithms as possible to each side of these equations. Write each resulting equation.

11. $R = KT^s$

12. $h = \left(\dfrac{r}{w}\right)^{1.8}$

13. $A = 800(1+i)^{20}$

Evaluate these formulas for the solved-for variable. Round as directed.

14. In $\boxed{M = GR^p}$, find M when G = 30, R = 120, and p = 0.25. Round to tenths.

15. In $\boxed{H = B(1+r)^n}$, find H when B = 2,000, r = 0.085, and n = 30.
    Round to the nearest whole number.

16. In $\boxed{f = \left(\dfrac{d}{w}\right)^{2.25}}$, find "f" when d = 36.8 and w = 22.9. Round to hundredths.

Evaluate these formulas for a non-solved-for variable. Round as directed.

17. In $\boxed{W = QR^a}$, find Q when W = 83.6, R = 3.19, and a = 1.72. Round to tenths.

18. In $\boxed{\dfrac{F_1}{F_2} = \left(\dfrac{A_1}{A_2}\right)^{0.72}}$, find F_2 when F_1 = 2,000, A_1 = 0.794, and A_2 = 0.152.
    Round to the nearest whole number.

19. In $\boxed{A = P(1+i)^n}$, find P when A = 500,000, i = 0.035, and n = 50. Round to hundreds.

In evaluating these formulas, use the "log principle for equations". Round as directed.

20. In $\boxed{H = KP^t}$, find "t" when H = 480, K = 73.6, and P = 1.95. Round to hundredths.

21. In $\boxed{F = BV^a}$, find V when F = 12.9, B = 0.604, and a = 0.538.
    Round to the nearest whole number.

22. In $\boxed{A = P(1+i)^n}$, find "i" when A = 60,000, P = 12,000, and n = 24.
    Round to ten-thousandths.

Chapter 8 POWERS OF "e" AND NATURAL LOGARITHMS

In this chapter, we will discuss powers of "e" and natural (base "e") logarithms. The graphs of functions and formulas containing powers of "e" are discussed. Evaluations are performed with formulas containing powers of "e" and "ln" formulas. Rearrangements of both types of formulas are also discussed. The relationship between equations and formulas containing powers of "e" and natural logarithms is emphasized.

8-1   POWERS OF "e"

Some formulas contain powers in which the base is a number called "e". The numerical value of "e" is 2.7182818.... In this section, we will convert powers of "e" to ordinary numbers and ordinary numbers to powers of "e".

1. To convert $e^{1.5}$ to an ordinary number, one of the two methods below is used. We rounded the answer to two decimal places.

    1) If your calculator has an $\boxed{e^x}$ key, enter 1.5 and press $\boxed{e^x}$.

Enter	Press	Display	
1.5	$\boxed{e^x}$	4.4816891	($e^{1.5}$ = 4.48)

    2) If your calculator does not have an $\boxed{e^x}$ key, enter 1.5 and press $\boxed{INV}$ $\boxed{\ln x}$.

Enter	Press	Display	
1.5	$\boxed{INV}$ $\boxed{\ln x}$	4.4816891	($e^{1.5}$ = 4.48)

    Convert to an ordinary number. Round to the indicated place.

    a) $e^{2.19}$ = _____   (Round to hundredths.)

    b) $e^{-3.75}$ = _____   (Round to four decimal places.)

2. Since $e = e^1$, we can confirm the fact that $e = 2.7182818...$ by entering "1" and pressing either $\boxed{e^x}$ or $\boxed{INV}$ $\boxed{\ln x}$. Do so.

    Convert to an ordinary number. Round to the indicated place.

    a) $e^{5.1}$ = _____   (Round to the nearest whole number.)

    b) e^{-1} = _____   (Round to thousandths.)

a) 8.94
b) 0.0235

353

3. Just as $10^0 = 1$, $e^0 = 1$. To confirm that fact, enter "0" and press either $\boxed{e^x}$ or $\boxed{INV}$ $\boxed{\ln x}$.

Any power of "e" with a positive exponent equals an ordinary number larger than "1". To confirm that fact, do these. Round to the indicated place.

    a) $e^{0.5}$ = _____ (Round to hundredths.)

    b) $e^{3.68}$ = _____ (Round to tenths.)

    c) $e^{9.757}$ = _____ (Round to hundreds.)

a) 1.64

b) 0.368

---

4. Any power of "e" with a negative exponent equals an ordinary number between 0 and 1. To confirm that fact, do these. Round to the indicated place.

    a) $e^{-0.545}$ = _____ (Round to thousandths.)

    b) $e^{-3.37}$ = _____ (Round to four decimal places.)

    c) $e^{-8.4}$ = _____ (Round to millionths.)

a) 1.65

b) 39.6

c) 17,300

---

5. To convert an ordinary number to a power of "e", we use the $\boxed{\ln x}$ key to find the exponent. We did so below for 67.8 and 0.045.

Enter	Press	Display
67.8	$\boxed{\ln x}$	4.2165622
0.045	$\boxed{\ln x}$	-3.1010928

Rounding each exponent to four decimal places, we get:

    67.8 = $e^{4.2166}$ 0.045 = _____

a) 0.580

b) 0.0344

c) 0.000225

---

6. Using the $\boxed{\ln x}$ key to find the exponent, convert each ordinary number to a power of "e". Round each exponent to two decimal places.

    a) 7.95 = _____   b) 0.00635 = _____

$e^{-3.1011}$

---

7. When "1" is converted to a power of "e", the exponent is "0". That is: $1 = e^0$. To confirm that fact, enter "1" and press $\boxed{\ln x}$.

When a number larger than "1" is converted to a power of "e", the exponent is a positive number. To confirm that fact, convert these numbers to powers of "e". Round each exponent to three decimal places.

    a) 1.25 = _____   b) 287 = _____   c) 45,900 = _____

a) $e^{2.07}$

b) $e^{-5.06}$

8. When a number between "0" and "1" is converted to a power of "e", the exponent is a negative number. To confirm that fact, convert these numbers to powers of "e". Round each exponent to two decimal places.

   a) 0.979 = _____   b) 0.000408 = _____

| a) $e^{0.223}$ |
| b) $e^{5.659}$ |
| c) $e^{10.734}$ |

| a) $e^{-0.02}$ b) $e^{-7.80}$ |

8-2 GRAPHING FUNCTIONS CONTAINING POWERS OF "e"

In this section, we will graph some of the basic functions containing powers of "e".

9. In $\boxed{y = e^x}$:

   a) when x = 6.2, y = _____ (Round to the nearest whole number.)

   b) when x = -0.25, y = _____ (Round to thousandths.)

10. In $\boxed{y = e^{-x}}$, the "-x" means "the opposite of x". Therefore, we can either mentally change the sign of the value of x or press $\boxed{+/-}$ after entering the value of x before pressing $\boxed{e^x}$ or $\boxed{INV}$ $\boxed{\ln x}$.

    In $\boxed{y = e^{-x}}$:

    a) when x = 1.7, y = _____ (Round to thousandths.)

    b) when x = -2.5, y = _____ (Round to tenths.)

| a) 493 |
| b) 0.779 |

11. Using the tables provided, we graphed $y = e^x$ and $y = e^{-x}$ at the right.

| a) y = 0.183   b) y = 12.2 |

$\boxed{y = e^x}$ $\boxed{y = e^{-x}}$

x	y	x	y
-2.2	0.11	-2.2	9.03
-2.0	0.14	-2.0	7.39
-1.6	0.20	-1.6	4.95
-1.3	0.27	-1.3	3.67
-1.0	0.37	-1.0	2.72
-0.6	0.55	-0.6	1.82
-0.2	0.82	-0.2	1.22
0	1.00	0	1.00
0.2	1.22	0.2	0.82
0.6	1.82	0.6	0.55
1.0	2.72	1.0	0.37
1.3	3.67	1.3	0.27
1.6	4.95	1.6	0.20
2.0	7.39	2.0	0.14
2.2	9.03	2.2	0.11

Continued on following page.

356 Powers of "e" and Natural Logarithms

11. Continued

   Notice these points about the graphs:

   1. The graph of $y = e^x$ is called an <u>ascending</u> exponential graph. The graph of $y = e^{-x}$ is called a <u>descending</u> exponential graph.

   2. The asymptote (that is, the line the graph approaches but never reaches) for each graph is the x-axis or $y = 0$.

   3. Since $y = 1$ when $x = 0$ because $e^0 = 1$, the y-intercept for each graph is $(0, 1)$.

---

12. To graph functions of the form $\boxed{y = ke^{cx}}$, we replace $\underline{k}$ and $\underline{c}$ with numerical constants. For example:

   If $k = 10$ and $c = 0.3$, we get: $y = 10e^{0.3x}$

   If $k = 20$ and $c = -0.5$, we get: $y = 20e^{-0.5x}$

   Let's find some pairs of values for $\boxed{y = 10e^{0.3x}}$.

   If $x = 1$, $y = 10e^{0.3(1)} = 10e^{0.3} = 10(1.35) = 13.5$

   a) If $x = 0$, $y = 10e^{0.3(0)} = 10e^0 = 10(1) =$ _____

   b) If $x = -1$, $y = 10e^{0.3(-1)} = 10e^{-0.3} = 10(0.741) =$ _____

---

13. Let's find some pairs of values for $\boxed{y = 20e^{-0.5x}}$.

   If $x = 2$, $y = 20e^{-0.5(2)} = 20e^{-1} = 20(0.368) = 7.36$

   a) If $x = 1.5$, $y = 20e^{-0.5(1.5)} = 20e^{-0.75} = 20(0.472) =$ _____

   b) If $x = -4$, $y = 20e^{-0.5(-4)} = 20e^2 = 20(7.39) =$ _____

a) 10
b) 7.41

a) 9.44
b) 147.8

Powers of "e" and Natural Logarithms   357

14. The following four specific instances of $\boxed{y = ke^{cx}}$ are graphed below.

$\boxed{y = 10e^{0.2x}}$ $\boxed{y = 10e^{-0.2x}}$ $\boxed{y = 20e^{0.5x}}$ $\boxed{y = 20e^{-0.5x}}$

Notice these points about the graphs:

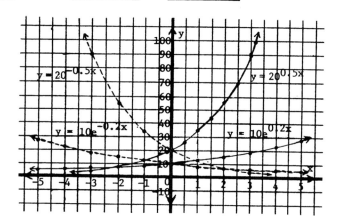

1. The constant $\underline{k}$ gives the point where the graph crosses the y-axis.

   For both $y = 10e^{0.2x}$ and $y = 10^{-0.2x}$, the y-intercept is $(0, 10)$.

   For both $y = 20e^{0.5x}$ and $y = 20^{-0.5x}$, the y-intercept is $(0, 20)$.

2. The constant $\underline{c}$ tells whether the graph ascends or descends.

   If $\underline{c}$ is positive (as in $y = 10e^{0.2x}$ and $y = 20e^{0.5x}$) the graph ascends.

   If $\underline{c}$ is negative (as in $y = 10^{-0.2x}$ and $y = 20e^{-0.5x}$) the graph descends.

3. The constant $\underline{c}$ tells how fast the graph ascends or descends. The larger the absolute value of $\underline{c}$, the faster the ascent or descent.

   $y = 20e^{0.5x}$ ascends faster than $y = 10e^{0.2x}$.

   $y = 20e^{-0.5x}$ descends faster than $y = 10e^{-0.2x}$.

4. The asymptote for each graph is the x-axis or $y = 0$.

---

15. If we graphed both $\boxed{y = 15e^{0.3x}}$ and $\boxed{y = 5e^{0.4x}}$: a) The y-intercept for $y = 15e^{0.3x}$ would be (,). b) The y-intercept for $y = 5e^{0.4x}$ would be (,). c) Which graph would ascend faster? _____	
16. $y = e^{0.5x}$ and $y = e^{-0.1x}$ are instances of $\boxed{y = ke^{cx}}$ in which $k = 1$. a) The y-intercept for both graphs is (,). b) The asymptote for $y = e^{-0.1x}$ is _____.	a) $(0, 15)$ b) $(0, 5)$ c) $y = 5e^{0.4x}$
17. $y = e^x$ and $y = e^{-x}$ are also instances of $\boxed{y = ke^{cx}}$ in which $k = 1$ and $c = 1$ or -1. a) Does $y = e^x$ ascend or descend? _____ b) The y-intercept for both graphs is (,).	a) $(0, 1)$ b) $y = 0$ (or the x-axis)

18. To graph functions of the form $y = k(1 - e^{-x})$, we replace k with a numerical constant. For example:

If $k = 5$, we get: $y = 5(1 - e^{-x})$

If $k = 10$, we get: $y = 10(1 - e^{-x})$

Let's find some pairs of values for $y = 10(1 - e^{-x})$.

a) If $x = 1$, $y = 10(1 - e^{-(1)})$
$= 10(1 - e^{-1})$
$= 10(1 - 0.368) = 10(0.632) = $ _____

b) If $x = 0$, $y = 10(1 - e^{-(0)})$
$= 10(1 - e^{0})$
$= 10(1 - 1) = 10(0) = $ _____

c) If $x = -1$, $y = 10(1 - e^{-(-1)})$
$= 10(1 - e^{1})$
$= 10(1 - 2.72) = 10(-1.72) = $ _____

a) ascends

b) (0, 1)

---

19. The following two specific instances of $y = k(1 - e^{-x})$ are graphed below.

a) 6.32   b) 0   c) -17.2

$y = 5(1 - e^{-x})$ $y = 10(1 - e^{-x})$

Notice these points:

1. The y-intercept for both graphs is at the origin (0, 0).

2. The constant k tells the asymptote of the graph.

For $y = 5(1 - e^{-x})$, the asymptote is the horizontal line $y = 5$.

For $y = 10(1 - e^{-x})$, the asymptote is the horizontal line $y = 10$.

For $y = 20(1 - e^{-x})$:

a) The y-intercept would be (,).

b) The equation of the asymptote would be _____.

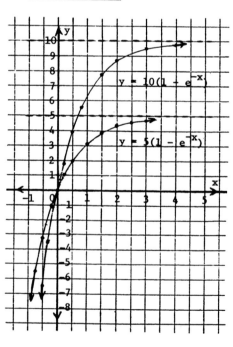

Powers of "e" and Natural Logarithms   359

20. Give the y-intercept for each function below.

   a) $y = 2.4e^{0.3x}$ b) $y = 25(1 - e^{-x})$

   (____ , ____)   (____ , ____)

a) (0, 0)
b) y = 20

---

21. Give the equation of the asymptote for each function below.

   a) $y = 15(1 - e^{-x})$ b) $y = 1.6e^{-0.5x}$

   _____   _____

a) (0, 2.4)
b) (0, 0)

---

a) y = 15   b) y = 0

8-3 EVALUATING FORMULAS CONTAINING POWERS OF "e"

In this section, we will show how a calculator can be used to evaluate formulas containing powers of "e".

22. We can use $\boxed{e^x}$ or $\boxed{\text{INV}}$ $\boxed{\ln x}$ for each evaluation below.

   a) In $\boxed{T = e^x}$, when x = 2.5, T = _____ (Round to tenths.)

   b) In $\boxed{R = e^p}$, when p = 9.15, R = _____
      (Round to the nearest whole number.)

---

23. In the formulas below, "-x" means "the <u>opposite</u> of x" and "-p" means "the <u>opposite</u> of p". Therefore, we can either mentally change the sign of the value of x or p or press $\boxed{+/-}$ after entering the value of x or p before pressing $\boxed{e^x}$ or $\boxed{\text{INV}}$ $\boxed{\ln x}$.

   a) In $\boxed{S = e^{-x}}$, when x = 1.7, S = _____
      (Round to thousandths.)

   b) In $\boxed{V = e^{-p}}$, when p = 4.1, V = _____
      (Round to four decimal places.)

a) 12.2
b) 9,414

---

a) 0.183
b) 0.0166

24. To perform the evaluation below, we press $\boxed{+/-}$ after entering 1.4 for "x" in e^{-x}. Notice that we press $\boxed{=}$ to complete the addition in the numerator before dividing. We rounded the answer to hundredths.

In $\boxed{H = \dfrac{e^x + e^{-x}}{2}}$, find H when $x = 1.4$.

Enter	Press	Display
1.4	$\boxed{e^x}$ $\boxed{+}$	4.0552
	or $\boxed{INV}$ $\boxed{\ln x}$ $\boxed{+}$	
1.4	$\boxed{+/-}$ $\boxed{e^x}$ $\boxed{=}$ $\boxed{\div}$	4.3017969
	or $\boxed{+/-}$ $\boxed{INV}$ $\boxed{\ln x}$ $\boxed{=}$ $\boxed{\div}$	
2	$\boxed{=}$	2.1508985 (H = 2.15)

Using the same steps, do this one. Round to hundredths.

In $\boxed{H = \dfrac{e^x + e^{-x}}{2}}$, when $x = 0.8$, H = _____

25. In the formula below, "-ft" means "the opposite of ft". Therefore, we press $\boxed{+/-}$ after completing the multiplication "ft" by pressing $\boxed{=}$. We rounded the answer to five decimal places.

In $\boxed{R = e^{-ft}}$, find R when $f = 0.5$ and $t = 10.6$.

Enter	Press	Display
0.5	$\boxed{\times}$	0.5
10.6	$\boxed{=}$ $\boxed{+/-}$ $\boxed{e^x}$	.00499159 (R = 0.00499)
	or $\boxed{INV}$ $\boxed{\ln x}$	

Use the same steps for this one. Round to thousandths.

In $\boxed{R = e^{-ft}}$, when $f = 0.1$ and $t = 24$, R = _____

1.34

0.091

26. In the formula below, we must multiply a power of "e" by 14.7 . Two methods are shown. We rounded the answer to tenths.

In $\boxed{P = 14.7e^{-0.2h}}$, find P when h = 1.8 .

1) If a calculator has parentheses symbols, we can use the parentheses to evaluate "-0.2h" before pressing $\boxed{e^x}$ or $\boxed{INV}\ \boxed{\ln x}$.

Enter	Press	Display
14.7	$\boxed{x}\ \boxed{(}$	14.7
0.2	$\boxed{+/-}\ \boxed{x}$	-0.2
1.8	$\boxed{)}\ \boxed{e^x}\ \boxed{=}$	10.255842 (P = 10.3)
	or $\boxed{INV}\ \boxed{\ln x}\ \boxed{=}$	

2) If a calculator does not have parentheses symbols, we have to evaluate the power of "e" first before multiplying by 14.7 .

Enter	Press	Display
0.2	$\boxed{+/-}\ \boxed{x}$	-0.2
1.8	$\boxed{=}\ \boxed{e^x}\ \boxed{x}$	.69767633
	or $\boxed{INV}\ \boxed{\ln x}\ \boxed{x}$	
14.7	$\boxed{=}$	10.255842 (P = 10.3)

Use a calculator for this one. Round to hundredths.

In $\boxed{P = 14.7e^{-0.2h}}$, when h = 2.1 , P = _____

---

27. Two methods for the evaluation below are shown. We rounded the answer to thousandths.

In $\boxed{A = ke^{-ct}}$, find A when k = 4 , c = 0.02 , and t = 140 .

1) If a calculator has parentheses symbols, we can use them to evaluate "-ct" before pressing $\boxed{e^x}$ or $\boxed{INV}\ \boxed{\ln x}$. Notice that we press $\boxed{+/-}$ after pressing $\boxed{)}$ to get "-ct".

Enter	Press	Display
4	$\boxed{x}\ \boxed{(}$	4.
0.02	$\boxed{x}$	0.02
140	$\boxed{)}\ \boxed{+/-}\ \boxed{e^x}\ \boxed{=}$	.24324025 (A = 0.243)
	or $\boxed{INV}\ \boxed{\ln x}\ \boxed{=}$	

Continued on following page.

9.66

27. Continued

    2) If a calculator does not have parentheses symbols, we must evaluate the power of "e" first before multiplying by the value of "k". Notice that we press $\boxed{+/-}$ to get "-ct" before pressing $\boxed{e^x}$ or $\boxed{INV}\ \boxed{\ln x}$.

Enter	Press	Display
0.02	$\boxed{x}$	0.02
140	$\boxed{=}\ \boxed{+/-}\ \boxed{e^x}\ \boxed{x}$	.06081006
	or $\boxed{INV}\ \boxed{\ln x}\ \boxed{x}$	
4	$\boxed{=}$	.24324025 (A = 0.243)

Use a calculator for this one. Round to thousandths.

In $\boxed{A = ke^{-ct}}$, when k = 5, c = 0.04, and t = 55, A = _____

                                                                              0.554

28. Two methods for the evaluation below are also shown. We rounded the answer to tenths.

In $\boxed{i = I_0 e^{-\frac{Rt}{L}}}$, find "i" when $I_0 = 70$, R = 15.7, t = 2.5, and L = 21.9.

    1) If a calculator has parentheses symbols, we can use them to evaluate "$-\frac{Rt}{L}$" before pressing $\boxed{e^x}$ or $\boxed{INV}\ \boxed{\ln x}$. Notice that we press $\boxed{+/-}$ after $\boxed{)}$ to get "$-\frac{Rt}{L}$" .

Enter	Press	Display
70	$\boxed{x}\ \boxed{(}$	70.
15.7	$\boxed{x}$	15.7
2.5	$\boxed{\div}$	39.25
21.9	$\boxed{)}\ \boxed{+/-}\ \boxed{e^x}\ \boxed{=}$	11.661092 (i = 11.7)
	or $\boxed{INV}\ \boxed{\ln x}\ \boxed{=}$	

Continued on following page.

Powers of "e" and Natural Logarithms 363

28. Continued

2) If a calculator does not have parentheses symbols, we must evaluate the power of "e" before multiplying by the value of I_o. Notice that we press $\boxed{+/-}$ to get $"-\frac{Rt}{L}"$ before pressing $\boxed{e^x}$ or $\boxed{INV}\ \boxed{\ln x}$.

Enter	Press	Display
15.7	$\boxed{x}$	15.7
2.5	$\boxed{\div}$	39.25
21.9	$\boxed{=}\ \boxed{+/-}\ \boxed{e^x}\ \boxed{x}$	.16658702
	or $\boxed{INV}\ \boxed{\ln x}\ \boxed{x}$	
70	$\boxed{=}$	11.661092 (i = 11.7)

Using the same method, do this one. Round to tenths.

In $\boxed{V = Ee^{-\frac{t}{RC}}}$, when E = 200, t = 0.75, R = 12, and C = 0.06, V = _____

---

29. To find <u>k</u> below, we substituted the values for A, c, and t and then solved the equation.

Find <u>k</u> when A = 0.75, c = 0.02, and t = 100. Round to hundredths.

$\boxed{A = ke^{-ct}}$

$0.75 = ke^{-(0.02)(100)}$

$0.75 = ke^{-2}$

$0.75 = k(0.1353)$

$k = \dfrac{0.75}{0.1353} = 5.54$

Find <u>k</u> when A = 0.85, c = 0.03, and t = 50. Round to hundredths.

$\boxed{A = ke^{-ct}}$

70.6

---

k = 3.81, from:

$0.85 = ke^{-1.5}$

$k = \dfrac{0.85}{0.223}$

30. In $\boxed{i = I_0 e^{-\frac{Rt}{L}}}$, find I_0 when: $i = 35$, $R = 2.9$, $t = 0.5$, $L = 0.62$

We get: $35 = I_0 e^{-\left(\frac{(2.9)(0.5)}{0.62}\right)}$

$35 = I_0 e^{-2.339}$

$35 = I_0(0.0964)$

$I_0 = \dfrac{35}{0.0964} =$ _____ (Round to the nearest whole number.)

$I_0 = 363$

31. In $\boxed{i = \dfrac{E}{R}\left(1 - e^{-\frac{Rt}{L}}\right)}$, find $\underline{i}$ when: $E = 14$, $R = 20$, $t = 0.04$, $L = 0.6$

We get: $i = \dfrac{14}{20}\left(1 - e^{-\left(\frac{(20)(0.04)}{0.6}\right)}\right)$

$i = 0.7(1 - e^{-1.333})$

$i = 0.7(1 - 0.2637)$

$i = 0.7(0.7363)$

$i =$ _____ (Round to thousandths.)

$i = 0.515$

32. In $\boxed{i = \dfrac{E}{R}\left(1 - e^{-\frac{Rt}{L}}\right)}$, find E when: $i = 3.84$, $R = 22.5$, $t = 2.8$, $L = 71.3$

We get: $3.84 = \dfrac{E}{22.5}\left(1 - e^{-\left(\frac{(22.5)(2.8)}{71.3}\right)}\right)$

$3.84 = \dfrac{E}{22.5}(1 - e^{-0.8836})$

$3.84 = \dfrac{E}{22.5}(1 - 0.4133)$

$3.84 = \dfrac{E}{22.5}(0.5867)$

$E = \dfrac{(3.84)(22.5)}{0.5867} =$ _____ (Round to the nearest whole number.)

$E = 147$

8-4 GRAPHING FORMULAS CONTAINING POWERS OF "e"

In this section, we will graph some basic formulas containing powers of "e".

33. Each formula below contains a power of "e" and several variables.

    $$i = I_0 e^{-\frac{Rt}{L}} \qquad v = V\left(1 - e^{-\frac{t}{RC}}\right)$$

    To graph either formula, we must substitute numerical constants for all but two variables. For example:

    If $I_0 = 100$, $R = 1.2$, and $L = 0.4$ in the left formula,

    we get: $i = 100e^{-\frac{1.2t}{0.4}}$ or $i = 100e^{-3t}$

    If $V = 50$, $R = 10$, and $C = 0.5$ in the right formula,

    we get: $v = 50\left(1 - e^{-\frac{t}{(10)(0.5)}}\right)$ or $v = $ _____

34. We graphed $\boxed{i = 100e^{-3t}}$ below. Since only positive values make sense for i and t, the graph appears only in Quadrant 1.

    $v = 50(1 - e^{-0.2t})$

    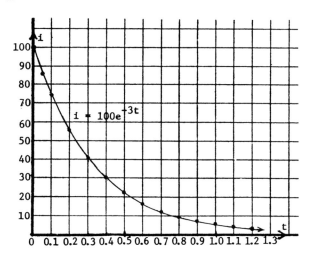

    a) The coordinates of the i-intercept are (,).

    b) Does the graph descend more rapidly between $t = 0.25$ and $t = 0.50$ or between $t = 1.00$ and $t = 1.25$? _____

    c. Write the equation of the asymptote for the graph. _____

366 Powers of "e" and Natural Logarithms

35. We graphed $v = 50(1 - e^{-0.4t})$ below. The graph appears only in Quadrant 1 since only positive values make sense for v and t.

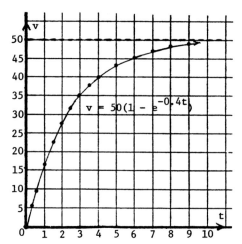

a) The coordinates of the v-intercept are (,   ).

b) Does the graph ascend more rapidly between t = 1 and t = 2 or between t = 5 and t = 6 ? _____

c) Write the equation of the asymptote for the graph. _____

a) (0, 100)

b) Between t = 0.25 and t = 0.50

c) i = 0 (the t-axis)

a) (0, 0)   b) Between t = 1 and t = 2   c) v = 50

SELF-TEST 25 (pages 353-367)

1. Convert $e^{7.83}$ to an ordinary number. Round to the nearest whole number.

   $e^{7.83}$ = _____

2. Convert 0.00495 to a power of "e". Round to thousandths.

   0.00495 = _____

3. Which of the following exponential functions have descending graphs? _____

   a) $y = 0.5e^x$ b) $y = 30e^{-x}$ c) $y = 12e^{-0.8x}$ d) $y = 200e^{0.1x}$

Write the coordinates of the y-intercept of each exponential function.	Write the equation of the asymptote of each exponential function.
4. $y = 40e^{-0.5x}$	6. $y = 6.4e^{-0.25x}$
5. $y = 150(1 - e^{-x})$	7. $y = 25(1 - e^{-x})$

Continued on following page.

Powers of "e" and Natural Logarithms   367

SELF-TEST 25 (Continued)

8. In $\boxed{T = Ce^{-pw}}$, find T when $C = 185$, $p = 0.28$, and $w = 6.75$. Round to tenths.

T = _____

9. In $\boxed{i = \dfrac{E}{R}\left(1 - e^{-\frac{Rt}{L}}\right)}$, find "i" when $E = 12$, $R = 30$, $t = 0.15$, and $L = 2.75$. Round to thousandths.

i = _____

10. In $\boxed{v = Ve^{-\frac{t}{RC}}}$, find V when $v = 34.8$, $t = 20$, $R = 5.20$, and $C = 1.36$. Round to the nearest whole number.

V = _____

11. In $\boxed{\dfrac{A_o}{A} = e^{-ap}}$, find A when $A_o = 11.4$, $a = 0.346$, and $p = 2.92$. Round to tenths.

A = _____

ANSWERS:
1. 2,515
2. $e^{-5.308}$
3. (b) and (c)
4. (0, 40)
5. (0, 0)
6. y = 0 (the x-axis)
7. y = 25
8. T = 27.9
9. i = 0.322
10. V = 589
11. A = 31.3

8-5   NATURAL LOGARITHMS

When a number is written in power-of-"e" form, the exponent of the "e" is called the <u>natural logarithm</u> of the number. In this section, we will discuss natural logarithms and contrast them with common logarithms. We will also contrast natural logarithmic notation with common logarithmic notation.

36. Any positive number can be written in either power-of-ten or power-of-"e" form. For example:

$$325 = 10^{2.5119} \qquad 325 = e^{5.7838}$$

a) The <u>common</u> logarithm of a number is the <u>exponent</u> when the number is written in <u>power-of-ten</u> form.

   The <u>common</u> logarithm of 325 is _____ .

b) The <u>natural</u> logarithm of a number is the <u>exponent</u> when the number is written in <u>power-of-"e"</u> form.

   The <u>natural</u> logarithm of 325 is _____ .

368 Powers of "e" and Natural Logarithms

37. Since $0.875 = e^{-0.1335}$ and $0.875 = 10^{-0.0580}$:

   a) The common logarithm of 0.875 is _____ .

   b) The natural logarithm of 0.875 is _____ .

   a) 2.5119
   b) 5.7838

38. a) To find the common logarithm of a number, we enter the number and press $\boxed{\log}$.

   To four decimal places, the common logarithm of 16.7 is _____ .

   b) To find the natural logarithm of a number, we enter the number and press $\boxed{\ln x}$.

   To four decimal places, the natural logarithm of 16.7 is _____ .

   a) -0.0580
   b) -0.1335

39. The common logarithm of 0.0189 is -1.7235 . The natural logarithm of 0.0189 is -3.9686 . Therefore:

   $0.0189 = 10^{-1.7235}$ $0.0189 = e^{\boxed{}}$

   a) 1.2227
   b) 2.8154

40. a) If the common logarithm of a number is 2.5639 , the number is $10^{2.5639}$ or _____ . (Round to the nearest whole number.)

   b) If the natural logarithm of a number is 2.5639 , the number is $e^{2.5639}$ or _____ . (Round to the nearest tenth.)

   $e^{\boxed{-3.9686}}$

41. a) If the common logarithm of a number is -1.5 , the number is _____ . (Round to four decimal places.)

   b) If the natural logarithm of a number is -1.5 , the number is _____ . (Round to three decimal places.)

   a) 366
   b) 13.0

42. The abbreviations "log" and "ln" are used for logarithms.

   "log 265" means "the common logarithm of 265".

   "ln 265" means "the natural logarithm of 265".

   Note: The abbreviation "ln" contains the first letters of the words "logarithm" and "natural".

   Using the $\boxed{\log}$ and $\boxed{\ln x}$ keys, complete these. Round to four decimal places.

   a) log 265 = _____   b) ln 265 = _____

   a) 0.0316 , from: $10^{-1.5}$
   b) 0.223 , from: $e^{-1.5}$

   a) 2.4232
   b) 5.5797

Powers of "e" and Natural Logarithms   369

43. Since $1 = e^0$, the natural logarithm of "1" is 0. That is:

$$\boxed{\ln 1 = 0}$$

The natural logarithm of any number greater than "1" is a <u>positive</u> number. To confirm that fact, do these:

a) $\ln 1.56 = $ _____   (Round to thousandths.)

b) $\ln 240 = $ _____   (Round to hundredths.)

The natural logarithm of any number between 0 and 1 is a <u>negative</u> number. To confirm that fact, do these:

c) $\ln 0.95 = $ _____   (Round to thousandths.)

d) $\ln 0.004 = $ _____   (Round to hundredths.)

44. Any power-of-ten or power-of-"e" equation can be written as a logarithmic equation. For example:

$27.5 = 10^{1.4393}$ can be written: $\log 27.5 = 1.4393$

$27.5 = e^{3.3142}$ can be written: $\ln 27.5 = 3.3142$

Write each equation below as a logarithmic equation.

a) $10^{-1.5288} = 0.0296$ b) $e^{-2.78} = 0.062$

a) 0.445
b) 5.48
c) -0.051
d) -5.52

45. When a power-of-ten or power-of-"e" equation contains a letter, it can also be written as a logarithmic equation. For example:

$t = 10^{-2.5684}$ can be written: $\log t = -2.5684$

$1.29 = e^x$ can be written: $\ln 1.29 = x$

Write each equation below as a logarithmic equation.

a) $10^y = 6.59$ b) $e^{-2.75} = h$

a) $-1.5288 = \log 0.0296$
b) $-2.78 = \ln 0.062$

a) $y = \log 6.59$
b) $-2.75 = \ln h$

46. Any logarithmic equation can be written as a power-of-ten or power-of-"e" equation. For example:

$\log 1.33 = 0.1239$ can be written: $1.33 = 10^{0.1239}$

$\ln 1.33 = 0.2852$ can be written: $1.33 = e^{0.2852}$

Write each logarithmic equation as a power-of-ten or power-of-"e" equation.

a) $2.8209 = \log 662$ b) $4.1759 = \ln 65.1$

---

47. When a logarithmic equation contains a letter, it can also be written as a power-of-ten or power-of-"e" equation. For example:

$\log 0.589 = d$ can be written: $0.589 = 10^d$

$\ln t = -3.55$ can be written: $t = e^{-3.55}$

a) $10^{2.8209} = 662$
b) $e^{4.1759} = 65.1$

Write each logarithmic equation as a power-of-ten or power-of-"e" equation.

a) $-0.2066 = \log x$ b) $y = \ln 0.265$

---

a) $10^{-0.2066} = x$ b) $e^y = 0.265$

8-6 SOLVING POWER-OF-"e" AND "ln" EQUATIONS

In this section, we will discuss the methods for solving power-of-"e" and "ln" equations that contain a letter.

48. In each power-of-"e" equation below, the letter stands for <u>an ordinary number</u>. Use $\boxed{e^x}$ or $\boxed{\text{INV}}\boxed{\ln x}$ to solve each equation.

a) Round to tenths.   b) Round to four decimal places.

$e^{4.25} = t$ $y = e^{-2.685}$

$t = \underline{\qquad}$ $y = \underline{\qquad}$

49. In each <u>power-of-"e"</u> equation below, the letter stands for <u>an exponent</u>. Use the $\boxed{\ln x}$ key to solve each equation. Round each exponent to three decimal places.

a) $2{,}750 = e^x$ b) $e^y = 0.0999$

$x = \underline{\qquad}$ $y = \underline{\qquad}$

a) $t = 70.1$
b) $y = 0.0682$

---

a) $x = 7.919$
b) $y = -2.304$

Powers of "e" and Natural Logarithms   371

50. When a "ln" equation contains a letter, we can solve it by writing the equation in power-of-"e" form. Let's use that method to solve the equation below.

$$\ln R = 1.8576$$

a) Write the equation in power-of-"e" form. _____

b) Therefore, rounded to hundredths, R = _____

---

51. Let's solve $\ln 0.193 = y$.

a) Write the equation in power-of-"e" form. _____

b) Therefore, rounded to four decimal places, y = _____ .

a) $R = e^{1.8576}$

b) $R = 6.41$

---

52. Let's solve $-2.56 = \ln V$.

a) Write the equation in power-of-"e" form. _____

b) Therefore, rounded to four decimal places, V = _____ .

a) $0.193 = e^y$

b) $y = -1.6451$

---

53. Let's solve $t = \ln 6,295$.

a) Write the equation in power-of-"e" form. _____

b) Therefore, rounded to hundredths, t = _____ .

a) $e^{-2.56} = V$

b) $V = 0.0773$

---

54. When the letter is on the opposite side of "ln" in an equation, it stands for the logarithm (or exponent). For example:

In $\ln 10.8 = v$, "v" stands for the logarithm since: $10.8 = e^v$

In $y = \ln 0.725$, "y" stands for the logarithm since: $e^y = 0.725$

In such cases, we can simply use the $\boxed{\ln x}$ key without converting to a power-of-"e" equation. Solve these. Round to three decimal places.

a) $\ln 0.087 = F$ b) $Q = \ln 13,500$

   F = _____   Q = _____

a) $e^t = 6,295$

b) $t = 8.75$

---

a) $F = -2.442$

b) $Q = 9.510$

372 Powers of "e" and Natural Logarithms

55. When the letter follows "ln" in an equation, it stands for the ordinary number. For example:

In ln P = -1.75 , "P" stands for the ordinary number since: $P = e^{-1.75}$

In 3.69 = ln H , "H" stands for the ordinary number since: $e^{3.69} = H$

In such cases, we can convert the "ln" equation to a power-of-"e" equation and use $\boxed{e^x}$ or $\boxed{INV}$ $\boxed{\ln x}$. Solve these.

    a) Round to tenths.   b) Round to thousandths.

      ln B = 4.488   -0.36 = ln G

      B = _____   G = _____

---

56. In each equation below, the natural logarithm is given.

    ln x = 2.43   -0.69 = ln T

Finding the value of the variable when the logarithm is given is called finding the "inverse logarithm" or finding the "antilogarithm". Therefore, we can solve each equation by simply using $\boxed{INV}$ $\boxed{\ln x}$ without converting to power-of-"e" form. Do so.

    a) Round to tenths.   b) Round to thousandths.

      ln x = 2.43   -0.69 = ln T

      x = _____   T = _____

a) B = 88.9 , from:
    B = $e^{4.488}$

b) G = 0.698 , from:
    $e^{-0.36} = G$

a) x = 11.4   b) T = 0.502

8-7 EVALUATING "ln" FORMULAS

In this section, we will show how a calculator can be used to evaluate formulas containing "ln" expressions.

57. The steps for the evaluation below are shown. Notice that we entered 50 and pressed $\boxed{\ln x}$ before pressing $\boxed{=}$. We rounded the answer to tenths.

In $\boxed{D = k \ln P}$, find D when k = 10 and P = 50 .

Enter	Press	Display	
10	$\boxed{\times}$	10.	
50	$\boxed{\ln x}$ $\boxed{=}$	39.12023	(D = 39.1)

Use a calculator for this one. Round to tenths.

In $\boxed{D = k \ln P}$, when k = 7.5 and P = 18.5 , D = _____

D = 21.9

58. The steps for the evaluation below are shown. Notice again that we entered the value for P and then pressed $\boxed{\ln x}$. We rounded the answer to thousandths.

In $\boxed{w = \dfrac{d}{\ln P}}$, find "w" when $d = 4.24$ and $P = 70$.

Enter	Press	Display	
4.24	$\boxed{\div}$	4.24	
70	$\boxed{\ln x}\ \boxed{=}$	.99800041	(w = 0.998)

Use a calculator for this one. Round to hundredths.

In $\boxed{w = \dfrac{d}{\ln P}}$, when $d = 3.99$ and $P = 80$, $w = $ _____

59. In the formula below, $\ln\left(\dfrac{A_0}{A}\right)$ is the natural logarithm of a fraction or division. Notice that we press $\boxed{=}$ to complete that division before pressing $\boxed{\ln x}$, and then divide by the value of "k". We rounded the answer to hundredths.

In $\boxed{t = \dfrac{\ln\left(\dfrac{A_0}{A}\right)}{k}}$, find "t" when $A_0 = 60$, $A = 20$, and $k = 0.31$.

Enter	Press	Display	
60	$\boxed{\div}$	60.	
20	$\boxed{=}\ \boxed{\ln x}\ \boxed{\div}$	1.0986123	
0.31	$\boxed{=}$	3.5439106	(t = 3.54)

Use a calculator for this one. Round to hundredths.

In $\boxed{k = \dfrac{\ln\left(\dfrac{A_0}{A}\right)}{t}}$, when $A_0 = 38.9$, $A = 9.45$, and $t = 1.29$,

$k = $ _____

$w = 0.91$

$k = 1.10$

60. Two calculator methods are shown for the evaluation below. We rounded the answer to hundredths.

In $\boxed{v = c \ln\left(\frac{M}{m}\right)}$, find "v" when $c = 1.88$, $M = 750$, and $m = 112$.

1) If a calculator has parentheses symbols, we can use them to evaluate $\left(\frac{M}{m}\right)$ before pressing $\boxed{\ln x}$.

Enter	Press	Display	
1.88	$\boxed{x}$ $\boxed{(}$	1.88	
750	$\boxed{\div}$	750.	
112	$\boxed{)}$ $\boxed{\ln x}$ $\boxed{=}$	3.5749598	(v = 3.57)

2) If a calculator does not have parentheses symbols, we must evaluate $\ln\left(\frac{M}{m}\right)$ first and then multiply by 1.88 . Notice that we pressed $\boxed{=}$ to complete the division before pressing $\boxed{\ln x}$.

Enter	Press	Display	
750	$\boxed{\div}$	750.	
112	$\boxed{=}$ $\boxed{\ln x}$ $\boxed{x}$	1.9015743	
1.88	$\boxed{=}$	3.5749598	(v = 3.57)

Use a calculator for this one. Round to hundredths.

In $\boxed{v = c \ln\left(\frac{M}{m}\right)}$, when $c = 1.86$, $M = 869$, and $m = 90$,

v = _____

61. In $\boxed{Q = -\ln B}$, "-ln B" means "the opposite of ln B". Therefore, after finding ln B , we press $\boxed{+/-}$ to get its opposite. An example is shown. We rounded the answer to hundredths.

In $\boxed{Q = -\ln B}$, find Q when $B = 3.5$.

Enter	Press	Display	
3.5	$\boxed{\ln x}$ $\boxed{+/-}$	-1.252763	(Q = -1.25)

Using the same steps, do this one. Round to hundredths.

In $\boxed{Q = -\ln B}$, when $B = 120$, Q = _____

v = 4.22

Q = -4.79

62. To complete the evaluation below, we used a calculator to find ln 40 and then solved the equation. Complete the other evaluation.

Find <u>k</u> when D = 60 and P = 40 . Round to tenths.

$$D = k \ln P$$

$60 = k \ln 40$

$60 = k(3.689)$

$k = \dfrac{60}{3.689} = 16.3$

Find <u>d</u> when w = 0.73 and P = 75 . Round to hundredths.

$$w = \dfrac{d}{\ln P}$$

d = 3.15 , from:

$0.73 = \dfrac{d}{4.317}$

$d = (0.73)(4.317)$

---

63. In the formula at the right, we want to find <u>k</u> when:

$A_o = 60$

$A = 15$

$t = 1.52$

$$\ln\left(\dfrac{A_o}{A}\right) = kt$$

$\ln\left(\dfrac{60}{15}\right) = k(1.52)$

$\ln 4 = 1.52k$

Complete the solution. Round to thousandths.

k = 0.912 , from:

$1.386 = 1.52k$

$k = \dfrac{1.386}{1.52}$

---

64. A two-step process is needed to find "A" in the evaluation below. We rounded the answer to tenths.

In $\boxed{\ln A = \dfrac{h}{t}}$, find A when h = 80 and t = 20 .

1) First we find "ln A" by substituting:

$\ln A = \dfrac{80}{20} = 4$

2) Then we use $\boxed{e^x}$ or $\boxed{\text{INV}}$ $\boxed{\ln x}$ to find A . The steps are:

Enter	Press	Display	
4	$\boxed{e^x}$ or $\boxed{\text{INV}}$ $\boxed{\ln x}$	54.59815	(A = 54.6)

Use the same steps for this one. Round to the nearest whole number.

In $\boxed{\ln A = \dfrac{h}{t}}$, when h = 150 and t = 30 , A = _____

A = 148

65. To solve for P below, we solved for ln P first. Complete the other evaluation.

Find P when D = 35 and k = 10. Round to tenths.

$$\boxed{D = k \ln P}$$

$35 = 10 \ln P$

$\ln P = \dfrac{35}{10} = 3.5$

$P = 33.1$

Find P when w = 0.93 and d = 5.07. Round to the nearest whole number.

$$\boxed{w = \dfrac{d}{\ln P}}$$

P = 233, from: $\ln P = \dfrac{5.07}{0.93}$

8-8 EVALUATIONS REQUIRING A CONVERSION TO POWER-OF-"e" FORM

In this section, we will discuss evaluations with "ln" formulas that require a conversion to power-of-"e" form.

66. In each equation below, the variable is part of the "ln" expression.

$\ln\left(\dfrac{V}{20}\right) = 1.5$ $\qquad$ $\ln\left(\dfrac{65}{P}\right) = -2.2$

To solve for the variable, we must convert the equation to power-of-"e" form. That is:

$\ln\left(\dfrac{V}{20}\right) = 1.5$ is converted to $\dfrac{V}{20} = e^{1.5}$

$\ln\left(\dfrac{65}{P}\right) = -2.2$ is converted to $\dfrac{65}{P} = e^{-2.2}$

Notice in each conversion that the whole quantity in parentheses equals a power of "e". Convert each equation below to power-of-"e" form.

a) $\ln\left(\dfrac{T}{100}\right) = 1.9$ $\qquad$ b) $\ln\left(\dfrac{50}{Q}\right) = -0.6$

a) $\dfrac{T}{100} = e^{1.9}$

b) $\dfrac{50}{Q} = e^{-0.6}$

67. We solved for S to the nearest whole number below by converting to power-of-"e" form. Using the same steps, solve for F in the other equation. Round to the nearest whole number.

$$\ln\left(\frac{S}{40}\right) = 1.4 \qquad \ln\left(\frac{80}{F}\right) = -2.1$$

$$\frac{S}{40} = e^{1.4}$$

$$\frac{S}{40} = 4.055$$

$$S = 40(4.055)$$

$$S = 162$$

---

68. In the formula at the right, we want to find A_0 when:

   $A = 3.75$
   $k = 0.529$
   $t = 4.67$

   We substituted and converted to power-of-"e" form. Complete the solution. Round to tenths.

$$\boxed{\ln\left(\frac{A_0}{A}\right) = kt}$$

$$\ln\left(\frac{A_0}{3.75}\right) = (0.529)(4.67)$$

$$\ln\left(\frac{A_0}{3.75}\right) = 2.470$$

$$\frac{A_0}{3.75} = e^{2.470}$$

$F = 653$, from:

$$\frac{80}{F} = e^{-2.1}$$

$$\frac{80}{F} = 0.1225$$

---

69. In the formula at the right, we want to find A_0 when:

   $A = 13.7$
   $k = 0.0529$
   $t = 17.5$

   We substituted and converted to power-of-"e" form. Complete the solution. Round to tenths.

$$\boxed{\ln\left(\frac{A}{A_0}\right) = -kt}$$

$$\ln\left(\frac{13.7}{A_0}\right) = -(0.0529)(17.5)$$

$$\ln\left(\frac{13.7}{A_0}\right) = -0.9258$$

$$\frac{13.7}{A_0} = e^{-0.9258}$$

$A_0 = 44.3$ from:

$$\frac{A_0}{3.75} = 11.83$$

$A_0 = 34.6$, from:

$$\frac{13.7}{A_0} = 0.3962$$

70. In the formula at the right, we want to find M when:

$$v = c \ln\left(\frac{M}{m}\right)$$

v = 3.89
c = 1.72
m = 102

$3.89 = 1.72 \ln\left(\frac{M}{102}\right)$

$\frac{3.89}{1.72} = \ln\left(\frac{M}{102}\right)$

We substituted, isolated the "ln" expression, and converted to power-of-"e" form. Complete the solution. Round to the nearest whole number.

$2.262 = \ln\left(\frac{M}{102}\right)$

$e^{2.262} = \frac{M}{102}$

M = 979, from: $9.602 = \frac{M}{102}$

8-9 EVALUATIONS REQUIRING A CONVERSION TO "ln" FORM

In this section, we will discuss evaluations with power-of-"e" formulas that require a conversion to "ln" form.

71. In each equation below, the variable is the exponent of "e" or part of the exponent of "e".

$2.5 = e^p$ $80 = e^{-q}$ $55 = e^{-0.4h}$

To solve for the variable, we convert the equation to "ln" form. That is:

$2.5 = e^p$ is converted to $\ln 2.5 = p$

$80 = e^{-q}$ is converted to $\ln 80 = -q$

$55 = e^{-0.4h}$ is converted to $\ln 55 = -0.4h$

Convert each equation below to "ln" form.

a) $1.8 = e^x$ b) $95 = e^{-t}$ c) $42.5 = e^{-0.1t}$

72. We solved for p below by converting to "ln" form. Solve for t in the other equation. Round to hundredths.

$1.2 = e^p$ $40 = e^t$

$\ln 1.2 = p$

$p = 0.182$

a) $\ln 1.8 = x$

b) $\ln 95 = -t$

c) $\ln 42.5 = -0.1t$

73. We converted to "ln" form to solve for q below. Notice that we took the opposite of both sides in the final step. Solve for x in the other equation. Round to hundredths.

$25 = e^{-q}$ $\qquad\qquad$ $70 = e^{-x}$

$\ln 25 = -q$

$3.22 = -q$

$q = -3.22$

$t = 3.69$, from:

$\ln 40 = t$

---

74. We converted to "ln" form to solve for y below. Solve for x in the other equation. Round to hundredths.

$2.65 = e^{-0.2y}$ $\qquad\qquad$ $6.13 = e^{-0.4x}$

$\ln 2.65 = -0.2y$

$0.9746 = -0.2y$

$y = \dfrac{0.9746}{-0.2}$

$y = -4.87$

$x = -4.25$, from:

$\ln 70 = -x$

$4.25 = -x$

---

75. To solve for x below, we isolated the power of "e" before converting to "ln" form. Solve for y in the other equation. Round to tenths.

$240 = 100e^{-30x}$ $\qquad\qquad$ $157 = 10e^{-0.2y}$

$\dfrac{240}{100} = e^{-30x}$

$2.4 = e^{-30x}$

$\ln 2.4 = -30x$

$0.8755 = -30x$

$x = \dfrac{0.8755}{-30}$

$x = -0.0292$

$x = -4.53$, from:

$\ln 6.13 = -0.4x$

$x = \dfrac{1.813}{-0.4}$

$y = -13.8$, from:

$\ln 15.7 = -0.2y$

$2.754 = -0.2y$

380 Powers of "e" and Natural Logarithms

76. We converted to "ln" notation to complete the evaluation below. Use the same method for the other evaluation. Round to thousandths.

 Find p when V = 37.6 .   Find t when R = 0.47 .
 Round to hundredths.   Round to thousandths.

$$\boxed{V = e^p}$$ $$\boxed{R = e^{-t}}$$

  $37.6 = e^p$

 $\ln 37.6 = p$

   $p = 3.63$

---

77. We converted to "ln" notation to complete the evaluation below. Use the same method for the other evaluation. Round to tenths.

 Find k when A = 2.66 ,   Find t when A = 0.417 ,
 A_0 = 17.4 , and t = 50 .   k = 3.61 , and c = 0.059 .
 Round to four decimal places.

$$\boxed{A = A_0 e^{-kt}}$$ $$\boxed{A = ke^{-ct}}$$

  $2.66 = 17.4 e^{-k(50)}$

  $\dfrac{2.66}{17.4} = e^{-50k}$

  $0.1529 = e^{-50k}$

 $\ln 0.1529 = -50k$

  $-1.878 = -50k$

   $k = \dfrac{-1.878}{-50}$

   $k = 0.0376$

---

t = 0.755 , from:

$\ln 0.47 = -t$

$-0.755 = -t$

---

t = 36.6

SELF-TEST 26 (pages 367-381)

1. Find the natural logarithm of 0.706 . Round to four decimal places. _____

2. Find the number whose natural logarithm is 3.1539 . Round to tenths. _____

3. Write this logarithmic equation as a power-of-"e" equation.

   $\ln 16.2 = 2.785$ _____

4. Write this power-of-"e" equation as a logarithmic equation.

   $e^{-0.263} = 0.769$ _____

5. Find "t". Round to thousandths.

   $e^t = 1.53$

   t = _____

6. Find W. Round to hundredths.

   $\ln 7{,}390 = W$

   W = _____

7. Find P . Round to the nearest whole number.

   $5.906 = \ln P$

   P = _____

8. In $\boxed{v = c \ln\left(\dfrac{M}{m}\right)}$, find "v" when $c = 1.78$, $M = 953$, and $m = 226$. Round to hundredths.

   v = _____

9. In $\boxed{r = k \ln Q}$, find Q when $r = 380$ and $k = 105$. Round to tenths.

   Q = _____

10. In $\boxed{\ln\left(\dfrac{P}{V}\right) = -cr}$, find V when $P = 45.9$, $c = 0.067$, and $r = 12.4$. Round to the nearest whole number.

    V = _____

11. In $\boxed{A = A_0 e^{-kt}}$, find "t" when $A = 36.1$, $A_0 = 50$, and $k = 1.25$. Round to thousandths.

    t = _____

ANSWERS:

1. -0.3481
2. 23.4
3. $16.2 = e^{2.785}$
4. $-0.263 = \ln 0.769$
5. $t = 0.425$
6. $W = 8.91$
7. $P = 367$
8. $v = 2.56$
9. $Q = 37.3$
10. $V = 105$
11. $t = 0.261$

382   Powers of "e" and Natural Logarithms

8-10 LAWS OF NATURAL LOGARITHMS

In this section, we will discuss the three basic laws of natural logarithms. We will also show how these laws can be reversed to write "ln" expressions in equivalent forms.

78. The law of common logarithms <u>for multiplication</u> is shown below.

$$\boxed{\log(ab) = \log a + \log b}$$

The same law applies to natural logarithms. That is:

$$\boxed{\ln(ab) = \ln a + \ln b}$$

Using the above law, write these in an equivalent form.

a) $\ln(pq) = $ _____   b) $\ln(RV) = $ _____

79. The law of common logarithms <u>for division (or a fraction)</u> is shown below.

$$\boxed{\log\left(\frac{a}{b}\right) = \log a - \log b}$$

The same law applies to natural logarithms. That is:

$$\boxed{\ln\left(\frac{a}{b}\right) = \ln a - \ln b}$$

Using the above law, write these in an equivalent form.

a) $\ln\left(\frac{p}{q}\right) = $ _____   b) $\ln\left(\frac{A_o}{A}\right) = $ _____

a) $\ln p + \ln q$
b) $\ln R + \ln V$

80. The law of common logarithms <u>for a power</u> is shown below.

$$\boxed{\log a^b = b \log a}$$

The same law applies to natural logarithms. That is:

$$\boxed{\ln a^b = b \ln a}$$

Using the above law, write these in an equivalent form.

a) $\ln p^q = $ _____   b) $\ln t^{-1} = $ _____

a) $\ln p - \ln q$
b) $\ln A_o - \ln A$

81. The law of logarithms for multiplication can be reversed. That is:

Just as: $\ln(ab) = \ln a + \ln b$,

$\ln a + \ln b = \ln(ab)$

Using the reversed law, write these in an equivalent form.

a) $\ln p + \ln q = $ _____   b) $\ln T + \ln V = $ _____

a) $q \ln p$
b) $-1 \ln t$ or $-\ln t$

82. The law of logarithms for division (or a fraction) can be reversed. That is:

$$\text{Just as:} \quad \ln\left(\frac{a}{b}\right) = \ln a - \ln b$$

$$\ln a - \ln b = \ln\left(\frac{a}{b}\right)$$

Using the reversed law, write these in an equivalent form.

a) $\ln p - \ln q = $ _____   b) $\ln A - \ln A_0 = $ _____

a) $\ln(pq)$

b) $\ln(TV)$

---

83. The law of logarithms for a power can also be reversed. That is:

$$\text{Just as:} \quad \ln a^b = b \ln a$$

$$b \ln a = \ln a^b$$

Using the reversed law, write these in an equivalent form.

a) $p \ln q = $ _____   b) $k \ln V = $ _____   c) $-1 \ln d = $ _____

a) $\ln\left(\frac{p}{q}\right)$

b) $\ln\left(\frac{A}{A_0}\right)$

---

84. Using the reversed law for a power, we can write $\boxed{P = -\ln A}$ in an equivalent form without a "$-$". The steps are:

1. Substitute -1 for the "$-$". $P = -1 \ln A$

2. Apply the reversed law for a power. $P = \ln A^{-1}$

3. Apply the definition of a power with a negative exponent. $P = \ln\left(\frac{1}{A}\right)$

Using the same steps, write each formula in an equivalent form without a "$-$".

a) $R = -\ln d$ b) $V = -\ln t$

a) $\ln q^p$

b) $\ln V^k$

c) $\ln d^{-1}$

---

a) $R = \ln\left(\frac{1}{d}\right)$

b) $V = \ln\left(\frac{1}{t}\right)$

384 Powers of "e" and Natural Logarithms

8-11 REARRANGING "ln" FORMULAS

In this section, we will show how "ln" formulas can be rearranged to solve for a variable or the logarithm of a variable.

85. We solved for <u>k</u> below. Solve for V in the other formula.

$$\boxed{D = k \ln P} \qquad \boxed{w = \frac{V}{\ln T}}$$

$$\frac{D}{\ln P} = k$$

or

$$k = \frac{D}{\ln P}$$

---

86. We solved for ln T below. Solve for ln P in the other formula.

$$\boxed{w = \frac{V}{\ln T}} \qquad \boxed{D = k \ln P}$$

$$w \ln T = V$$

$$\ln T = \frac{V}{w}$$

$V = w \ln T$

---

87. We solved for $\ln\left(\frac{M}{N}\right)$ below. Solve for $\ln\left(\frac{p}{q}\right)$ in the other formula.

$$\boxed{v = c \ln\left(\frac{M}{N}\right)} \qquad \boxed{R = k \ln\left(\frac{p}{q}\right)}$$

$$\frac{v}{c} = \ln\left(\frac{M}{N}\right)$$

or

$$\ln\left(\frac{M}{N}\right) = \frac{v}{c}$$

$\ln P = \frac{D}{k}$

---

88. To solve for $\ln A_0$ below, we used the law of logarithms for division. Solve for ln p in the other formula.

$$\boxed{\ln\left(\frac{A_0}{A}\right) = kt} \qquad \boxed{cv = \ln\left(\frac{p}{q}\right)}$$

$$\ln A_0 - \ln A = kt$$

$$\ln A_0 = kt + \ln A$$

$\ln\left(\frac{p}{q}\right) = \frac{R}{k}$

$\ln p = cv + \ln q$

89. To solve for ln t below, we had to take the opposite of both sides in the final step. Solve for ln A in the other formula.

$$\boxed{\ln\left(\frac{c}{t}\right) = b}$$ $$\boxed{kt = \ln\left(\frac{A_o}{A}\right)}$$

$\ln c - \ln t = b$

$-\ln t = b - \ln c$

$\ln t = \ln c - b$

---

90. To solve for ln a below, we isolated the "ln" expression first. Solve for ln N in the other formula.

$$\boxed{D = k\ln\left(\frac{a}{b}\right)}$$ $$\boxed{v = c\ln\left(\frac{M}{N}\right)}$$

$\frac{D}{k} = \ln\left(\frac{a}{b}\right)$

$\frac{D}{k} = \ln a - \ln b$

$\frac{D}{k} + \ln b = \ln a$

or

$\ln a = \frac{D}{k} + \ln b$

$\ln A = \ln A_o - kt$

---

91. In the evaluation below, <u>k</u> is a non-solved-for variable.

In $\boxed{D = k \ln P}$, find <u>k</u> when D = 25 and P = 40.

We can use either of the methods below for the evaluation.

Substitute into the formula as it stands and then solve for <u>k</u>.	Rearrange the formula to solve for <u>k</u> and then substitute.
$\boxed{D = k \ln P}$	$\boxed{D = k \ln P}$
$25 = k \ln 40$	$k = \frac{D}{\ln P}$
$25 = k(3.689)$	$k = \frac{25}{\ln 40}$
$k = \frac{25}{3.689}$	$k = \frac{25}{3.689}$
$k = 6.78$	$k = 6.78$

Did we get the same value of <u>k</u> with both methods? _____

$\ln N = \ln M - \frac{v}{c}$

Yes

386 Powers of "e" and Natural Logarithms

8-12 REARRANGEMENTS REQUIRING A CONVERSION TO POWER-OF-"e" FORM

To solve for a variable in an "ln" formula, we sometimes have to convert to power-of-"e" form. We will discuss rearrangements of that type in this section.

92. We converted two "ln" formulas to power-of-"e" formulas below.

$$\boxed{\ln Q = -t}$$
$$Q = e^{-t}$$

$$\boxed{\ln\left(\frac{p}{q}\right) = ab}$$
$$\frac{p}{q} = e^{ab}$$

Following the examples, convert these to power-of-"e" formulas.

a) $\ln A = -bv$

b) $\ln\left(\frac{A_o}{A}\right) = \frac{c}{k}$

93. To convert the formula below to power-of-"e" form, we had to isolate the "ln" expression first. Convert the other formula to power-of-"e" form.

$$\boxed{v = c \ln\left(\frac{M}{N}\right)}$$
$$\frac{v}{c} = \ln\left(\frac{M}{N}\right)$$
$$e^{\frac{v}{c}} = \frac{M}{N}$$
or
$$\frac{M}{N} = e^{\frac{v}{c}}$$

$$\boxed{t = \frac{\ln\left(\frac{D_o}{D}\right)}{k}}$$

a) $A = e^{-bv}$

b) $\frac{A_o}{A} = e^{\frac{c}{k}}$

94. To solve for P below, we isolated ln P and then converted to power-of-"e" form. Solve for T in the other formula.

$$\boxed{D = k \ln P}$$
$$\ln P = \frac{D}{k}$$
$$P = e^{\frac{D}{k}}$$

$$\boxed{w = \frac{S}{\ln T}}$$

$\frac{D_o}{D} = e^{kt}$

$T = e^{\frac{S}{w}}$

95. We solved for A_o below. Solve for a in the other formula.

$$\boxed{\ln\left(\frac{A_o}{A}\right) = kt} \qquad \boxed{D = k\ln\left(\frac{a}{b}\right)}$$

$$\frac{A_o}{A} = e^{kt}$$

$$A_o = Ae^{kt}$$

96. We solved for t below. Solve for b in the other formula.

$$\boxed{\ln\left(\frac{c}{t}\right) = b} \qquad \boxed{D = k\ln\left(\frac{a}{b}\right)}$$

$$\frac{c}{t} = e^b$$

$$c = te^b$$

$$t = \frac{c}{e^b}$$

$a = be^{\frac{D}{k}}$

97. In the last frame, we got these two solutions.

$$t = \frac{c}{e^b} \qquad\qquad b = \frac{a}{e^{\frac{D}{k}}}$$

To put the two solutions in the preferred form, we move the power of "e" to the numerator by changing the sign of its exponent. We get:

$$t = \frac{c}{e^b} = ce^{-b} \qquad\qquad b = \frac{a}{e^{\frac{D}{k}}} = ae^{-\frac{D}{k}}$$

Write each solution in the preferred form.

a) $V_1 = \dfrac{V_2}{e^t} = $ _____ b) $R = \dfrac{T}{\frac{a}{e^c}} = $ _____

$b = \dfrac{a}{e^{\frac{D}{k}}}$

a) $V_1 = V_2 e^{-t}$

b) $R = Te^{-\frac{a}{c}}$

388 Powers of "e" and Natural Logarithms

98. Three methods are shown for the evaluation below. Any of the three methods can be used.

In $\boxed{D = k \ln P}$, find $\underline{P}$ when $D = 45$ and $k = 10$.

Substituting, solving for $\ln P$, and then using a calculator.	Solving for $\ln P$, then substituting and using a calculator.	Solving for P, then substituting and using a calculator.
$\boxed{D = k \ln P}$	$\boxed{D = k \ln P}$	$\boxed{D = k \ln P}$
$45 = 10 \ln P$	$\ln P = \dfrac{D}{k}$	$\ln P = \dfrac{D}{k}$
$\ln P = \dfrac{45}{10} = 4.5$	$\ln P = \dfrac{45}{10} = 4.5$	$P = e^{\frac{D}{k}}$
$P = 90.0$	$P = 90.0$	$P = e^{\frac{45}{10}} = e^{4.5}$
		$P = 90.0$

8-13 REARRANGING FORMULAS CONTAINING POWERS OF "e"

In this section, we will discuss the rearrangement of formulas containing powers of "e". Emphasis is given to solving for a variable in the exponent of "e".

99. We solved for $\underline{k}$ below and wrote the solution in the preferred form.
Solve for I_0 in the other formula.

$\boxed{R = ke^{-ay}}$ $\boxed{i = I_0 e^{-\frac{Rt}{L}}}$

$k = \dfrac{R}{e^{-ay}}$

$k = Re^{ay}$

100. We converted to "ln" form to solve for $\underline{y}$ below. Solve for $\underline{p}$ in the other formula.

$\boxed{T = e^y}$ $\boxed{V = e^p}$

$\ln T = y$

or

$y = \ln T$

$I_0 = ie^{\frac{Rt}{L}}$

$p = \ln V$

101. We converted to "ln" form to solve for t below. Solve for b in the other formula.

$$\boxed{P = e^{kt}} \qquad \boxed{H = e^{bm}}$$

$$\ln P = kt$$

$$t = \frac{\ln P}{k}$$

---

102. To solve for m below, we used the oppositing principle in the final step. Solve for x in the other formula.

$$\boxed{Q = e^{-m}} \qquad \boxed{V = e^{-x}}$$

$$\ln Q = -m$$

$$-\ln Q = m$$

or

$$m = -\ln Q$$

$b = \dfrac{\ln H}{m}$

---

103. We got these two solutions in the last frame.

$$m = -\ln Q \qquad x = -\ln V$$

To get rid of the "-" and get the solution in the preferred form, we use the reverse law of logarithms for a power and then use the definition of a power with a negative exponent. We get:

$$m = -\ln Q = \ln Q^{-1} = \ln\left(\frac{1}{Q}\right)$$

$$x = -\ln V = \ln V^{-1} = \ln\left(\frac{1}{V}\right)$$

Write each solution below in the preferred form.

a) $t = -\ln D =$ _____ b) $y = -\ln P =$ _____

$x = -\ln V$

---

104. We solved for x below. Solve for c in the other formula.

$$\boxed{S = e^{-bx}} \qquad \boxed{H = e^{-ct}}$$

$$\ln S = -bx$$

$$-\ln S = bx$$

$$x = -\frac{\ln S}{b}$$

a) $\ln\left(\dfrac{1}{D}\right)$

b) $\ln\left(\dfrac{1}{P}\right)$

Powers of "e" and Natural Logarithms 389

390 Powers of "e" and Natural Logarithms

105. We got these two solutions in the last frame.

$$x = -\frac{\ln S}{b} \qquad c = -\frac{\ln H}{t}$$

The solutions are not in the preferred form because of the "-". To put them in the preferred form, we use these steps:

$$x = -\frac{\ln S}{b} = \frac{-\ln S}{b} = \frac{\ln S^{-1}}{b} = \frac{\ln\left(\frac{1}{S}\right)}{b} = \frac{1}{b}\ln\left(\frac{1}{S}\right)$$

$$c = -\frac{\ln H}{t} = \frac{-\ln H}{t} = \frac{\ln H^{-1}}{t} = \frac{\ln\left(\frac{1}{H}\right)}{t} = \frac{1}{t}\ln\left(\frac{1}{H}\right)$$

Write each solution below in the preferred form.

a) $m = -\frac{\ln P}{d} = $ _____ b) $k = -\frac{\ln T}{a} = $ _____

$c = -\frac{\ln H}{t}$

106. Before converting to "ln" form, we must isolate the power of "e" first. Two examples are shown.

$$\boxed{T = be^{-y}}$$
$$\frac{T}{b} = e^{-y}$$
$$\ln\left(\frac{T}{b}\right) = -y$$

$$\boxed{i = I_0 e^{-\frac{Rt}{L}}}$$
$$\frac{i}{I_0} = e^{-\frac{Rt}{L}}$$
$$\ln\left(\frac{i}{I_0}\right) = -\frac{Rt}{L}$$

To solve for p below, we isolated the power of "e" first. Solve for x in the other formula.

$$\boxed{V = ke^p}$$
$$\frac{V}{k} = e^p$$
$$\ln\left(\frac{V}{k}\right) = p$$
or
$$p = \ln\left(\frac{V}{k}\right)$$

$$\boxed{R = ce^x}$$

a) $\frac{1}{d}\ln\left(\frac{1}{P}\right)$

b) $\frac{1}{a}\ln\left(\frac{1}{T}\right)$

$x = \ln\left(\frac{R}{c}\right)$

107. To solve for y below, we used the opposing principle in the last step. Solve for t in the other formula.

$$T = be^{-y}$$

$$R = ke^{-t}$$

$$\frac{T}{b} = e^{-y}$$

$$\ln\left(\frac{T}{b}\right) = -y$$

$$-\ln\left(\frac{T}{b}\right) = y$$

or

$$y = -\ln\left(\frac{T}{b}\right)$$

---

108. We got these two solutions in the last frame.

$$y = -\ln\left(\frac{T}{b}\right) \qquad t = -\ln\left(\frac{R}{k}\right)$$

The preferred form for each solution is shown below. Notice that the "-" is gone and the fraction is replaced by its reciprocal.

$$y = \ln\left(\frac{b}{T}\right) \qquad t = \ln\left(\frac{k}{R}\right)$$

The steps needed to convert one solution to the preferred form are shown below.

$$y = -\ln\left(\frac{T}{b}\right) = (-1)\ln\left(\frac{T}{b}\right) = \ln\left(\frac{T}{b}\right)^{-1} = \ln\left(\frac{1}{\frac{T}{b}}\right) = \ln\left(\frac{b}{T}\right)$$

Write each solution in the preferred form.

a) $V = -\ln\left(\frac{a}{b}\right) = $ _____

b) $F = -\ln\left(\frac{P}{S}\right) = $ _____

$t = -\ln\left(\frac{R}{k}\right)$

---

109. We solved for x in the formula below.

$$P = ke^{-bx}$$

$$\frac{P}{k} = e^{-bx}$$

$$\ln\left(\frac{P}{k}\right) = -bx$$

$$-\ln\left(\frac{P}{k}\right) = bx$$

$$x = \frac{-\ln\left(\frac{P}{k}\right)}{b}$$

Continued on following page.

a) $\ln\left(\frac{b}{a}\right)$

b) $\ln\left(\frac{S}{P}\right)$

109. Continued

To put the solution in a preferred form, we must get rid of the "-" in the numerator. To do so, the steps are:

$$x = \frac{-\ln\left(\frac{P}{k}\right)}{b} = \frac{\ln\left(\frac{k}{P}\right)}{b} = \frac{1}{b}\ln\left(\frac{k}{P}\right)$$

Write each solution in the preferred form.

a) $y = \dfrac{-\ln\left(\frac{R}{c}\right)}{t} =$ _____

b) $v = \dfrac{-\ln\left(\frac{M}{d}\right)}{q} =$ _____

a) $\dfrac{1}{t}\ln\left(\dfrac{c}{R}\right)$

b) $\dfrac{1}{q}\ln\left(\dfrac{d}{M}\right)$

110. We solved for $\underline{x}$ in the formula below.

$$\boxed{p = P_1 e^{-\frac{ax}{C}}}$$

$$\frac{p}{P_1} = e^{-\frac{ax}{C}}$$

$$\ln\left(\frac{p}{P_1}\right) = -\frac{ax}{C}$$

$$-\ln\left(\frac{p}{P_1}\right) = \frac{ax}{C}$$

$$-C\ln\left(\frac{p}{P_1}\right) = ax$$

$$x = \frac{-C\ln\left(\frac{p}{P_1}\right)}{a}$$

The steps needed to get rid of the "-" in the numerator of the solution are shown below.

$$x = \frac{-C\ln\left(\frac{p}{P_1}\right)}{a} = \frac{C\left[-\ln\left(\frac{p}{P_1}\right)\right]}{a} = \frac{C\ln\left(\frac{P_1}{p}\right)}{a} = \frac{C}{a}\ln\left(\frac{P_1}{p}\right)$$

Write each solution below in the preferred form.

a) $v = \dfrac{-L\ln\left(\frac{A_0}{A}\right)}{t} =$ _____

b) $m = \dfrac{-R\ln\left(\frac{c}{d}\right)}{h} =$ _____

a) $\dfrac{L}{t}\ln\left(\dfrac{A}{A_0}\right)$

b) $\dfrac{R}{h}\ln\left(\dfrac{d}{c}\right)$

111. We solved for L in the formula below.

$$\boxed{i = I_0 e^{-\frac{Rt}{L}}}$$

$$\frac{i}{I_0} = e^{-\frac{Rt}{L}}$$

$$\ln\left(\frac{i}{I_0}\right) = -\frac{Rt}{L}$$

$$L \ln\left(\frac{i}{I_0}\right) = -Rt$$

$$L = \frac{-Rt}{\ln\left(\frac{i}{I_0}\right)} \quad \text{or} \quad -\frac{Rt}{\ln\left(\frac{i}{I_0}\right)}$$

To get rid of the "-" in front of the solution, we use the steps below.

$$L = -\frac{Rt}{\ln\left(\frac{i}{I_0}\right)} = \frac{Rt}{-\ln\left(\frac{i}{I_0}\right)} = \frac{Rt}{\ln\left(\frac{I_0}{i}\right)}$$

Write each solution in the preferred form.

a) $M = -\dfrac{Sv}{\ln\left(\frac{a}{b}\right)} = $ _____

b) $T = -\dfrac{Fv}{\ln\left(\frac{A_0}{A}\right)} = $ _____

---

112. The steps needed to solve for R are shown below.

$$\boxed{b = B(1 - e^{-R})}$$

$$\frac{b}{B} = 1 - e^{-R}$$

$$\frac{b}{B} - 1 = -e^{-R}$$

$$1 - \frac{b}{B} = e^{-R}$$

$$\frac{B - b}{B} = e^{-R}$$

$$\ln\left(\frac{B - b}{B}\right) = -R$$

$$-\ln\left(\frac{B - b}{B}\right) = R$$

$$\text{or} \quad R = -\ln\left(\frac{B - b}{B}\right)$$

Write the solution in the preferred form.

$R = $ _____

---

a) $\dfrac{Sv}{\ln\left(\frac{b}{a}\right)}$

b) $\dfrac{Fv}{\ln\left(\frac{A}{A_0}\right)}$

113. Two methods are shown for the evaluation below.

In $\boxed{P = ke^{-x}}$, find $\underline{x}$ when $P = 30$ and $k = 15$.

Substituting and then solving for $\underline{x}$.

$\boxed{P = ke^{-x}}$

$30 = 15e^{-x}$

$\dfrac{30}{15} = e^{-x}$

$2 = e^{-x}$

$\ln 2 = -x$

$x = -\ln 2$

$x = -0.693$

Rearranging to solve for $\underline{x}$ and then substituting.

$\boxed{P = ke^{-x}}$

$x = \ln\left(\dfrac{k}{P}\right)$

$x = \ln\left(\dfrac{15}{30}\right)$

$x = \ln 0.5$

$x = -0.693$

Did we get the same solution both ways? _____

$R = \ln\left(\dfrac{B}{B-b}\right)$

Yes

8-14 "exp" NOTATION FOR POWER-OF-"e" FORMULAS

A notation involving the abbreviation "exp" is sometimes used to write formulas containing powers of "e". We will discuss "exp" notation in this section.

114. Instead of writing the base "e" in a formula, the abbreviation "exp" is sometimes used. For example:

$S = e^{-ct}$ is written $S = \exp(-ct)$

$A = 40e^{-0.4t}$ is written $A = 40\exp(-0.4t)$

Notice these points about "exp" notation.

1. The base "e" is not shown.
2. There is no period after "exp".
3. The exponent is written in parentheses after the "exp" and on the same line as the "exp".

Write each formula in "exp" notation.

a) $T = e^{-p}$ b) $P = ke^{-bm}$

a) $T = \exp(-p)$

b) $P = k\exp(-bm)$

115. "exp" notation is especially useful with complicated exponents. For example:

$$p = P_1 e^{-\frac{t}{RC}} \quad \text{is written} \quad p = P_1 \exp\left(-\frac{t}{RC}\right)$$

Write each formula in "exp" notation.

a) $M = Te^{-\frac{Rt}{L}}$

b) $i = k(1 - e^{-R})$

---

116. The abbreviation "exp" stands for the word "exponential". In this context, the word "exponential" means a base-e exponential. The following words are used to read a formula in "exp" notation.

$A = \exp(-cm)$ is read: A equals exponential (-cm).

$R = k \exp(-bt)$ is read: R equals k times _____ (-bt)

a) $M = T \exp\left(-\frac{Rt}{L}\right)$

b) $i = k(1 - \exp(-R))$

---

117. Any formula in "exp" notation can be written in power-of-"e" form. For example:

$H = \exp(-0.2t)$ is the same as $H = e^{-0.2t}$

$R = 20 \exp(-5y)$ is the same as $R = 20e^{-5y}$

Write each formula below in power-of-"e" form.

a) $T = 4.5 \exp(-0.1t)$

b) $V = E \exp\left(-\frac{t}{RC}\right)$

exponential

---

a) $T = 4.5e^{-0.1t}$

b) $V = Ee^{-\frac{t}{RC}}$

SELF-TEST 27 (pages 382-396)

1. Using a law of logarithms, write this expression in an equivalent form.

    $$\ln\left(\frac{r}{t}\right)$$

2. Using a reversed law of logarithms, write this expression in an equivalent form.

    $$a \ln P$$

3. Solve for "ln w".

    $$d = p \ln\left(\frac{w}{a}\right)$$

4. Solve for F_2.

    $$\ln\left(\frac{F_1}{F_2}\right) = -kp$$

5. Solve for "s".

    $$R = e^{-ms}$$

6. Solve for "t".

    $$G = Ke^{-t}$$

7. Solve for "w".

    $$B = Ae^{-\frac{w}{r}}$$

8. Solve for "p".

    $$h = m(1 - e^{-p})$$

9. Write this formula in "exp" notation.

    $$I = I_0 e^{-kt}$$

10. Write this formula in power-of-"e" form.

    $$d = k \exp\left(-\frac{t}{s}\right)$$

ANSWERS:

1. $\ln r - \ln t$
2. $\ln P^a$
3. $\ln w = \dfrac{d}{p} + \ln a$
4. $F_2 = F_1 e^{kp}$
5. $s = \dfrac{1}{m} \ln\left(\dfrac{1}{R}\right)$
6. $t = \ln\left(\dfrac{K}{G}\right)$
7. $w = r \ln\left(\dfrac{A}{B}\right)$
8. $p = \ln\left(\dfrac{m}{m-h}\right)$
9. $I = I_0 \exp(-kt)$
10. $d = ke^{-\frac{t}{s}}$

Powers of "e" and Natural Logarithms   397

SUPPLEMENTARY PROBLEMS - CHAPTER 8

Assignment 25

Convert each power of "e" to an ordinary number. Round as directed.

1. Round to hundredths.
   $e^{2.02}$
2. Round to thousands.
   $e^{12.8}$
3. Round to thousandths.
   $e^{-1.937}$

Convert each ordinary number to a power of "e". Round each exponent to four decimal places.

4. 5.47
5. 2,630
6. 1.18
7. 0.00703

For the exponential function $\boxed{y = 80e^{-0.6x}}$:

8. Write the coordinates of its y-intercept.
9. Is its graph an ascending or descending exponential graph?
10. Write the equation of its asymptote.
11. If $x = -2$, $y = ?$ (Round to tenths.)
12. If $x = 2$, $y = ?$ (Round to hundredths.)

For the exponential function $\boxed{y = 200(1 - e^{-0.3x})}$:

13. Write the coordinates of its y-intercept.
14. Is its graph an ascending or descending exponential graph?
15. Write the equation of its asymptote.
16. If $x = 0.5$, $y = ?$ (Round to tenths.)
17. If $x = 5$, $y = ?$ (Round to the nearest whole number.)

Evaluate these exponential formulas.

18. In $\boxed{S = \dfrac{e^x - e^{-x}}{2}}$, find S when $x = 1.38$. Round to hundredths.

19. In $\boxed{G = Ke^{-mt}}$, find G when $K = 125$, $m = 1.18$, and $t = 2$. Round to tenths.

20. In $\boxed{i = I(1 - e^{-ct})}$, find "i" when $I = 37.4$, $c = 0.463$, and $t = 0.75$. Round to tenths.

21. In $\boxed{m = Me^{-kr}}$, find M when $m = 395$, $k = 2.18$, and $r = 1.77$. Round to hundreds.

22. In $\boxed{i = \dfrac{E}{R} e^{-\frac{t}{RC}}}$, find E when $i = 2.58$, $R = 10$, $t = 0.55$, and $C = 0.0414$. Round to tenths.

23. In $\boxed{v = V\left(1 - e^{-\frac{Rt}{L}}\right)}$, find V when $v = 6.27$, $R = 4.7$, $L = 12.5$, and $t = 1$. Round to tenths.

Powers of "e" and Natural Logarithms

These applied problems require evaluation of exponential formulas.

24. If a principal of P dollars is invested at <u>r</u> percent per year compounded continuously, its value A after <u>t</u> years is found by the formula $\boxed{A = Pe^{rt}}$. Find A when P = $1,000, r = 0.08, and t = 10. Round to the nearest cent.

25. The temperature T_o of a tank of hot water is 100°C. The temperature T_m of the surrounding air is 20°C. The water is allowed to cool. Its temperature T after <u>t</u> minutes can be calculated from the formula $\boxed{T = T_m + (T_o - T_m)e^{-At}}$ where A = 0.049. Find temperature T when t = 20. Round to the nearest whole number.

26. The formula $\boxed{v = V\left(1 - e^{-\frac{t}{RC}}\right)}$ applies to the charging of a capacitor in an electrical circuit where V is the applied voltage, <u>v</u> is the capacitor voltage after <u>t</u> seconds, and R and C are circuit constants. Find <u>v</u> when V = 400, t = 5, R = 2.3, and C = 4.5. Round to the nearest whole number.

27. The formula $\boxed{N = N_o e^{-Lt}}$ applies to the radioactive disintegration of radium where N_o is the initial amount, N is the amount remaining after <u>t</u> years, and L = 0.0004278. If N_o = 5 grams of radium, find the amount N remaining after 100 years. Round to hundredths.

Assignment 26

Find each natural logarithm. Round to four decimal places.

1. ln 41.9
2. ln 1,000
3. ln 1
4. ln 0.0738

Find the number whose natural logarithm is:

5. 2.1704 (Round to hundredths.)
6. 13 (Round to thousands.)
7. -1.578 (Round to thousandths.)

Write each power-of-"e" equation as a logarithmic equation.

8. $40 = e^{3.6889}$
9. $e^{-1.63} = 0.196$
10. $P = e^t$

Write each logarithmic equation as a power-of-"e" equation.

11. ln 0.00674 = -5
12. 0 = ln 1
13. ln r = a

Solve each equation. Round as directed.

14. Round to tenths.
    ln R = 3.65

15. Round to hundredths.
    ln 14.2 = h

16. Round to thousandths.
    -1.84 = ln V

17. Round to thousandths.
    $e^x = 2.09$

18. Round to four decimal places.
    $w = e^{-0.27}$

19. Round to hundredths.
    $0.004608 = e^y$

Evaluate each formula. Round as directed.

20. In $\boxed{P = A \ln V}$, find A when P = 13.8 and V = 4.43 . Round to hundredths.

21. In $\boxed{\ln\left(\dfrac{F_1}{F_2}\right) = -ar}$, find "r" when $F_1 = 25.6$, $F_2 = 72.7$, and a = 0.0419 . Round to tenths.

22. In $\boxed{s = K \ln N}$, find N when s = 226 and K = 309 . Round to thousandths.

23. In $\boxed{p = b \ln\left(\dfrac{A}{G}\right)}$, find A when p = 9.16 , b = 4.72 , and G = 1,850 . Round to hundreds.

24. In $\boxed{W = e^{-s}}$, find "s" when W = 42.5 . Round to thousandths.

25. In $\boxed{B = Ke^{-cv}}$, find "v" when B = 135 , K = 542 , and c = 0.0337. Round to tenths.

These applied problems require evaluation of logarithmic or exponential formulas.

26. The formula $\boxed{v = c \ln\left(\dfrac{M}{m}\right)}$ is used to find the velocity $\underline{v}$ of a rocket whose lift-off weight is M , whose weight at engine shutdown is $\underline{m}$, and whose exhaust gases have velocity $\underline{c}$. Find $\underline{v}$ when M = 1,550 , m = 800 , and c = 4,900 . Round to hundreds.

27. When discharging a capacitor charged to V volts, the capacitor voltage $\underline{v}$ after $\underline{t}$ seconds is calculated by the formula $\boxed{v = Ve^{-\frac{t}{RC}}}$. Find $\underline{t}$ when v = 40 , V = 80 , R = 1.8 , and C = 4.5 . Round to hundredths.

28. If the initial number of bacteria in a nutrient solution is N , the formula $\boxed{n = Ne^{0.82t}}$ defines the number of bacteria $\underline{n}$ present $\underline{t}$ hours later. Find $\underline{t}$ when n = 50,000 and N = 100 . Round to hundredths.

29. The formula $\boxed{t = -2337 \ln\left(\dfrac{N}{N_0}\right)}$ applies to the radioactive disintegration of radium where N_0 is the initial amount and N is the amount remaining after $\underline{t}$ years. Find $\underline{t}$ when N = 12 and N_0 = 24 . Round to the nearest whole number.

Assignment 27

Using a law of logarithms, write each expression in an equivalent form.

1. $\ln R^b$
2. $\ln\left(\dfrac{K}{N}\right)$
3. $\ln(2w)$

Using a reversed law of logarithms, write each expression in an equivalent form.

4. $\ln t - \ln p$
5. $k \ln Q$
6. $-\ln G$

State whether each of the following is "true" or "false".

7. $\ln(F - H) = \ln F - \ln H$
8. $-\ln\left(\dfrac{w}{v}\right) = \ln\left(\dfrac{v}{w}\right)$
9. $\ln A + \ln B = \ln(A + B)$

Powers of "e" and Natural Logarithms

Rearrange these formulas. Use the laws of logarithms where necessary.

10. Solve for P.

    $$H = P \ln Q$$

11. Solve for r.

    $$\ln A = br$$

12. Solve for "$\ln T$".

    $$t = -\frac{1}{k}\ln T$$

13. Solve for "$\ln A$".

    $$\ln\left(\frac{A}{P}\right) = kt$$

14. Solve for "$\ln R$".

    $$a = -d \ln\left(\frac{V}{R}\right)$$

15. Solve for t.

    $$w = \ln B^t$$

Rearrange these formulas. Conversion to power-of-"e" form is necessary.

16. Solve for T.

    $$w = -\ln T$$

17. Solve for A.

    $$p = b \ln A$$

18. Solve for K.

    $$\ln\left(\frac{K}{N}\right) = c$$

19. Solve for G.

    $$r = \ln\left(\frac{H}{G}\right)$$

20. Solve for R.

    $$\ln\left(\frac{D}{R}\right) = -kt$$

21. Solve for n.

    $$s = a \ln\left(\frac{m}{n}\right)$$

Rearrange these formulas containing powers of "e".

22. Solve for p.

    $$M = e^{ap}$$

23. Solve for v.

    $$F = ke^v$$

24. Solve for r.

    $$A = Pe^{rt}$$

25. Solve for s.

    $$H = Be^{-cs}$$

26. Solve for t.

    $$v = Ve^{-\frac{t}{RC}}$$

27. Solve for t.

    $$i = I\left(1 - e^{-\frac{Rt}{L}}\right)$$

Write each formula in power-of-"e" form.

28. $R = A \exp(kw)$

29. $d = k \exp(-cp)$

30. $v = V \exp\left(-\frac{Rt}{L}\right)$

Write each formula in "exp" notation.

31. $y = e^{-aw}$

32. $p = Pe^{\frac{v}{2n}}$

33. $v = V\left(1 - e^{-\frac{t}{RC}}\right)$

Chapter 9 SYSTEMS OF THREE EQUATIONS AND DETERMINANTS

In this chapter, we will discuss the determinant method for solving systems of two equations. We will discuss the addition method, the equivalence method, and the determinant method for solving systems of three equations. Systems of three formulas are also discussed.

9-1 A REVIEW OF THREE METHODS

In this section, we will review three methods for solving systems of two equations.

1. We used the <u>addition method</u> below to solve the system at the right. The solution is: $x = 3$, $y = 1$.

   $$\begin{array}{r} x + 2y = 5 \\ 3x - 2y = 7 \end{array}$$

   1) <u>Finding the value of "x"</u>.

      By adding the equations, we can eliminate "y" and solve for "x".

      $$\begin{array}{r} x + 2y = 5 \\ 3x - 2y = 7 \\ \hline 4x = 12 \\ x = 3 \end{array}$$

   2) <u>Finding the value of "y"</u>.

      To find the corresponding value of "y", we can substitute 3 for "x" in either of the original equations. We substituted 3 in the top equation.

      $$\begin{array}{r} x + 2y = 5 \\ 3 + 2y = 5 \\ 2y = 2 \\ y = 1 \end{array}$$

   Use the addition method to solve this system.

   $$\begin{array}{r} 2p + q = 10 \\ 2p - q = 6 \end{array}$$

   a) Find the value of "p".

   b) Find the value of "q" by substitution.

---

a) $p = 4$
b) $q = 2$

2. We used the equivalence method below to solve the system at the right. The solution is: V = 10, T = -2

$$V - 3T = 16$$
$$V + T = 8$$

   1) Solve for the same variable in each equation. (We solved for V.)

$$V = 3T + 16$$
$$V = 8 - T$$

   2) Use the equivalence principle to eliminate the solved-for variable V.

$$3T + 16 = 8 - T$$

   3) Find the value of T.

$$4T = -8$$
$$T = -2$$

   4) Find the corresponding value of V. (We substituted -2 for T in one of the equations in which V is solved for.)

$$V = 3T + 16$$
$$V = 3(-2) + 16$$
$$V = -6 + 16$$
$$V = 10$$

Use the equivalence method to solve the system at the right.

$$a + 3b = 10$$
$$a + b = 6$$

   a) Solve for "a" in each equation.

$$a =$$
$$a =$$

   b) Use the equivalence principle to eliminate "a".

$$\underline{\hspace{2cm}} = \underline{\hspace{2cm}}$$

   c) Find the value of "b".

   d) Find the corresponding value of "a".

---

3. We used the substitution method below to solve the system at the right. The solution is: x = 19, y = 16

$$y = x - 3$$
$$4y - 3x = 7$$

a) $$a = 10 - 3b$$
$$a = 6 - b$$

b) $10 - 3b = 6 - b$

c) $b = 2$

d) $a = 4$

   1) Eliminate "y" by substituting (x - 3) for "y" in the bottom equation.

$$4(x - 3) - 3x = 7$$
$$4x - 12 - 3x = 7$$

   2) Solve for "x".

$$x - 12 = 7$$
$$x = 19$$

   3) Find the corresponding value of "y". (We substituted in the top equation.)

$$y = x - 3$$
$$y = 19 - 3$$
$$y = 16$$

Continued on following page.

3. Continued

Use the substitution method to solve the system at the right.

$$q = p + 4$$
$$2p + q = 10$$

a) Eliminate "q" by substituting (p + 4) for "q" in the bottom equation.

b) Solve for "p".

c) Solve for "q" by substituting in one of the original equations.

a) $2p + p + 4 = 10$ b) $p = 2$ c) $q = 6$

9-2 SECOND ORDER DETERMINANTS

In this section, we will discuss second order or two-by-two determinants.

4. A determinant is a square array of numbers enclosed by vertical lines. Two examples are shown.

$$\begin{vmatrix} 3 & 5 \\ 2 & 1 \end{vmatrix} \qquad \begin{vmatrix} -2 & 6 \\ 4 & -3 \end{vmatrix}$$

The general form for the above determinants is:

$$\begin{vmatrix} a_1 & b_1 \\ a_2 & b_2 \end{vmatrix}$$

The rows and columns of the general form are shown below. Since there are 2 rows and 2 columns, the determinant is called a "second order" or "two-by-two" determinant.

a) Circle the bottom row in $\begin{vmatrix} 3 & 5 \\ 2 & 1 \end{vmatrix}$

b) Circle the rightmost column in $\begin{vmatrix} -2 & 6 \\ 4 & -3 \end{vmatrix}$

5. Any determinant equals one number. To find that number, we use the definition below.

$$\begin{vmatrix} a_1 & b_1 \\ a_2 & b_2 \end{vmatrix} \text{ is equal to: } a_1b_2 - a_2b_1$$

That is, a determinant equals the difference of the product of the diagonals. Using the above definition, we get:

$$\begin{vmatrix} 4 & 2 \\ 3 & 5 \end{vmatrix} = (4)(5) - (3)(2) = 20 - 6 = 14$$

$$\begin{vmatrix} 3 & -5 \\ 2 & 4 \end{vmatrix} = (3)(4) - (2)(-5) = 12 - (-10) = \underline{\qquad}$$

a) $\begin{vmatrix} 3 & 5 \\ 2 & 1 \end{vmatrix}$

b) $\begin{vmatrix} -2 & 6 \\ 4 & -3 \end{vmatrix}$

---

6. Using the definition in the last frame, find the value of each determinant.

a) $\begin{vmatrix} 5 & -2 \\ 3 & 4 \end{vmatrix} = ()() - ()() = \underline{\qquad}$

b) $\begin{vmatrix} -7 & 3 \\ 2 & 2 \end{vmatrix} = ()() - ()() = \underline{\qquad}$

c) $\begin{vmatrix} 6 & -5 \\ -5 & -6 \end{vmatrix} = ()() - ()() = \underline{\qquad}$

22

---

7. Using the same method, evaluate these.

a) $\begin{vmatrix} 4 & -1 \\ 0 & 2 \end{vmatrix} = ()() - ()() = \underline{\qquad}$

b) $\begin{vmatrix} 1 & -3 \\ -2 & 0 \end{vmatrix} = ()() - ()() = \underline{\qquad}$

a) $(5)(4) - (3)(-2)$
$= 20 - (-6) = 26$

b) $(-7)(2) - (2)(3)$
$= (-14) - 6 = -20$

c) $(6)(-6) - (-5)(-5)$
$= (-36) - 25 = -61$

---

8. The determinant below contains decimal numbers.

$$\begin{vmatrix} 2.3 & 3.1 \\ 1.7 & 8.6 \end{vmatrix} = (2.3)(8.6) - (1.7)(3.1)$$

We can use a calculator to find the value of the determinant. (The value is 14.51.) The steps are:

Enter	Press	Display
2.3	☒	2.3
8.6	⊟	19.78
1.7	☒	1.7
3.1	⊟	14.51

a) $(4)(2) - (0)(-1)$
$= 8 - 0 = 8$

b) $(1)(0) - (-2)(-3)$
$= 0 - 6 = -6$

Continued on following page.

8. Continued

Use a calculator to evaluate these. Don't forget to press [+/-] after entering a negative number. Round to tenths.

a) $\begin{vmatrix} 2.4 & 1.5 \\ -1.7 & 3.7 \end{vmatrix}$ = _____

b) $\begin{vmatrix} 1.3 & 3.2 \\ 6.9 & 2.7 \end{vmatrix}$ = _____

a) 11.4   b) -18.6

9-3 USING DETERMINANTS TO SOLVE SYSTEMS OF TWO EQUATIONS

In this section, we will discuss the determinant method for solving systems of two equations. The determinant method is called <u>Cramer's Rule</u>.

9. The standard form for a system of two equations is given below.

$$\begin{array}{l} a_1x + b_1y = c_1 \\ a_2x + b_2y = c_2 \end{array}$$

The a's and b's are coefficients and the c's are number terms. We can use the a's, b's, and c's to set up determinants to solve the system. The three determinants used are D, D_x, and D_y. They are shown below.

$$D = \begin{vmatrix} a_1 & b_1 \\ a_2 & b_2 \end{vmatrix} \qquad D_x = \begin{vmatrix} c_1 & b_1 \\ c_2 & b_2 \end{vmatrix} \qquad D_y = \begin{vmatrix} a_1 & c_1 \\ a_2 & c_2 \end{vmatrix}$$

Determinant D is formed by detaching the coefficients of the x's and y's. Notice these points about D_x and D_y.

1) To get D_x from D, we replace the a's (the coefficients of the x's) with the c's (see the arrow).

2) To get D_y from D, we replace the b's (the coefficients of the y's) with the c's (see the arrow).

Using the determinants above, we can set up the following rules to solve for <u>x</u> and <u>y</u>.

$$x = \frac{D_x}{D} = \frac{\begin{vmatrix} c_1 & b_1 \\ c_2 & b_2 \end{vmatrix}}{\begin{vmatrix} a_1 & b_1 \\ a_2 & b_2 \end{vmatrix}} \qquad y = \frac{D_y}{D} = \frac{\begin{vmatrix} a_1 & c_1 \\ a_2 & c_2 \end{vmatrix}}{\begin{vmatrix} a_1 & b_1 \\ a_2 & b_2 \end{vmatrix}}$$

10. Using the definition, we set up D, D_x, and D_y for the system below.

$$3x + y = 7$$
$$4x + 2y = 9$$

$$D = \begin{vmatrix} 3 & 1 \\ 4 & 2 \end{vmatrix} \qquad D_x = \begin{vmatrix} 7 & 1 \\ 9 & 2 \end{vmatrix} \qquad D_y = \begin{vmatrix} 3 & 7 \\ 4 & 9 \end{vmatrix}$$

Following the example, set up D, D_x, and D_y for the system below.

$$5x + 3y = 10$$
$$x + 6y = 3$$

D = | | D_x = | | D_y = | |

---

11. We set up D, D_x, and D_y for the system below.

$$4p - q = 5$$
$$-p + 3q = 8$$

$$D = \begin{vmatrix} 4 & -1 \\ -1 & 3 \end{vmatrix} \qquad D_p = \begin{vmatrix} 5 & -1 \\ 8 & 3 \end{vmatrix} \qquad D_q = \begin{vmatrix} 4 & 5 \\ -1 & 8 \end{vmatrix}$$

$$D = \begin{vmatrix} 5 & 3 \\ 1 & 6 \end{vmatrix}$$

$$D_x = \begin{vmatrix} 10 & 3 \\ 3 & 6 \end{vmatrix}$$

$$D_y = \begin{vmatrix} 5 & 10 \\ 1 & 3 \end{vmatrix}$$

Following the example, set up D, D_{F_1}, and D_{F_2} for this system.

$$-1.5F_1 + 2.7F_2 = 12.7$$
$$3.6F_1 - 8.5F_2 = 0$$

D = | | D_{F_1} = | | D_{F_2} = | |

---

12. The rules for solving for $\underline{x}$ and $\underline{y}$ are shown below.

$$x = \frac{D_x}{D} \qquad\qquad y = \frac{D_y}{D}$$

Using those rules, we set up the solution for the system below.

$$x - 3y = 4$$
$$5x + 2y = 9$$

$$x = \frac{D_x}{D} = \frac{\begin{vmatrix} 4 & -3 \\ 9 & 2 \end{vmatrix}}{\begin{vmatrix} 1 & -3 \\ 5 & 2 \end{vmatrix}} \qquad y = \frac{D_y}{D} = \frac{\begin{vmatrix} 1 & 4 \\ 5 & 9 \end{vmatrix}}{\begin{vmatrix} 1 & -3 \\ 5 & 2 \end{vmatrix}}$$

$$D = \begin{vmatrix} -1.5 & 2.7 \\ 3.6 & -8.5 \end{vmatrix}$$

$$D_{F_1} = \begin{vmatrix} 12.7 & 2.7 \\ 0 & -8.5 \end{vmatrix}$$

$$D_{F_2} = \begin{vmatrix} -1.5 & 12.7 \\ 3.6 & 0 \end{vmatrix}$$

Continued on following page.

12. Continued

Following the example, set up the solution for this system.

$$\begin{array}{r} 3m - t = 10 \\ m + 8t = 11 \end{array}$$

a) $m = \dfrac{D_m}{D} = \dfrac{\begin{vmatrix} & \\ & \end{vmatrix}}{\begin{vmatrix} & \\ & \end{vmatrix}}$ b) $t = \dfrac{D_t}{D} = \dfrac{\begin{vmatrix} & \\ & \end{vmatrix}}{\begin{vmatrix} & \\ & \end{vmatrix}}$

---

13. We used the determinant method to solve the system below.

$$\begin{array}{r} 5x - 2y = 19 \\ 7x + 3y = 15 \end{array}$$

$$x = \frac{D_x}{D} = \frac{\begin{vmatrix} 19 & -2 \\ 15 & 3 \end{vmatrix}}{\begin{vmatrix} 5 & -2 \\ 7 & 3 \end{vmatrix}} = \frac{(19)(3) - (15)(-2)}{(5)(3) - (7)(-2)} = \frac{87}{29} = 3$$

$$y = \frac{D_y}{D} = \frac{\begin{vmatrix} 5 & 19 \\ 7 & 15 \end{vmatrix}}{\begin{vmatrix} 5 & -2 \\ 7 & 3 \end{vmatrix}} = \frac{(5)(15) - (7)(19)}{(5)(3) - (7)(-2)} = \frac{-58}{29} = -2$$

The solution is: $x = 3$, $y = -2$. Check the solution in each original equation below.

a) $5x - 2y = 19$ b) $7x + 3y = 15$

Answers:

a) $m = \dfrac{\begin{vmatrix} 10 & -1 \\ 11 & 8 \end{vmatrix}}{\begin{vmatrix} 3 & -1 \\ 1 & 8 \end{vmatrix}}$

b) $t = \dfrac{\begin{vmatrix} 3 & 10 \\ 1 & 11 \end{vmatrix}}{\begin{vmatrix} 3 & -1 \\ 1 & 8 \end{vmatrix}}$

---

14. Use the determinant method to solve this system.

$$\begin{array}{r} 4V + 3t = 1 \\ 3V - 2t = 22 \end{array}$$

a) $V = \dfrac{D_V}{D} = \dfrac{\begin{vmatrix} & \\ & \end{vmatrix}}{\begin{vmatrix} & \\ & \end{vmatrix}} = \underline{} = \underline{}$

b) $t = \dfrac{D_t}{D} = \dfrac{\begin{vmatrix} & \\ & \end{vmatrix}}{\begin{vmatrix} & \\ & \end{vmatrix}} = \underline{} = \underline{}$

Answers:

a) $5(3) - 2(-2) = 19$
   $15 - (-4) = 19$
   $19 = 19$

b) $7(3) + 3(-2) = 15$
   $21 + (-6) = 15$
   $15 = 15$

408   Systems Of Three Equations And Determinants

15. Use the determinant method to solve this system. Use a calculator to evaluate the determinants. Round F_1 to thousandths and F_2 to hundredths.

$$6.3F_1 - 1.9F_2 = -17$$
$$3.6F_1 + 2.7F_2 = 23$$

a) $F_1 = \dfrac{D_{F_1}}{D} = \underline{} = \underline{} = \underline{}$

b) $F_2 = \dfrac{D_{F_2}}{D} = \underline{} = \underline{} = \underline{}$

a) $V = \dfrac{\begin{vmatrix} 1 & 3 \\ 22 & -2 \end{vmatrix}}{\begin{vmatrix} 4 & 3 \\ 3 & -2 \end{vmatrix}} = \dfrac{-68}{-17} = 4$

b) $t = \dfrac{\begin{vmatrix} 4 & 1 \\ 3 & 22 \end{vmatrix}}{\begin{vmatrix} 4 & 3 \\ 3 & -2 \end{vmatrix}} = \dfrac{85}{-17} = -5$

a) $F_1 = \dfrac{\begin{vmatrix} -17 & -1.9 \\ 23 & 2.7 \end{vmatrix}}{\begin{vmatrix} 6.3 & -1.9 \\ 3.6 & 2.7 \end{vmatrix}} = \dfrac{-2.2}{23.85} = -0.092$ b) $F_2 = \dfrac{\begin{vmatrix} 6.3 & -17 \\ 3.6 & 23 \end{vmatrix}}{\begin{vmatrix} 6.3 & -1.9 \\ 3.6 & 2.7 \end{vmatrix}} = \dfrac{206.1}{23.85} = 8.64$

9-4 CONVERTING TO STANDARD FORM FOR THE DETERMINANT METHOD

To solve a system by the determinant method, both equations must be in standard form. When one or both equations are not in standard form, we must convert them to standard form first. We will discuss solutions of that type in this section.

16. The standard form for a system of two equations is shown below. Notice that both variables are on the left side, the number term is on the right side, and the terms containing the same variable are lined up.

$$a_1x + b_1y = c_1$$
$$a_2x + b_2y = c_2$$

Which of the following systems are in standard form? _____

a) $10x - 17 = 12y$
    $9x + 25 = 7y$

b) $1.5F_1 - 2.6F_2 = 35$
    $4.7F_1 - 6.2F_2 = 40$

Only (b)

17. In each original system below, one equation is not in standard form. The standard form for each system is shown.

   Original System
   $$2x - 3y - 10 = 0$$
   $$4x + y = 5$$

   Original System
   $$a = 5b - 9$$
   $$2a + 7b = 11$$

   Standard System
   $$2x - 3y = 10$$
   $$4x + y = 5$$

   Standard System
   $$a - 5b = -9$$
   $$2a + 7b = 11$$

   Convert each system below to standard form.

   a)
   $$9m - 8p = 0$$
   $$10m - 7p + 11 = 0$$

   b)
   $$2p + 5q = 9$$
   $$q = 8p - 12$$

18. In each original system below, neither equation is in standard form. The standard form for each system is shown.

   Original System
   $$3x - 11 = 2y$$
   $$x = 5y$$

   Original System
   $$4V_2 = V_1 + 13$$
   $$7V_1 + 27 = 5V_2$$

   Standard System
   $$3x - 2y = 11$$
   $$x - 5y = 0$$

   Standard System
   $$-V_1 + 4V_2 = 13$$
   $$7V_1 - 5V_2 = -27$$

   Convert each system below to standard form.

   a)
   $$3a = 7b$$
   $$8a + 12 = 3b$$

   b)
   $$1.2F_1 - 40 = 2.3F_2$$
   $$4.1F_1 = 3.3F_2 - 25$$

a) $9m - 8p = 0$
    $10m - 7p = -11$

b) $2p + 5q = 9$
    $-8p + q = -12$

a) $3a - 7b = 0$
    $8a - 3b = -12$

b) $1.2F_1 - 2.3F_2 = 40$
    $4.1F_1 - 3.3F_2 = -25$

410   Systems Of Three Equations And Determinants

19. The system below is not in standard form. To use the determinant method, we must put it in standard form first.

$$2.3x + 1.2y = 17$$
$$1.8x - 28 = 2.0y$$

a) Convert the system to standard form.

b) $x = \dfrac{D_x}{D} = \dfrac{\begin{vmatrix} & \\ & \end{vmatrix}}{\begin{vmatrix} & \\ & \end{vmatrix}} = \underline{} = \underline{}$

c) $y = \dfrac{D_y}{D} = \dfrac{\begin{vmatrix} & \\ & \end{vmatrix}}{\begin{vmatrix} & \\ & \end{vmatrix}} = \underline{} = \underline{}$

---

a) $2.3x + 1.2y = 17$
   $1.8x - 2.0y = 28$

b) $x = \dfrac{\begin{vmatrix} 17 & 1.2 \\ 28 & -2.0 \end{vmatrix}}{\begin{vmatrix} 2.3 & 1.2 \\ 1.8 & -2.0 \end{vmatrix}} = \dfrac{-67.6}{-6.76} = 10$

c) $y = \dfrac{\begin{vmatrix} 2.3 & 17 \\ 1.8 & 28 \end{vmatrix}}{\begin{vmatrix} 2.3 & 1.2 \\ 1.8 & -2.0 \end{vmatrix}} = \dfrac{33.8}{-6.76} = -5$

SELF-TEST 28 (pages 401-411)

Solve each system by the determinant method.

1. $3x - 2y = 18$
   $2x + y = 5$

2. $4F_2 = 20 - F_1$
   $3F_1 - 8F_2 = 0$

Continued on following page.

SELF-TEST 28 - Continued

3. Round each answer to hundredths.

$$2.9p + 1.4r = 10$$
$$3.2p - 4.5r = 8$$

4. Round each answer to thousandths.

$$0.7i_1 - 1.5 = 3i_2$$
$$0.4i_2 = 2i_1 + 2.7$$

ANSWERS: 1. $x = 4$ 2. $F_1 = 8$ 3. $p = 3.21$ 4. $i_1 = -1.521$
              $y = -3$ $F_2 = 3$ $r = 0.50$ $i_2 = -0.855$

9-5 SYSTEMS OF THREE EQUATIONS - THE ADDITION METHOD

In this section, we will show how the addition method can be used to solve systems of three equations.

20. The system below contains three equations and three variables.

(1) $x + y + z = 4$
(2) $x - 2y - z = 1$
(3) $2x - y - 2z = -1$

The solution of the system is the set of values for x, y, and z that satisfies each of the three equations.

a) Is $x = 1$, $y = 5$, and $z = -2$ the solution of the system?

b) Is $x = 2$, $y = -1$, and $z = 3$ the solution of the system?

21. The system below contains three equations and three variables.

$$\begin{array}{ll}(1) & 4m - p + q = 6 \\ (2) & -3m + 2p - q = -3 \\ (3) & 2m + p + 2q = 3\end{array}$$

The four steps needed to solve the system by the addition method are discussed below.

1) Eliminate <u>one</u> variable to get a system of two equations with two variables.

   a) Add a pair of equations to eliminate one variable. We added (1) and (2) below to eliminate q.

   $$\begin{array}{r}4m - p + q = 6 \\ -3m + 2p - q = -3 \\ \hline m + p = 3\end{array}$$

   b) Add a different pair of equations to eliminate <u>the same variable</u>. We added (2) and (3) below to eliminate q. Notice that we multiplied (2) by 2 to to get the opposites "-2q" and "2q".

   $$\begin{array}{ll}-3m + 2p - q = -3 & -6m + 4p - 2q = -6 \\ 2m + p + 2q = 3 & \underline{2m + p + 2q = 3} \\ & -4m + 5p = -3\end{array}$$

2) Solve the system of two equations to find the values of the two variables "m" and "p".

   a) We multiplied the top equation by 4 to get the opposites "4m" and "-4m", and then solved for p.

   $$\begin{array}{ll}m + p = 3 & 4m + 4p = 12 \\ -4m + 5p = -3 & \underline{-4m + 5p = -3} \\ & 9p = 9 \\ & p = 1\end{array}$$

   b) We substituted "1" for p in the top equation to solve for m.

   $$\begin{array}{l}m + p = 3 \\ m + 1 = 3 \\ m = 2\end{array}$$

3) Substitute m = 2 and p = 1 in one of the three original equations to solve for q. We substituted in equation (1).

   $$\begin{array}{l}4m - p + q = 6 \\ 4(2) - 1 + q = 6 \\ 8 - 1 + q = 6 \\ 7 + q = 6 \\ q = -1\end{array}$$

Continued on following page.

a) No. It satisfies (1), but not (2) and (3).

b) Yes. It satisfies (1), (2), and (3).

21. Continued

4) Check the solution (m = 2, p = 1, q = -1) in the three original equations.

4m - p + q = 6	-3m + 2p - q = -3	2m + p + 2q = 3
4(2) - 1 + (-1) = 6	-3(2) + 2(1) - (-1) = -3	2(2) + 1 + 2(-1) = 3
8 - 1 + (-1) = 6	(-6) + 2 + 1 = -3	4 + 1 - 2 = 3
6 = 6	-3 = -3	3 = 3

22. Let's use the addition method to solve the system below.

(1) d + k + t = 6
(2) 2d - k + 3t = 9
(3) -d + 2k + 2t = 9

a) Write the system of two equations obtained if k is eliminated by adding (1) and (2) and then adding (2) and (3).

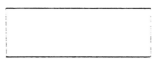

b) Solve the system of two equations to find the values of d and t.

c) Substitute those values for d and t in one of the three original equations to find the corresponding value of k.

d) The solution of the system is: d = _____, k = _____, t = _____

a) 3d + 4t = 15
   3d + 8t = 27
b) d = 1, t = 3
c) k = 2
d) d = 1, k = 2, t = 3

23. In the system below, the variable z does not appear in equation (1). Therefore the system is easier to solve.

$$\begin{array}{ll}(1) & x - y = 4 \\ (2) & x - 2y + z = 5 \\ (3) & 2x + 3y - 2z = 3\end{array}$$

    a) Let's eliminate z to get a system of two equations. One of the equations is (1) above. Find the other equation by adding (2) and (3) to eliminate z.

$$x - y = 4$$

    b) Solve the system of two equations to find the values of x and y.

    c) Substitute those values for x and y in one of the three original equations to find the corresponding value of z.

    d) The solution of the system is: x = _____, y = _____, z = _____

a) $x - y = 4$
$4x - y = 13$
   b) $x = 3, y = -1$
   c) $z = 0$
   d) $x = 3, y = -1, z = 0$

9-6 CONVERTING TO STANDARD FORM FOR THE ADDITION METHOD

To use the addition method, a system of three equations must be in standard form. If a system is not in standard form, we must convert it to standard form first. We will discuss the method in this section.

24. The standard form for a system of three equations with three variables is shown below.

$$\begin{array}{l}a_1x + b_1y + c_1z = d_1 \\ a_2x + b_2y + c_2z = d_2 \\ a_3x + b_3y + c_3z = d_3\end{array}$$

Continued on following page.

24. Continued

In the standard form, the a's, b's, and c's are coefficients of the variables x, y, and z, and the d's are the number terms. The variables are lined up on the left side with the d's on the right side.

Which of the following systems are in standard form? _____

a) $p + q + r = 7$
    $2p - 3q - r = 1$
    $3p - 2q + 4r = 0$

b) $2d - m + t = 0$
    $d + m - 9 = 4t$
    $3d - 2t + 8 = 5m$

25. We converted the original system below to standard form.

Original System:
$x - 2y + 5 = 3z$
$2x - z - 1 = 3y$
$y = x + 2z$

Standard Form:
$x - 2y - 3z = -5$
$2x - 3y - z = 1$
$-x + y - 2z = 0$

Only (a)

Following the example, convert this system to standard form.

$2b - d - 7 = c$
$b - 5c + 3 = 4d$
$c - 3b = 2d + 5$

26. The system below is not in standard form. To solve it by the addition method, we must begin by converting it to standard form.

(1) $2x + 2y - 3 = 3z$
(2) $x - 3z - 9 = 2y$
(3) $7x = 2y - 9z - 39$

$2b - c - d = 7$
$b - 5c - 4d = -3$
$-3b + c - 2d = 5$

a) Convert the system to standard form.

(1)

(2)

(3)

Continued on following page.

416 Systems Of Three Equations And Determinants

26. Continued

    b) Write the system of two equations obtained if we eliminate y by adding (1) and (2) and then adding (1) and (3).

    c) Solve the system of two equations to find the values of x and z.

    d) Find the value of y by substituting those values in one of the three original equations.

    e) The solution of the system is: x = _____, y = _____, z = _____

a) $2x + 2y - 3z = 3$
   $x - 2y - 3z = 9$
   $7x - 2y + 9z = -39$

b) $3x - 6z = 12$
   $9x + 6z = -36$

c) $x = -2$, $z = -3$

d) $y = -1$

e) $x = -2$, $y = -1$, $z = -3$

9-7 SYSTEMS OF THREE EQUATIONS - THE EQUIVALENCE METHOD

In this section, we will show how the equivalence method can be used to solve systems of three equations.

27. The system below contains three equations and three variables.

(1) $x + y + z = 9$
(2) $x + 2y - z = 4$
(3) $x - y + z = 3$

Continued on following page.

27. Continued

The four steps needed to solve the system by the equivalence method are discussed below.

1) Eliminate <u>one</u> variable to get a system of two equations with two variables.

   a) Using one pair of equations, solve for the same variable and equate the solutions to eliminate that variable. We used (1) and (2) to eliminate $\underline{x}$.

   $$x = 9 - y - z$$
   $$x = 4 - 2y + z$$
   $$9 - y - z = 4 - 2y + z$$
   or
   $$y - 2z = -5$$

   b) Using a different pair of equations, solve for the same variable in each and equate the solutions to eliminate that variable. We used (2) and (3) to eliminate $\underline{x}$.

   $$x = 4 - 2y + z$$
   $$x = 3 + y - z$$
   $$4 - 2y + z = 3 + y - z$$
   or
   $$-3y + 2z = -1$$

2) Solve the system of two equations to find the values of $\underline{y}$ and $\underline{z}$.

   $$\boxed{\begin{array}{l} y - 2z = -5 \\ -3y + 2z = -1 \end{array}}$$

   The solution is: $y = 3$ and $z = 4$

3) Substitute those values in one of the three original equations to find the corresponding value of $\underline{x}$. We used equation (1) below.

   $$x + y + z = 9$$
   $$x + 3 + 4 = 9$$
   $$x = 2$$

4) The solution of the system is: $x = 2$, $y = 3$, $z = 4$

418   Systems Of Three Equations And Determinants

28.   Let's use the equivalence method to solve the system below.

(1) $2k + p + q = 2$
(2) $-k + p - 4q = 0$
(3) $3k + p - q = -1$

a) Write the system of equations obtained if we use the equivalence principle to eliminate p from (1) and (2) and then from (2) and (3).

b) Solve the system of two equations to find the values of k and q.

c) Substitute those values in one of the three original equations to find the corresponding value of p.

d) The solution of the system is: k = _____, p = _____, q = _____

a) $3k + 5q = 2$
   $4k + 3q = -1$
b) $k = -1$, $q = 1$
c) $p = 3$
d) $k = -1$, $p = 3$, $q = 1$

29. In the system below, the variable $\underline{t}$ does not appear in equation (1). Therefore, the system is easier to solve.

(1) $d + m = -1$
(2) $2d - m - t = 3$
(3) $d + 3m + 2t = 5$

a) Let's eliminate $\underline{t}$ to get a system of two equations. One of the equations is (1) above. Use the equivalence principle to eliminate $\underline{t}$ from equations (2) and (3).

$d + m = -1$

b) Solve the system of two equations to find the values of $\underline{d}$ and $\underline{m}$.

c) Substitute those values in one of the three original equations to find the corresponding value of $\underline{t}$.

d) The solution of the system is: $d = $ _____, $m = $ _____, $t = $ _____

a) $d + m = -1$
   $5d + m = 11$
b) $d = 3$, $m = -4$
c) $t = 7$
d) $d = 3$, $m = -4$, $t = 7$

9-8 SYSTEMS OF THREE FORMULAS AND FORMULA DERIVATION

In this section, we will define a system of three formulas and show how we can derive a new formula from a system of that type by eliminating at least two variables.

30. Three formulas form a system if two of the formulas contain <u>at least one common variable</u> and the third formula contains <u>at least one variable</u> in common with <u>one</u> of the other two.

The three formulas at the right form a system because:

(1) $F = ma$
(2) $a = \dfrac{v^2}{r}$
(3) $v = wr$

1) Formulas (1) and (2) contain an $\underline{a}$ in common.

2) Formula (3) contains an $\underline{r}$ and a $\underline{v}$ in common with formula (2).

Continued on following page.

30. Continued

The three formulas at the right form a system because:

(1) $B = \dfrac{m}{a}$

(2) $p = m + n$

(3) $q = a - t$

a) Formulas (1) and (2) contain an _____ in common.

b) Formula (3) contains an _____ in common with formula (1).

---

31. To get a new relationship, <u>at least</u> <u>two</u> <u>variables</u> can be eliminated from any system of three formulas. For example, we can eliminate <u>a</u> and <u>r</u> from the system below and then solve for F. The steps are discussed.

a) m

b) a

(1) $F = ma$

(2) $a = \dfrac{v^2}{r}$

(3) $v = wr$

1) To eliminate <u>a</u>, find two formulas that contain <u>a</u>. The two formulas are (1) and (2).

    a) Solve for <u>a</u> in formulas (1) and (2). (<u>Note</u>: <u>a</u> is already solved-for in formula (2).)

$$F = ma \qquad\qquad a = \dfrac{v^2}{r}$$

$$a = \dfrac{F}{m}$$

    b) Use the equivalence principle to eliminate <u>a</u>. That is, equate the two solutions above.

$$\dfrac{F}{m} = \dfrac{v^2}{r} \qquad \text{(This is a new formula.)}$$

2) To eliminate <u>r</u>, use the new formula above and the original formula that has not been used. That is, use formula (3).

    a) Solve for <u>r</u> in formula (3) and the new formula above.

$$v = wr \qquad\qquad \dfrac{F}{m} = \dfrac{v^2}{r}$$

$$r = \dfrac{v}{w} \qquad\qquad Fr = mv^2$$

$$\qquad\qquad\qquad r = \dfrac{mv^2}{F}$$

    b) Use the equivalence principle to eliminate "r".

$$\dfrac{v}{w} = \dfrac{mv^2}{F} \qquad \text{(This is a new formula. \underline{a} and \underline{r} are eliminated.)}$$

Continued on following page.

31. Continued

   3) Solve for F in the new formula in which a and r are eliminated.

   $$\frac{v}{w} = \frac{mv^2}{F}$$

   $$Fv = mv^2w$$

   $$F = \frac{mv^2w}{v}$$

   $$F = mvw\left(\frac{v}{v}\right) \quad \text{or} \quad mvw(1)$$

   $$F = mvw \quad \underline{\text{Answer}}$$

32. Let's eliminate a and m from the system below and then solve for B.

   (1) $B = \dfrac{m}{a}$
   (2) $p = m + n$
   (3) $q = a - t$

   a) Solve for m in formulas (1) and (2) and equate the two solutions.

   b) Solve for a in the new formula and in formula (3), and equate the two solutions.

   c) Now solve for B in this new formula in which a and m are eliminated.

   a) $aB = p - n$

   b) $\dfrac{p-n}{B} = q + t$

   c) $B = \dfrac{p-n}{q+t}$

422 Systems Of Three Equations And Determinants

33. Eliminate e_o and i from the system below and then solve for A.
(Note: When simplifying at the end, e_i is also eliminated.)

(1) $\quad A = \dfrac{e_o}{e_i}$

(2) $\quad e_o = iR$

(3) $\quad i = \dfrac{me_i}{R + r_p}$

$A = \dfrac{mR}{R + r_p}$

SELF-TEST 29 (pages 411-423)

1. Solve this system by the <u>addition</u> <u>method</u>.

$P + Q + R = 7$
$2P + 2Q - R = 5$
$P - Q + R = 9$

Continued on following page.

SELF-TEST 29 - Continued

2. Solve this system by the equivalence method.

$$\begin{aligned} 2x + y + z &= 6 \\ x - 2y - 2x &= 8 \\ 2x + 3z &= 2 \end{aligned}$$

3. In this system of formulas, eliminate $\underline{s}$ and $\underline{d}$ and then solve for $\underline{r}$.

$$\begin{aligned} p &= ks \\ s &= \frac{d}{r} \\ k &= d - r \end{aligned}$$

ANSWERS: 1. P = 5   2. x = 4   3. $r = \dfrac{k^2}{p - k}$
          Q = -1   y = 0
          R = 3   z = -2

9-9 THIRD ORDER DETERMINANTS

In this section, we will discuss third order or three-by-three determinants.

34. The general form for a third order determinant is:

$$\begin{vmatrix} a_1 & b_1 & c_1 \\ a_2 & b_2 & c_2 \\ a_3 & b_3 & c_3 \end{vmatrix}$$

The rows and columns of the general form are shown below. Since there are 3 rows and 3 columns, the determinant is called a "<u>third order</u>" or "<u>three-by-three</u>" determinant.

Rows
$$\begin{vmatrix} a_1 & b_1 & c_1 \\ a_2 & b_2 & c_2 \\ a_3 & b_3 & c_3 \end{vmatrix}$$

Continued on following page.

424   Systems Of Three Equations And Determinants

34. Continued

a) Circle the middle row in this determinant.

$$\begin{vmatrix} 3 & 2 & 8 \\ 5 & 0 & 1 \\ -1 & 6 & 4 \end{vmatrix}$$

b) Circle the third column in this determinant.

$$\begin{vmatrix} 9 & -3 & 4 \\ 1 & -1 & -3 \\ 2 & -2 & 0 \end{vmatrix}$$

a)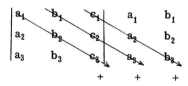

b)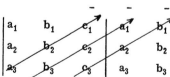

35. The general form of a third order determinant D is shown again below.

$$D = \begin{vmatrix} a_1 & b_1 & c_1 \\ a_2 & b_2 & c_2 \\ a_3 & b_3 & c_3 \end{vmatrix}$$

Any third order determinant equals one number. To find that number, we use the definition below.

$$D = a_1 b_2 c_3 + b_1 c_2 a_3 + c_1 a_2 b_3 - a_3 b_2 c_1 - b_3 c_2 a_1 - c_3 a_2 b_1$$

To find the six terms in the definition, we copy the first two columns to the right of the third column.

$$\begin{vmatrix} a_1 & b_1 & c_1 \\ a_2 & b_2 & c_2 \\ a_3 & b_3 & c_3 \end{vmatrix} \begin{matrix} a_1 & b_1 \\ a_2 & b_2 \\ a_3 & b_3 \end{matrix}$$

The first three terms are the diagonals from left to right starting from a_1, b_1, and c_1. Notice that all three terms ($a_1 b_2 c_3 + b_1 c_2 a_3 + c_1 a_2 b_3$) are <u>positive</u>.

The last three terms are the diagonals from left to right starting from a_3, b_3, and c_3. Notice that all three terms ($-a_3 b_2 c_1 - b_3 c_2 a_1 - c_3 a_2 b_1$) are <u>negative</u>.

Continued on following page.

35. Continued

Let's use the preceding definition to find the value of the determinant D below. We recopied the determinant and wrote the first two columns on the right side.

$$D = \begin{vmatrix} 1 & -2 & -3 \\ 4 & -1 & 1 \\ 2 & 0 & 2 \end{vmatrix}$$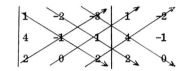

D = (1)(-1)(2) + (-2)(1)(2) + (-3)(4)(0) - (2)(-1)(-3) - (0)(1)(1) - (2)(4)(-2)

= (-2) + (-4) + 0 - 6 - 0 - (-16)

= 4

36. Using the definition in the last frame, we found the value of the determinant below.

$$D = \begin{vmatrix} 3 & 2 & -1 \\ 1 & 0 & -3 \\ 4 & -2 & 5 \end{vmatrix}$$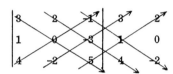

D = 0 + (-24) + 2 - 0 - 18 - 10 = -50

Following the example, find the value of this determinant.

$$D = \begin{vmatrix} 2 & 0 & 1 \\ 0 & 1 & 1 \\ 2 & -1 & 1 \end{vmatrix}$$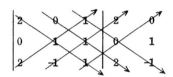

D = ___ + ___ + ___ - ___ - ___ - ___ = ___

37. Using the same method, find the value of each determinant.

a) $\begin{vmatrix} 3 & -2 & -1 \\ 1 & 3 & -3 \\ 2 & -1 & 2 \end{vmatrix}$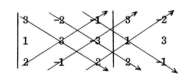

D = ___ + ___ + ___ - ___ - ___ - ___ = ___

b) $\begin{vmatrix} 2 & 3 & -5 \\ 1 & 4 & 3 \\ 3 & -2 & 1 \end{vmatrix}$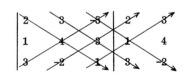

D = ___ + ___ + ___ - ___ - ___ - ___ = ___

D = 2 + 0 + 0 - 2 - (-2) - 0 = 2

38. Using the same method, find the value of each determinant.

a) $$D = \begin{vmatrix} 0 & 2 & 0 \\ 3 & -1 & 1 \\ 1 & -2 & 2 \end{vmatrix}$$

b) $$D = \begin{vmatrix} 1 & 4 & 1 \\ 2 & -1 & -2 \\ 3 & -2 & 1 \end{vmatrix}$$

a) D = 18 + 12 + 1
    − (−6) − 9 − (−4) = 32

b) D = 8 + 27 + 10 − (−60)
    − (−12) − 3 = 114

---

39. The determinant below contains decimal numbers.

$$D = \begin{vmatrix} 1.2 & 2.1 & 0.6 \\ 0.7 & -1.5 & 0.9 \\ 1.4 & -2.2 & -0.1 \end{vmatrix}$$

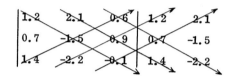

We can use a calculator to find the value of D. The value is 5.685. The steps for the first two terms are:

	Enter	Press	Display
First Term	1.2	[x]	1.2
	1.5	[+/−] [x]	−1.8
	0.1	[+/−] [+]	0.18
Second Term	2.1	[x]	2.1
	0.9	[x]	1.89
	1.4	[+]	2.826

Use a calculator to complete the evaluation. <u>Remember that the last three terms are subtracted.</u>

D = _____

a) D = −10, from:
    0 + 2 + 0 − 0 − 0 − 12

b) D = −38, from:
    (−1) + (−24) + (−4)
    − (−3) − 4 − 8

---

40. Use a calculator to find the value of each determinant.

a) $$D = \begin{vmatrix} 0.8 & 0.7 & 1.7 \\ 1.3 & 2.4 & 1.4 \\ -1.1 & -3.1 & 0.3 \end{vmatrix}$$

D = _____

D = 5.685

Continued on following page.

40. Continued

b) $$D = \begin{vmatrix} 4.3 & 0.8 & -5.1 \\ 0.1 & 6.2 & -0.9 \\ 0.4 & -0.1 & -1.3 \end{vmatrix}$$

   $D = _____$

a) $D = 0.334$

b) $D = -22.53$

9-10 USING DETERMINANTS TO SOLVE SYSTEMS OF THREE EQUATIONS

In this section, we will discuss the determinant method for solving systems of three equations. The determinant method for systems of three equations is also called <u>Cramer's Rule</u>.

41. The standard form for a system of three equations is given below.

$$\boxed{\begin{aligned} a_1x + b_1y + c_1z &= d_1 \\ a_2x + b_2y + c_2z &= d_2 \\ a_3x + b_3y + c_3z &= d_3 \end{aligned}}$$

The a's, b's, and c's are coefficients and the d's are number terms. We can use the a's, b's, c's, and d's to set up determinants to solve the system. The four determinants used are D, D_x, D_y, D_z. They are shown below.

$$D = \begin{vmatrix} a_1 & b_1 & c_1 \\ a_2 & b_2 & c_2 \\ a_3 & b_3 & c_3 \end{vmatrix} \qquad D_x = \begin{vmatrix} d_1 & b_1 & c_1 \\ d_2 & b_2 & c_2 \\ d_3 & b_3 & c_3 \end{vmatrix}$$

$$D_y = \begin{vmatrix} a_1 & d_1 & c_1 \\ a_2 & d_2 & c_2 \\ a_3 & d_3 & c_3 \end{vmatrix} \qquad D_z = \begin{vmatrix} a_1 & b_1 & d_1 \\ a_2 & b_2 & d_2 \\ a_3 & b_3 & d_3 \end{vmatrix}$$

Determinant D is formed by detaching the coefficients of the x's, y's, and z's. Notice these points about D_x, D_y, and D_z.

1) To get D_x from D, we replace the a's (the coefficients of the x's) with the d's (see the arrow).

2) To get D_y from D, we replace the b's (the coefficients of the y's) with the d's (see the arrow).

3) To get D_z from D, we replace the c's (the coefficients of the z's) with the d's (see the arrow).

Continued on following page.

41. Continued

Using the preceding determinants, we can set up the following rules to solve for x, y, and z.

$$x = \frac{D_x}{D} \qquad y = \frac{D_y}{D} \qquad z = \frac{D_z}{D}$$

42.

Using the rules from the last frame, we solved the system below.

$$\begin{array}{rcl} x - 2y + 3z &=& 6 \\ 2x - y - z &=& -3 \\ x + y + z &=& 6 \end{array}$$

$$x = \frac{D_x}{D} = \frac{\begin{vmatrix} 6 & -2 & 3 \\ -3 & -1 & -1 \\ 6 & 1 & 1 \end{vmatrix}}{\begin{vmatrix} 1 & -2 & 3 \\ 2 & -1 & -1 \\ 1 & 1 & 1 \end{vmatrix}} = \frac{15}{15} = 1$$

$$y = \frac{D_y}{D} = \frac{\begin{vmatrix} 1 & 6 & 3 \\ 2 & -3 & -1 \\ 1 & 6 & 1 \end{vmatrix}}{\begin{vmatrix} 1 & -2 & 3 \\ 2 & -1 & -1 \\ 1 & 1 & 1 \end{vmatrix}} = \frac{30}{15} = 2$$

$$z = \frac{D_z}{D} = \frac{\begin{vmatrix} 1 & -2 & 6 \\ 2 & -1 & -3 \\ 1 & 1 & 6 \end{vmatrix}}{\begin{vmatrix} 1 & -2 & 3 \\ 2 & -1 & -1 \\ 1 & 1 & 1 \end{vmatrix}} = \frac{45}{15} = 3$$

The solution is: $x = 1$, $y = 2$, and $z = 3$. Check the solution in each original equation below.

a) $x - 2y + 3z = 6$ b) $2x - y - z = -3$ c) $x + y + z = 6$

43. Use the determinant method to solve this system.

$$2x - y + z = 2$$
$$x + 3y + 2z = 7$$
$$5x + y - z = -9$$

$$x = \frac{D_x}{D} = \underline{\qquad\qquad} = \underline{\qquad} = \underline{\qquad}$$

$$y = \frac{D_y}{D} = \underline{\qquad\qquad} = \underline{\qquad} = \underline{\qquad}$$

$$z = \frac{D_z}{D} = \underline{\qquad\qquad} = \underline{\qquad} = \underline{\qquad}$$

a) $1 - 2(2) + 3(3) = 6$
   $1 - 4 + 9 = 6$
   $6 = 6$

b) $2(1) - 2 - 3 = -3$
   $2 - 2 - 3 = -3$
   $-3 = -3$

c) $1 + 2 + 3 = 6$
   $6 = 6$

---

44. Use the determinant method to solve this system. Use a calculator to evaluate the determinants.

$$1.1F_1 + 0.6F_2 + 1.3F_3 = 7.00$$
$$0.9F_1 + 0.7F_2 + 1.5F_3 = 7.65$$
$$2.1F_1 + 0.1F_2 + 0.8F_3 = 5.45$$

$$F_1 = \frac{D_{F_1}}{D} = \underline{\qquad\qquad} = \underline{\qquad} = \underline{\qquad}$$

$$x = \frac{\begin{vmatrix} 2 & -1 & 1 \\ 7 & 3 & 2 \\ -9 & 1 & -1 \end{vmatrix}}{\begin{vmatrix} 2 & -1 & 1 \\ 1 & 3 & 2 \\ 5 & 1 & -1 \end{vmatrix}} = \frac{35}{-35} = -1$$

$$y = \frac{\begin{vmatrix} 2 & 2 & 1 \\ 1 & 7 & 2 \\ 5 & -9 & -1 \end{vmatrix}}{\begin{vmatrix} 2 & -1 & 1 \\ 1 & 3 & 2 \\ 5 & 1 & -1 \end{vmatrix}} = \frac{0}{-35} = 0$$

$$z = \frac{\begin{vmatrix} 2 & -1 & 2 \\ 1 & 3 & 7 \\ 5 & 1 & -9 \end{vmatrix}}{\begin{vmatrix} 2 & -1 & 1 \\ 1 & 3 & 2 \\ 5 & 1 & -1 \end{vmatrix}} = \frac{-140}{-35} = 4$$

Continued on following page.

430 Systems Of Three Equations And Determinants

44. Continued

$$F_2 = \frac{D_{F_2}}{D} = \underline{} = \underline{} = \underline{}$$

$$F_3 = \frac{D_{F_3}}{D} = \underline{} = \underline{} = \underline{}$$

$$F_1 = \frac{\begin{vmatrix} 7.00 & 0.6 & 1.3 \\ 7.65 & 0.7 & 1.5 \\ 5.45 & 0.1 & 0.8 \end{vmatrix}}{\begin{vmatrix} 1.1 & 0.6 & 1.3 \\ 0.9 & 0.7 & 1.5 \\ 2.1 & 0.1 & 0.8 \end{vmatrix}} = \frac{0.138}{0.115} = 1.2$$

$$F_2 = \frac{\begin{vmatrix} 1.1 & 7.00 & 1.3 \\ 0.9 & 7.65 & 1.5 \\ 2.1 & 5.45 & 0.8 \end{vmatrix}}{\begin{vmatrix} 1.1 & 0.6 & 1.3 \\ 0.9 & 0.7 & 1.5 \\ 2.1 & 0.1 & 0.8 \end{vmatrix}} = \frac{0.2415}{0.115} = 2.1$$

$$F_3 = \frac{\begin{vmatrix} 1.1 & 0.6 & 7.00 \\ 0.9 & 0.7 & 7.65 \\ 2.1 & 0.1 & 5.45 \end{vmatrix}}{\begin{vmatrix} 1.1 & 0.6 & 1.3 \\ 0.9 & 0.7 & 1.5 \\ 2.1 & 0.1 & 0.8 \end{vmatrix}} = \frac{0.391}{0.115} = 3.4$$

Solving systems of three equations is a lengthly process, even when a calculator is used. Fortunately, programmable calculators can be used to solve systems of that type. When a programmable calculator is used, the calculator does all the calculations and prints out the solution. The operator only has to enter the coefficients and number terms.

SELF-TEST 30 (pages 423-431)

Evaluation each determinant.

1. $\begin{vmatrix} 1 & -3 & 2 \\ -1 & 4 & -3 \\ 2 & -1 & 1 \end{vmatrix}$

2. $\begin{vmatrix} 5 & 0 & 2 \\ 1 & -3 & 0 \\ 0 & 4 & -2 \end{vmatrix}$

3. Use the determinant method to solve this system of three equations.

$$\begin{aligned} x + 2y - z &= 7 \\ 4x + y + 2z &= 9 \\ 2x - 3y + z &= 1 \end{aligned}$$

4. Use the determinant method to solve this system of three equations.

$$\begin{aligned} 1.5r + 0.2s + 0.5t &= 6 \\ 0.4r + s + 0.3t &= 7 \\ 0.6r + 0.6s + 0.2t &= 5 \end{aligned}$$

ANSWERS: 1. 2
2. 38
3. $x = 3$, $y = 1$, $z = -2$
4. $r = 2$, $s = 5$, $t = 4$

432   Systems Of Three Equations And Determinants

SUPPLEMENTARY PROGRAMS - CHAPTER 9

Assignment 28

Solve each system of equations using the method listed.

1. Addition Method
$$3x - 5y = 22$$
$$4x + 5y = 6$$

2. Equivalence Method
$$2P + R = 13$$
$$3P + R = 16$$

3. Substitution Method
$$h = 3 - 3k$$
$$2h + 5k = 4$$

Find the numerical value of each determinant.

4. $\begin{vmatrix} 2 & 3 \\ 7 & 5 \end{vmatrix}$
5. $\begin{vmatrix} -3 & -5 \\ 6 & 4 \end{vmatrix}$
6. $\begin{vmatrix} 8 & 12 \\ -7 & 0 \end{vmatrix}$
7. $\begin{vmatrix} 8.7 & 3.9 \\ -1.4 & -2.6 \end{vmatrix}$

Solve each system of equations by the determinant method. Record the numerical values of the <u>three</u> determinants used in each solution.

8. $4x + 5y = 10$
   $3x + 2y = 11$
   a) $D = ?$ d) $x = ?$
   b) $D_x = ?$ e) $y = ?$
   c) $D_y = ?$

9. $3t - 2w = 6$
   $4t - 3w = 10$
   a) $D = ?$ d) $t = ?$
   b) $D_t = ?$ e) $w = ?$
   c) $D_w = ?$

10. $8Q = 6 - 5P$
    $2P = 3Q + 21$
    a) $D = ?$ d) $P = ?$
    b) $D_P = ?$ e) $Q = ?$
    c) $D_Q = ?$

11. $a = 2.5b + 3$
    $b = 1.5a - 10$
    a) $D = ?$ d) $a = ?$
    b) $D_a = ?$ e) $b = ?$
    c) $D_b = ?$

12. $9r - 1 = 5t$
    $27 - 6t = 8r$
    a) $D = ?$ d) $r = ?$
    b) $D_r = ?$ e) $t = ?$
    c) $D_t = ?$

13. $3.7x - 2.3y = 5$
    $1.9x + 5.2y = 8$
    (Round x and y to hundredths.)
    a) $D = ?$ d) $x = ?$
    b) $D_x = ?$ e) $y = ?$
    c) $D_y = ?$

14. Using electrical principles, the following system was set up to calculate currents i_1 and i_2 in a circuit. Find i_1 and i_2. Round each to hundredths.
$$2.8i_1 + 3.4i_2 = 6.3$$
$$4.3i_1 - 1.5i_2 = 6.3$$

15. Using principles of equilibrium, the following system was set up to calculate forces F_1 and F_2 in a structure. Find F_1 and F_2. Round to the nearest whole number.
$$85F_1 - 54F_2 = 8,800$$
$$62F_1 - 15F_2 = 9,400$$

Assignment 29

Solve each system of equations by the addition method.

1. $2x - y + z = 9$
   $x - 3y - z = 4$
   $3x + y + z = 10$

2. $r + s - 2t = 8$
   $2r - s + 2t = 4$
   $r + s = 2$

3. $a + c + 6 = b$
   $2a + b + c = 2$
   $a + 2b = c + 7$

Solve each system of equations by the equivalence method.

4. $h + k + p = 5$
   $h + 2k - p = 13$
   $h - k - 2p = 3$

5. $t + v + 2w = 4$
   $t - 2v + w = 5$
   $v + w = 1$

6. $F_1 + F_2 = F_3 + 9$
   $F_2 = F_1 + 2$
   $F_1 = 8 - F_3$

Continued on following page.

Assignment 29 - Continued

Do each formula derivation.

7. Eliminate A and P and then solve for V.

   $$\begin{array}{|l|} \hline V = P - A \\ B = A + F \\ P = KV \\ \hline \end{array}$$

8. Eliminate w and r and then solve for a.

   $$\begin{array}{|l|} \hline t = \dfrac{w}{d} \\ r = w + p \\ d = ar \\ \hline \end{array}$$

9. Eliminate b and r and then solve for h.

   $$\begin{array}{|l|} \hline s = bt \\ b = \dfrac{h^2}{r} \\ h = kr \\ \hline \end{array}$$

Assignment 30

Evaluate each determinant.

1. $\begin{vmatrix} 2 & 1 & 1 \\ 1 & 4 & 2 \\ 3 & 0 & 1 \end{vmatrix}$

2. $\begin{vmatrix} 1 & -2 & -1 \\ 3 & 0 & 2 \\ 0 & 5 & 4 \end{vmatrix}$

3. $\begin{vmatrix} 0.2 & 1 & -0.5 \\ -1 & 1.5 & 0.1 \\ 2 & -2.4 & 3 \end{vmatrix}$

Solve each system of equations by the determinant method. Record the numerical values of the <u>four</u> determinants used in each solution.

4. $\begin{array}{|l|} \hline x + y + z = 3 \\ 2x + y - z = 0 \\ x + 2y + z = 6 \\ \hline \end{array}$

   a) D = ?   e) x = ?
   b) D_x = ?   f) y = ?
   c) D_y = ?   g) z = ?
   d) D_z = ?

5. $\begin{array}{|l|} \hline t + 2v + w = 8 \\ 2t - v - w = 6 \\ t - 2v + w = 0 \\ \hline \end{array}$

   a) D = ?   e) t = ?
   b) D_t = ?   f) v = ?
   c) D_v = ?   g) w = ?
   d) D_w = ?

6. $\begin{array}{|l|} \hline 2a + b + c = 5 \\ a - 3b - c = 2 \\ 4a - b - c = 1 \\ \hline \end{array}$

   a) D = ?   e) a = ?
   b) D_a = ?   f) b = ?
   c) D_b = ?   g) c = ?
   d) D_c = ?

7. The following system was set up to find the forces on a steel beam. Solve the system and find F_1, F_2, and F_3.

   $$\begin{array}{|l|} \hline 12F_1 + 25F_2 - 20F_3 = 3,200 \\ 15F_1 - 30F_2 + 30F_3 = 0 \\ 10F_1 + 15F_2 - 12F_3 = 2,200 \\ \hline \end{array}$$

8. The following system was set up to find three currents in an electronics circuit. Solve the system and find i_1, i_2, and i_3. Round to hundredths.

   $$\begin{array}{|l|} \hline 1.4i_1 - i_2 + 0.2i_3 = 3 \\ i_1 + 0.5i_2 + 1.2i_3 = 5 \\ 0.8i_1 - 1.6i_2 - i_3 = 0 \\ \hline \end{array}$$

ANSWERS FOR SUPPLEMENTARY PROBLEMS

CHAPTER 1 - POLYNOMIALS

Assignment 1

1. t^5 2. m^7 3. a^5b^3 4. p^2r^6 5. x^4y^5 6. $s^4v^3w^5$ 7. $d^4h^2k^3$ 8. $a^5c^2d^3$ 9. p^4
10. xy^3 11. r^3t 12. h^2 13. a^2dk 14. t^3vw 15. hr^2 16. m^3s^2 17. x^3y^6 18. r^8t^6
19. 1 20. $d^8h^{12}p^4$ 21. $2y^3 - 3y^2 + 4y - 1$ 22. $a^2y^3 + a^3y^2 + ay + 2a$ 23. $ry^5 - by^3 + ky - 5t$
24. 1 25. 2 26. 3 27. 3 28. 6 29. binomial 30. monomial 31. trinomial
32. binomial

Assignment 2

1. $2x^2 + 1$ 2. $3x^3 - x$ 3. $4x^4 + x^2 - 1$ 4. $6x^2 - 5$ 5. $kx^4 + ax^2 - b$ 6. $2y^2 + 4$ 7. $-y^3 + 3y + 2$
8. $y^2 - y + 3$ 9. $xy^4 - 3$ 10. $2by^2 + dy$ 11. $ty + 2w$ 12. $-12r^4t^3$ 13. abx^3y^2 14. $10d^4k^6$
15. $-18a^3p^4s^2$ 16. $16c^2d^2$ 17. $4x^6$ 18. $s^4t^2w^8$ 19. $25h^6p^2$ 20. $4x^3 - 4x$ 21. $2a^3 + a^2b$
22. $3my^4 - 2m^2y$ 23. $2t^3 - 2t^2 + 10t$ 24. $8cd^3 + 8d^2k + 8dp$ 25. $x^3y + x^2y^2 + xy^3$
26. $2r^5w + 3r^4w^2 + 5r^3w$ 27. $3m^3v^3 - 6mv^4 + 3m^2v^2$ 28. $15h^3s^5 - 5h^4s^4 - 10h^3s^4$

Assignment 3

1. $br + bt + dr + dt$ 2. $xy - 2x + 3y - 6$ 3. $3ap + aw - 12cp - 4cw$ 4. $h^2s - 2h^2t - p^2s + 2p^2t$
5. $6x^2 - xy - 2y^2$ 6. $20h^2 + 3hk - 2k^2$ 7. $c^4 - 3c^2d + 2d^2$ 8. $8r^4 + 10r^2t^2 + 3t^4$ 9. $m^2 - n^2$
10. $25p^2 - 9w^2$ 11. $r^4 - t^4$ 12. $x^2y^2 - 4$ 13. $4s^2 + 4sv + v^2$ 14. $9x^2 - 24xy + 16y^2$
15. $a^4 - 10a^2b^2 + 25b^4$ 16. $25r^4 + 30r^2w + 9w^2$ 17. $2d$ 18. $3t^2$ 19. k^2p 20. $6b^3$
21. $3(3y^2 - 5)$ 22. $2t(5t + 3)$ 23. $xy(x - 2 + y)$ 24. $4a^2b^2(3a^2 + a - 2b^2)$ 25. $x^2(2x^2 + 3x - 1)$
26. $5dh(d - 3h - 2)$ 27. $(R + r)(R - r)$ 28. $(p + 4t)(p - 4t)$ 29. $(1 + bc)(1 - bc)$
30. $(3s^2 + t^2)(3s^2 - t^2)$ 31. $(5ad^2 + 8p)(5ad^2 - 8p)$ 32. $(k + 3)^2$ 33. $(2y - 5)^2$ 34. $(m + 2n)^2$
35. $(3P - 2)^2$ 36. $(4dh - 1)^2$

Assignment 4

1. $(x + 1)(x + 3)$ 2. $(t + 1)(t - 7)$ 3. $(w - 2)(w - 5)$ 4. $(d - 3)(d + 4)$ 5. $(a + b)(a - 2b)$
6. $(x + 2y)(x - 6y)$ 7. $(r^2 + 1)(r^2 + 4)$ 8. $(pt + 1)(pt + 6)$ 9. $(3y + 2)(y + 1)$ 10. $(5h - 1)(h + 2)$
11. $(3x - 2)(2x - 3)$ 12. $(2t - 5)(2t + 3)$ 13. $(2r - s)(r + 3s)$ 14. $(5a + b)(a + 2b)$
15. $(3tw + 2)(tw - 3)$ 16. $(4x - 3y)(x + 2y)$ 17. $2(p + 3)(p - 3)$ 18. $5(y + 2)(y + 3)$
19. $8(d + k)(d - k)$ 20. $h(r + 2s)^2$ 21. $3x^2$ 22. $-h$ 23. $-4ac$ 24. $4t^2 - 3$ 25. $3rs^2 + 2s$
26. $5y^2 - 2y + 4$ 27. $3p^2w + px - 2$ 28. $m^3t^2 - m^2t - m$ 29. $x + 6$ 30. $d - 5$ 31. $a - 6$
32. $3y - 4$ 33. $2r + 3$ 34. $5t - 2$

CHAPTER 2 - LINEAR EQUATIONS

Assignment 5

1. non-linear 2. linear 3. linear 4. non-linear 5. linear 6. (-3, 0) 7. (5, 0)
8. $\left(\frac{1}{2}, 0\right)$ 9. $\left(-\frac{8}{5}, 0\right)$ 10. (0, -9) 11. (0, 2) 12. $\left(0, \frac{1}{3}\right)$ 13. $\left(0, -\frac{5}{2}\right)$

14. 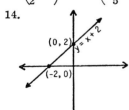 15. 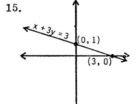 16.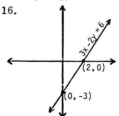

17. Slope = 1 18. Slope = $-\frac{1}{2}$ 19. Slope = -4 20. Slope = $\frac{1}{6}$ 21. A, D 22. A 23. B, C, E

435

Assignment 5 (Continued)
24. C   25. C   26. 32 units   27. 3 units   28. 3.95 units   29. $\Delta y = -8$ 30. $\Delta y = 15$
31. $\Delta y = -1.5$ 32. Slope = 5; y-intercept (0, 2)   33. Slope = 1; y-intercept (0, -3)
34. Slope = $-\frac{2}{5}$; y-intercept $(0, \frac{3}{2})$ 35. Slope = -4.2; y-intercept (0, -7.5)   36. $y = -x + 2$
37. $y = 3x - 5$ 38. $y = \frac{1}{2}x - 1$ 39. $y = \frac{8}{3}x + \frac{7}{4}$

Assignment 6
1. $y = -4x + 6$; $m = -4$; (0, 6)   2. $y = 3x + 2$; $m = 3$; (0, 2)   3. $y = x - 5$; $m = 1$; (0, -5)
4. $y = \frac{1}{4}x - 2$; $m = \frac{1}{4}$; (0, -2)   5. $y = -3x + \frac{3}{2}$; $m = -3$; $(0, \frac{3}{2})$ 6. $y = \frac{1}{2}x - \frac{5}{3}$; $m = \frac{1}{2}$; $(0, -\frac{5}{3})$
7. $m = 2$ 8. $m = -1$ 9. $m = -\frac{3}{5}$ 10. $m = \frac{1}{4}$ 11. $m = 1$ 12. $m = -\frac{3}{5}$ 13. $y = 3x - 6$
14. $y = -x + 2$ 15. $y = \frac{1}{3}x + 5$ 16. $y = \frac{5}{2}x + 8$ 17. $y = -4x + 6$ 18. $y = -\frac{1}{2}x + \frac{7}{2}$
19. $y = 2x - 1$ 20. $y = -3x + 6$ 21. $y = -x + 1$ 22. $y = \frac{1}{2}x - 2$ 23. $y = -\frac{2}{3}x + \frac{2}{3}$
24. $y = \frac{5}{2}x - 20$ 25. a, d, e   26. $y = \frac{3}{4}x$ 27. $y = 2x$ 28. $y = -3x$ 29. $y = \frac{1}{2}x$
30. c, d, and e   31. $y = 6$ 32. $x = -3$ 33. $x = 5$ 34. $y = -8$ 35. 0   36. undefined

Assignment 7
1. (6, 0) and (0, 9)   2. (30, 0) and (0, -6)   3. (6, 0) and (0, -4)   4. (-7, 0) and (0, 2)
5. a) (See graph)
   b) Slope = $\frac{2}{3}$
6. a) (See graph)
   b) Slope = -2

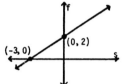
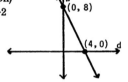

7. a) $v = 3t + 5$
   b) Slope = 3
   c) (0, 5)
8. a) $p = \frac{3}{4}h - 5$
   b) Slope = $\frac{3}{4}$
   c) (0, -5)
9. a) $s = -\frac{1}{2}r + 1$
   b) Slope = $-\frac{1}{2}$
   c) (0, 1)
10. a) $T = -\frac{4}{5}V + 10$
    b) Slope = $-\frac{4}{5}$
    c) (0, 10)
11. a) 5-unit increase
    b) 20-unit increase
12. a) $\frac{4}{3}$-unit increase
    b) 16-unit increase
13. $d = 8w$ 14. $P = \frac{2}{5}R$ 15. 90 meters
16. $F = \frac{1}{2}W$ 17. $p = 8t$ 18. $V = 25G$ 19. $y = 2x - 2$ 20. $d = -\frac{1}{2}s + 3$

CHAPTER 3 - SQUARE ROOT RADICALS
Assignment 8
1. a, b, e   2. 3 and -3   3. 6 and -6   4. 9 and -9   5. 20 and -20   6. 17   7. 380
8. 15.3   9. .024   10. $W = 10$ 11. $h = 24$ 12. $A = 20$ 13. $p = 6$ 14. $F = 8$ 15. $s = 5$
16. $\sqrt{15xy}$ 17. $4pr\sqrt{2dt}$ 18. $\sqrt{21aw}$ 19. $\sqrt{75}$ 20. $\sqrt{bd^2}$ 21. $\sqrt{12x}$ 22. $\sqrt{8r^2}$ 23. $3\sqrt{5}$
24. $10\sqrt{3x}$ 25. $2r\sqrt{7w}$ 26. $2h^2p\sqrt{hp}$ 27. $6\sqrt{y}$ 28. $4a^2\sqrt{3b}$ 29. $10tw\sqrt{2w}$ 30. $2h^2k^2\sqrt{3h}$
31. $2\sqrt{2}$ 32. $3\sqrt{x}$ 33. $p^2\sqrt{d}$ 34. $5ab\sqrt{a}$

Assignment 9
1. $\frac{2}{3}$ 2. $\frac{p^3}{a^2}$ 3. $\frac{2t\sqrt{2t}}{5d}$ 4. $\frac{4x\sqrt{y}}{3w}$ 5. $b\sqrt{\frac{a}{2}}$ 6. $3\sqrt{rv}$ 7. $a\sqrt{\frac{b}{3d}}$ 8. $\frac{1}{2h}$ 9. $\sqrt{\frac{3k}{r}}$
10. $\frac{1}{t}\sqrt{\frac{p}{5}}$ 11. $h\sqrt{2bh}$ 12. $\frac{1}{r}\sqrt{\frac{1}{t}}$ 13. $4x$ 14. $2x$ 15. $\frac{9p}{4r}$ 16. $\frac{3p}{2r}$ 17. 8   18. 3   19. $\frac{a}{d}$
20. $\frac{9s}{w}$ 21. (b) and (d)   22. $\frac{\sqrt{3}}{2}$ 23. $\frac{a\sqrt{k}}{2dk}$ 24. $\frac{v\sqrt{mp}}{pt}$ 25. $\frac{\sqrt{2rw}}{w}$ 26. $\frac{2}{\sqrt{5}}$ 27. $\frac{x}{\sqrt{y}}$
28. $\frac{ad}{p\sqrt{dh}}$ 29. $\frac{v}{2\sqrt{sv}}$ 30. $9\sqrt{3}$ 31. $4\sqrt{F}$ 32. $(p + r)\sqrt{y}$ 33. $\sqrt{10}$ 34. $-4\sqrt{x}$
35. $(K - W)\sqrt{A}$ 36. Only (a)

Assignment 10
1. (a) and (d)   2. Not possible   3. $a+b$ 4. Not possible   5. cd 6. $\sqrt{2r^2-2r}$
7. $p\sqrt{dw+2d}$ 8. $2b\sqrt{a^2+4a}$ 9. $h-p$ 10. $3\sqrt{x^2+4y^2}$ 11. $a\sqrt{t-w}$
12. $\dfrac{h\sqrt{2p-r}}{3p}$ 13. $\dfrac{\sqrt{m^2+r^2}}{4}$ 14. $\dfrac{2\sqrt{r-1}}{r-1}$ 15. $3\sqrt{y+2}$ 16. $\dfrac{a+b}{c\sqrt{a+b}}$ 17. $\dfrac{k(t-1)}{2t\sqrt{t-1}}$
18. $11\sqrt{x+y}$ 19. $3\sqrt{m-1}$ 20. $2\sqrt{h+4}$ 21. $\dfrac{8}{3}$ 22. $\dfrac{8b}{a^2}$ 23. $\dfrac{r}{w}$ 24. $4x(x-y)$
25. $9-6\sqrt{F}+F$ 26. $\dfrac{x+8\sqrt{x}+16}{16}$ 27. $9+4\sqrt{t+5}+t$ 28. $10-6\sqrt{1-R}-R$ 29. 22   30. $a-4$
31. $1-w$ 32. $\dfrac{5+\sqrt{3}}{11}$ 33. $\dfrac{3(\sqrt{x}-1)}{x-1}$ 34. $\dfrac{1}{2(2+\sqrt{2})}$ 35. $\dfrac{r-25}{2(\sqrt{r}-5)}$

Assignment 11
1. $4i$ 2. $3i$ 3. $10i$ 4. $-7i$ 5. $-i$ 6. $12i$ 7. 0 8. -4 9. $-42i$ 10. 10
11. $-2+i$ 12. $0-4i$ or $-4i$ 13. $5+0i$ or 5 14. $-4+5i$ 15. $-12+0i$ or -12
16. $0+10i$ or $10i$ 17. $18-i$ 18. $9+5i$ 19. $-30-24i$ 20. 29 21. 10 22. 100
23. $1-2i$ 24. $-2+3i$ 25. $-\dfrac{6}{5}+\dfrac{7i}{5}$ or $-1.2+1.4i$ 26. $\dfrac{3}{13}-\dfrac{11i}{13}$ 27. $x=3i$ and $-3i$
28. $y=7i$ and $-7i$ 29. $w=2i$ and $-2i$ 30. $x=2+2i$ and $2-2i$ 31. $t=1+3i$ and $1-3i$
32. $y=2+3i$ and $2-3i$

CHAPTER 4 - RADICAL EQUATIONS AND FORMULAS

Assignment 12
1. $x=7$ 2. $t=18$ 3. $y=6$ 4. $w=\dfrac{1}{10}$ 5. $P=14$ 6. $r=\dfrac{5}{4}$ 7. $x=\pm 3$ 8. $y=\pm\dfrac{2}{3}$
9. $t=2$ 10. $x=\dfrac{9}{20}$ 11. $s=3$ 12. $w=4$ 13. $d=12$ 14. $y=\dfrac{2}{25}$ 15. $A=3$ 16. $m=\dfrac{19}{9}$
17. $x=\dfrac{1}{16}$ 18. $R=14$ 19. $d=\dfrac{4}{9}$ 20. $s=15$ 21. $w=16$ 22. $r=2$ 23. $v=1$ 24. $p_1=7$
25. $k=5$ 26. $B_1=4$ 27. $V_2=9$ 28. $t=18$ 29. $y=5$ (not $y=2$)   30. $x=2$ (not $x=-2$)
31. $t=6$ (not $t=0$)   32. $w=1$ and $w=2$

Assignment 13
1. $A=s^2$ 2. $r=\dfrac{d}{a^2}$ 3. $m=\dfrac{p^2+w}{2}$ 4. $N=\dfrac{BH^2}{G}$ 5. $P=\dfrac{E^2}{R}$ 6. $h=s^2t$ 7. $a=\dfrac{bp}{h^2}$
8. $E=\dfrac{P^2}{F}$ 9. $w=\dfrac{d^2-b^2}{2}$ 10. $K=\dfrac{1}{AT^2}$ 11. $V=\dfrac{G^2}{G^2+1}$ 12. $t=\dfrac{1}{1-v^2}$ 13. $R=\left(\dfrac{W}{P}\right)^2$
14. $t=4d^2$ 15. $h=\left(\dfrac{rs}{b}\right)^2$ 16. $x=\dfrac{v^2}{2p^2}$ 17. $A=\dfrac{1}{BF^2K^2}$ 18. $C=(HT)^2-N$ 19. $p=\dfrac{dt^2}{4\pi^2}$
20. $r=\dfrac{a^2s}{bw^2}$ 21. $d=(v-t)^2$ 22. $m=(a-b)^2$ 23. $P=\left(\dfrac{C+H}{B}\right)^2$ 24. $W=\left(\dfrac{G-F}{R}\right)^2$
25. $B=P\sqrt{G}$ 26. $A=\dfrac{DF}{\sqrt{R}}$ 27. $c=h-k\sqrt{p}$ 28. $m=\dfrac{t-r}{\sqrt{a}}$ 29. $V_1=V_2\sqrt{\dfrac{A_1}{A_2}}$ 30. $b=a\sqrt{\dfrac{t}{r}}$
31. $h=\dfrac{1}{p}\sqrt{\dfrac{s_1}{s_2}}$ 32. $K=P\sqrt{\dfrac{F}{P}}$ or $\sqrt{FP}$ 33. $v=\sqrt{d^2-w^2}$ 34. $P=\sqrt{B^2-G^2}$ 35. $A=\sqrt{1-\dfrac{R^2}{D^2}}$
36. $d=\sqrt{\dfrac{1}{t^2}-1}$

Assignment 14
1. $d=2\sqrt{A}$ 2. $r_1=r_2\sqrt{\dfrac{k}{p}}$ 3. $R=V\sqrt{T}$ 4. $F=B\sqrt{\dfrac{1}{P}}$ 5. $t_1=t_2\sqrt{\dfrac{W_1}{W_2}}$ 6. $m=r\sqrt{\dfrac{s}{a}}$ 7. $P=G\sqrt{\dfrac{B}{2K}}$
8. $w=\dfrac{1}{2\sqrt{pt}}$ 9. $V=\sqrt{KR}$ 10. $m=\dfrac{1}{p}\sqrt{\dfrac{h}{t}}$ 11. $T=\dfrac{H}{A^2G^2}$ 12. $a=\dfrac{s^2}{s+1}$ 13. $V=\dfrac{F^2}{W}$
14. $r=k\sqrt{b^2+p^2}$ 15. $N=\sqrt{\dfrac{B}{F}}$ 16. $w=\sqrt{2kt}$ 17. $P=R\sqrt{\dfrac{F}{V}}$ 18. $w=\dfrac{1}{\sqrt{bp}}$ 19. $S=\dfrac{\sqrt{R^2-T^2}}{K}$
20. $a=\sqrt{\dfrac{d^2-b^2}{2}}$ 21. $F=\sqrt{\dfrac{V}{A}}$ 22. $d=\dfrac{r}{a}$ 23. $t=\dfrac{c}{4w^2}$ 24. $V=\sqrt{\dfrac{2E}{M}}$ 25. $P=\dfrac{h}{k}$
26. $t=\dfrac{p\pm\sqrt{p^2-4dn}}{2d}$ 27. $F=-m\pm\sqrt{m^2-h}$ 28. $w=\dfrac{s\pm\sqrt{s^2+4r}}{2}$ 29. $V=\dfrac{-p\pm\sqrt{p^2+k^2}}{k}$

438   Answers

CHAPTER 5 - FUNCTIONS, NON-LINEAR GRAPHS, VARIATION

Assignment 15
1. y   2. t   3. (b) and (c)   4. (a) and (c)   5. h   6. horizontal axis   7. $f(x) = x^3 + 4$
8. $f(t) = 5t^2 - 2t$ 9. $f(-1) = 10$ 10. $f(0) = -3$ 11. $f(a - 1) = a^2 - 1$ 12. $f(2h) = \dfrac{2h}{h - 1}$
13. $f(3, 4) = \dfrac{3}{2}$ 14. $f(1, 0) = \dfrac{1}{3}$ 15. (a) and (b)   16. 7 ... -4   17. 80 ... 0   18. $y = x + 1$
19. $w = 5t$ 20. $P = -3T + 60$ 21. (a) and (d)   22. (a) (0, -3)   (c) (0, -3)
            (b) (-2, 0) and (2, 0)   (d) -3
23. (a) (2, 4)   (c) (0, 0)
    (b) (0, 0) and (4, 0)   (d) 4

Assignment 16
1. $x^2 + y^2 = 4$ 2. $(x + 1)^2 + (y - 1)^2 = 16$ 3. $x^2 + (y - 2)^2 = 9$ 4. $\dfrac{x^2}{9} + \dfrac{y^2}{1} = 1$ 5. $\dfrac{x^2}{4} + \dfrac{y^2}{16} = 1$
6. $y = \dfrac{2}{3}x$ and $y = -\dfrac{2}{3}x$ 7. x-axis and y-axis   8. (a) and (b)   9. A decreases   10. d decreases
11. (a) $k = 3$ 12. (a) $k = 200$ 13. (a) and (d)   14. $G = 100$ 15. $w = 40$
    (b) $P = 3V$ (b) $f = \dfrac{200}{t}$
16. $d = 375$ kilometers   17. $X_L = 120$ ohms   18. $B = 3$ 19. $h = 60$ 20. $X_C = 32$ ohms
21. $V = 60$ liters

Assignment 17
1. $A = 0.7854d^2$ 2. $t = 250$ 3. $H = 18$ 4. $R = \dfrac{50}{I^2}$ 5. $d = 2$ 6. $W = 40$ 7. $E = 200$ joules
8. $P = 36$ watts   9. $F = 8$ newtons   10. $I = 3$ microwatts   11. $T = 240$ 12. $w = 490$
13. $A = \dfrac{bh}{r^2}$ 14. $t = \dfrac{p^2}{a+b}$ 15. $P = \dfrac{m(w-r)}{a^2}$ 16. A increases   17. N decreases   18. F increases

CHAPTER 6 - EXPONENTS, POWERS, AND ROOTS

Assignment 18
1. 16   2. 9   3. 64   4. 1   5. 243   6. $\dfrac{1}{8}$ 7. $\dfrac{1}{25}$ 8. $\dfrac{1}{10}$ 9. $\dfrac{1}{x^4}$ 10. $\dfrac{1}{w}$ 11. y^{-3}
12. R^{-1} 13. 5^{-1} 14. 4^{-2} 15. a^2 16. h^4 17. y^{-1} 18. $40x$ 19. d^3t^{-2} 20. p^{-3}
21. $63x^{-1}y^2$ 22. $24a^3b^{-4}$ 23. x^{-2} 24. P^2 25. r^{-1} 26. $5ab^4$ 27. $5y^3$ 28. $\dfrac{3}{4m^2}$
29. $\dfrac{p}{2}$ 30. $\dfrac{2h^3}{3}$ 31. t^{-3} 32. 10^8 33. a^2b^{-5} 34. $x^{-1}y^{-2}$ 35. s^2w^{-3} 36. $kp^{-2}t^{-1}$
37. ax^4y^2 38. $2dh^{-1}m$ 39. $\dfrac{x}{y^2}$ 40. $\dfrac{T}{R}$ 41. $\dfrac{1}{m^3n^3}$ 42. $\dfrac{a^5c^4}{b^2d^3}$

Assignment 19
1. w^{20} 2. 1   3. y^{-6} 4. P^2 5. a^3 6. x^8y^4 7. $a^{-3}b^6$ 8. $w^{-2}v^2$ 9. 1   10. $125r^3w^{-3}$
11. $32m^{-10}n^{15}$ 12. $d^{-6}h^4p^2$ 13. $s^3v^{-12}w^6$ 14. $\dfrac{1}{16}$ 15. $\dfrac{x^3}{64}$ 16. $\dfrac{a^{-3}}{b^{12}}$ 17. $\dfrac{p^{-4}r^2}{d^6h^{-2}}$ 18. $\left(\dfrac{k}{w}\right)^2$
19. $\left(\dfrac{2}{m}\right)^1$ or $\dfrac{2}{m}$ 20. $\left(\dfrac{c}{ab}\right)^3$ 21. $\left(\dfrac{2t}{s}\right)^5$ 22. $\sqrt[3]{50}$ 23. $\sqrt[5]{x}$ 24. $\sqrt{A}$ 25. 14   26. 4
27. 0.042   28. 2.4   29. 2.49   30. 0.986   31. 41.7   32. P^5 33. a^{-4} 34. t^{-2} 35. b^5
36. x^1 or x 37. m^{-1} 38. $(3y)^{-5}$ 39. $\left(\dfrac{v}{p}\right)^2$

Assignment 20
1. $\sqrt[4]{8^3}$ 2. $\sqrt[3]{x}$ 3. $\sqrt{a^{-3}}$ 4. $\sqrt[5]{10^{-2}}$ 5. $\sqrt{p}$ 6. $R^{\frac{2}{3}}$ 7. $b^{\frac{1}{8}}$ 8. $7^{\frac{1}{2}}$ 9. $k^{-\frac{5}{4}}$ 10. $y^{-\frac{1}{2}}$
11. $t^{\frac{5}{4}}$ 12. $A^{\frac{1}{6}}$ 13. $y^{\frac{2}{3}}$ 14. $P^{\frac{3}{2}}$ 15. $X^{1.5}$ 16. $N^{0.25}$ 17. $r^{-0.5}$ 18. $p^{-1.6}$ 19. $\sqrt{P}$
20. $\sqrt[3]{y^1}$ 21. $\sqrt[4]{b^{-3}}$ 22. $\sqrt{V^{-5}}$ 23. $y^{3.6}$ 24. $w^{0.6}$ 25. $x^{0.60}$ 26. $M^{-0.5}$ 27. 262,144
28. 0.07776   29. 0.000125   30. 0.0625   31. 240   32. 4.42   33. 0.135   34. $P = 73.9$
35. $Q = 1.77$ 36. $F = 34.5$ 37. $A = 232$ 38. $w = 57.6$

CHAPTER 7 - COMMON LOGARITHMS

Assignment 21
1. 1,000   2. 0.00001   3. 10   4. 1   5. 10^2 6. 10^6 7. 10^{-3} 8. 10^{-1} 9. 63,100
10. 0.00655   11. 1.24   12. $10^{1.7642}$ 13. $10^{7.7952}$ 14. $10^{-4.0809}$ 15. $10^{-0.5287}$ 16. 1
17. 1,000   18. 0.01   19. 0.000151   20. 26.1   21. 7   22. 1.2175   23. -4   24. -0.0375
25. 10^4 26. $10^{-0.5172}$ 27. T = 642   28. D = 2.25   29. R = 0.0376   30. A = 0.114
31. (d) Between 100 and 1,000   32. 2 = log 100   33. log 0.1 = -1   34. log 59.2 = 1.7723
35. -2.68 = log 0.00209   36. $1 = 10^0$ 37. $10^{-3} = 0.001$ 38. $10^{4.13} = 13,490$ 39. $0.875 = 10^{-0.058}$
40. y = 5   41. y = -2   42. No value for "y", since negative numbers have no logarithms.

Assignment 22
1. x = 1.9069   2. t = -1.2874   3. A = 4.8627   4. r = -3.8356   5. P = 2,240   6. V = 0.192
7. w = 79.4   8. d = 0.00308   9. S = 32.6   10. P = 4.96   11. v = 2.26   12. H = 54
13. M = 15,500   14. V = 0.000302   15. k = 156   16. A = 73.3   17. a = 3.19   18. E_2 = 40.8
19. P_1 = 0.202   20. a = 2.72   21. R = 358   22. $\log P = \frac{a}{k}$ 23. log A = -H
24. $m = \frac{s}{\log w}$ 25. $c = \frac{p - t}{\log F}$ 26. H = B log R - V   27. $\log t = \frac{r - a}{d}$

Assignment 23
1. logarithmic scale   2. uniform scale   3. three cycles   4. semi-log graph   5. A: (0.05, 4)
6. B: (0.3, 6)   7. C: (7, 7)   8. D: (0.03, 10)   9. E: (0.9, 16)   10. F: (4, 17)   11. log G - log N
12. p log V   13. log 3 + log R   14. Not possible   15. log t = v - log h   16. log G = log A - P
17. $d = \frac{N}{\log R}$ 18. $\log M = \frac{S}{K} + \log W$ 19. $r = \frac{f}{A \log T}$ 20. $\log P = \frac{w}{b} - \log 2$ 21. R = 29.4
22. a = 4.27   23. Q = 248   24. P_1 = 0.119

Assignment 24
1. y = 14.4   2. x = 2.35   3. V = 131   4. t = 8.97   5. w = 20.8   6. s = 0.828
7. log x + 3 log y   8. 4(log A - log B)   9. log R + t log (d + 1)   10. 2 log w - log p
11. log R = log K + s log T   12. log h = 1.8 (log r - log w)   13. log A = log 800 + 20 log (1 + i)
14. M = 99.3   15. H = 23,117   16. f = 2.91   17. Q = 11.4   18. F_2 = 608   19. P = 89,500
20. t = 2.81   21. V = 296   22. i = 0.0694

CHAPTER 8 - POWERS OF "e" AND NATURAL LOGARITHMS

Assignment 25
1. 7.54   2. 362,000   3. 0.144   4. $e^{1.6993}$ 5. $e^{7.8747}$ 6. $e^{0.1655}$ 7. $e^{-4.9576}$ 8. (0, 80)
9. descending   10. y = 0 (the x-axis)   11. y = 265.6   12. y = 24.10   13. (0, 0)
14. ascending   15. y = 200   16. y = 27.9   17. y = 155   18. S = 1.86   19. G = 11.8
20. i = 11.0   21. M = 18,700   22. E = 97.4   23. V = 20.0   24. A = \$2,225.54   25. T = 50°C
26. v = 153 volts   27. N = 4.79 grams

Assignment 26
1. 3.7353   2. 6.9078   3. 0   4. -2.6064   5. 8.76   6. 442,000   7. 0.206
8. ln 40 = 3.6889   9. -1.63 = ln 0.196   10. ln P = t   11. $0.00674 = e^{-5}$ 12. $e^0 = 1$
13. $r = e^a$ 14. R = 38.5   15. h = 2.65   16. V = 0.159   17. x = 0.737   18. w = 0.7634
19. y = -5.38   20. A = 9.27   21. r = 24.9   22. N = 2.078   23. A = 12,900   24. s = -3.750
25. v = 41.2   26. v = 3,200   27. t = 5.61 seconds   28. t = 7.58 hours   29. t = 1,620 years

Assignment 27
1. b ln R   2. ln K - ln N   3. ln 2 + ln w   4. $\ln\left(\frac{t}{p}\right)$ 5. $\ln Q^k$ 6. $\ln\left(\frac{1}{G}\right)$ 7. False
8. True   9. False   10. $P = \frac{H}{\ln Q}$ 11. $r = \frac{\ln A}{b}$ 12. ln T = -kt   13. ln A = kt + ln P
14. $\ln R = \frac{a}{d} + \ln V$ 15. $t = \frac{w}{\ln B}$ 16. $T = e^{-w}$ 17. $A = e^{\frac{p}{b}}$ 18. $K = Ne^c$ 19. $G = He^{-r}$

Continued on following page.

Assignment 27 - Continued

20. $R = De^{kt}$ 21. $n = me^{-\frac{s}{a}}$ 22. $p = \frac{1}{a}\ln M$ 23. $v = \ln\left(\frac{F}{k}\right)$ 24. $r = \frac{1}{t}\ln\left(\frac{A}{P}\right)$

25. $s = \frac{1}{c}\ln\left(\frac{B}{H}\right)$ 26. $t = RC \ln\left(\frac{V}{v}\right)$ 27. $t = \frac{L}{R}\ln\left(\frac{I}{I-i}\right)$ 28. $R = Ae^{kw}$ 29. $d = ke^{-cp}$

30. $v = Ve^{-\frac{Rt}{L}}$ 31. $y = \exp(-aw)$ 32. $p = P \exp\left(\frac{v}{2n}\right)$ 33. $v = V\left[1 - \exp\left(-\frac{t}{RC}\right)\right]$

CHAPTER 9 - SYSTEMS OF THREE EQUATIONS AND DETERMINANTS

Assignment 28
1. $x = 4$, $y = -2$ 2. $P = 3$, $R = 7$ 3. $h = -3$, $k = 2$ 4. -11 5. 18 6. 84 7. -17.16
8. a) $D = -7$ b) $D_x = -35$ c) $D_y = 14$ d) $x = 5$ e) $y = -2$ 9. a) $D = -1$ b) $D_t = 2$ c) $D_w = 6$
   d) $t = -2$ e) $w = -6$ 10. a) $D = -31$ b) $D_P = -186$ c) $D_Q = 93$ d) $P = 6$ e) $Q = -3$
11. a) $D = 2.75$ b) $D_a = 22$ c) $D_b = 5.5$ d) $a = 8$ e) $b = 2$
12. a) $D = 94$ b) $D_r = 141$ c) $D_t = 235$ d) $r = 1.5$ e) $t = 2.5$
13. a) $D = 23.61$ b) $D_x = 44.4$ c) $D_y = 20.1$ d) $x = 1.88$ e) $y = 0.85$
14. $i_1 = 1.64$, $i_2 = 0.50$ 15. $F_1 = 181$, $F_2 = 122$

Assignment 29
1. $x = 3$, $y = -1$, $z = 2$ 2. $r = 4$, $s = -2$, $t = -3$ 3. $a = -2$, $b = 5$, $c = 1$ 4. $h = 3$, $k = 4$, $p = -2$
5. $t = 1$, $v = -1$, $w = 2$ 6. $F_1 = 5$, $F_2 = 7$, $F_3 = 3$
7. $V = \frac{B-F}{K-1}$ or $V = \frac{F-B}{1-K}$ 8. $a = \frac{d}{dt+p}$ 9. $h = \frac{s}{kt}$

Assignment 30
1. 1 2. -1 3. 4.448
4. a) $D = 3$ b) $D_x = -3$ c) $D_y = 9$ d) $D_z = 3$ e) $x = -1$ f) $y = 3$ g) $z = 1$
5. a) $D = -12$ b) $D_t = -48$ c) $D_v = -24$ d) $D_w = 0$ e) $t = 4$ f) $v = 2$ g) $w = 0$
6. a) $D = 12$ b) $D_a = 12$ c) $D_b = -24$ d) $D_c = 60$ e) $a = 1$ f) $b = -2$ g) $c = 5$
7. $F_1 = 100$, $F_2 = 200$, $F_3 = 150$ 8. $i_1 = 6.29$, $i_2 = 5.16$, $i_3 = -3.23$

INDEX

Addition
    in solving systems 401, 411-416
    of complex numbers 156-158
    of polynomials 19-20
    of radicals 134-136, 143
Asymptote 234, 358
Axis of symmetry 227

Base 1
Binomial(s) 11
    as radicands 137-143
    division by 53-55
    factoring of 35-39
    FOIL method 28
    multiplication of 27-33
    squaring 31-33, 146-149

Calculators
    and determinants 404, 426
    and "e" 353-355, 359-363, 370-372
    and exponential equations 338-339
    and exponents 304-306
    and formulas 293-295
    and logarithms 311-313, 315-320, 372-375
    and square roots 110-111, 280
    powers 290-295
    roots 281, 286
Circle 230-232
Combined Variation 254-256
Combining Like Terms 15-20
Common Factors 35-36
Common Logarithms 308-313
Complete Complex Numbers 156
    as solutions to equations 163-165
Complex numbers 155-156
    addition of 156-158
    complete 156
    division of 161-163
    incomplete 156
    multiplication of 158-161
    subtraction of 157-158
Conics
    circle 230-232
    ellipse 232-233
    hyperbola 234-235
    parabola 225-229
Conjugates
    of complex numbers 161
    of radicals 150

Constant 215
    of variation 237, 240, 244, 248
Cramer's Rule 405-408, 427-430
Cube Root 278

Decimal Exponents 288-290, 304-306
Degree
    of a term 13
Dependent Variable 215-217, 238, 242, 246
Determinants
    second order 403-405
    third order 423-427
Direct Square Variation 244-247
Direct Variation 237-240
Division
    by binomials 53-55
    by monomials 51-53
    law of exponents 5, 266
    of complex numbers 161-163
    of radicals 122-123, 141
Domain 217

"e" 353
Ellipse 232-233
Equations
    exponential 338-343, 370
    logarithmic 315-317, 371-372
    radical 169-183
    see also: Functions, Formulas
Equivalence Method for Solving Systems 402, 416-419
Estimating Linear Equations 99-102
"exp" Notation 394-395
Exponent(s) 1
    decimal 288-290, 304-306
    fractional 283-288
    in division 266-268
    in multiplication 264-265
    laws of 4, 5, 9, 263-278, 287-288
    negative 263, 270-272, 303-304
    powers of ten 302-306
    raising to a power 9, 273-274
Exponential Functions
    evaluation of 343-347
    "exp" notation 394-395
    graph of 295-297, 306-308, 355-359, 365-366
    solution of 338-341, 370
Extraneous Roots 182-183

Factoring
    binomials 35-39
    completely 50-51
    difference of two perfect squares 37-39
    monomials 33-35
    perfect square trinomials 40-42
    trinomials 40-51
Focus 228-229
    property of a parabola 229
FOIL Method 28, 148, 159
Formula Derivation
    the equivalence method 201-205
    the substitution method 206-208
    with three equations 419-422
Formula Evaluation
    and calculators 293-295
    exponential 343-347, 359-364
    involving radicals 111-113, 179-181, 184-197
    logarithmic 317-325, 372-380
Formulas
    and logarithms 317-325, 334-337
    and natural logarithms 372-380, 384-394
    exponential 343-347
    graphs of 365-366
    in quadratic form 208-210
Fractional Exponents 283-288
Fractions
    involving radicals 124-127, 130-134
    powers of 276-278
Function 215-229
    exponential 295-297, 306-308, 338-343
    linear 224-225
    logarithmic 313-314
    notation 219-222
    of "e" 355-359
    of several variables 223
    Quadratic 225-229

Graphs
    of exponential functions 295-297, 306-308, 355-359, 365-366
    of functions 218-219
    of linear equations 62, 90-91
    of logarithmic functions 313-314
    on log paper 330-332

Horizontal Intercept 60-61, 90-91
Horizontal Lines 85-87
Hyperbola 234-235

Imaginary Numbers 153-155, 216-217
    as solutions to equations 163-165
Incomplete Complex Numbers 156

Independent Variable 215-217, 238, 242, 246
Intercepts
    horizontal 60-61, 90-91
    used in graphing 62, 90-91
    vertical 60-61, 90-91
    x-intercept 60-61
Inverse Square Variation 248-250
Inverse Variation 240-243

Joint Variation 252-253

Laws of Exponents 4, 5, 9, 263-278, 287-288
Laws of Logarithms 332-337, 382-383
Like Terms 15-16
    combining 15-20, 22-23
Linear Equations (s) 58-59
    estimation of 99-102
    slope-intercept form 71-77, 94-99
    using point-point 80-81
    using point-slope 79-80
Linear Functions 224-225
    see also: Line(s)
Line(s)
    equations of 60-81, 94-102
    estimating the graph of 99-102
    horizontal 85-87
    parallel 84
    slope of 63-67
    through the origin 81-83, 92-99
    vertical 85-87
Logarithmic Formulas 317-325
    and laws of logarithms 334-337
Logarithmic Function
    graph of 313-314
    solution of 315-317, 371-372
Logarithmic Scales 327-332
Logarithms
    common 308-313
    laws of 332-337, 382-383
    natural 367-370
Log Principle for Equations 338-341

Monomial 11
    division by 51-53
    factoring 33-36
    multiplication by 24-26
    powers of 274-275
Multiplication
    by FOIL method 28
    by monomials 24-26
    law of exponents 4, 264
    of binomials 27-33
    of complex numbers 158-161
    of radicals 113-116, 140

Natural Logarithms 367-370
Negative Exponent 263, 270-272, 303-304
Non-linear Equation 58-59

Opposites
   of complex numbers 157-158
   of polynomials 21
Origin
   lines through 81-83, 97-99, 238

Parabola 225-229
Parallel Lines 84
Perfect Squares 116-121, 139, 198-201
Perfect Square Trinomials 40-42
Point-point Form of a Linear Equation 80-81
Point-slope Form of a Linear Equation 79-80
Polynomials
   in one variable 10-12
   in several variables 12-14
   addition of 19-20
   opposites of 21
   subtraction of 22-23
   factoring completely 50-51
Powers 2, 262
   of "e" 353-355
   of fractions 276-278
   of ten 302-306
   on calculators 290-295
   raising to a power 9, 273-274
   zero exponent 7
Principles
   squaring for radical equations 169-170
   squaring for radical formulas 184-197
Principle Square Root 109
Proportional
   see: Variation

Quadratic equations
   with complex solutions 163-165
Quadratic Formula 164, 210
Quadratic Functions 225-229

Radical 110
Radical Equations 169-179, 182-183
Radicals
   addition and subtraction of 134-136, 143
   containing binomials 137-143
   division of 122-123, 141
   multiplication of 113-116, 140
   simplifying 116-128
   squaring 129, 143-149
Radicand 110
Range 217

Rationalizing Denominators 130-132, 142, 150-151
Rationalizing Numerators 132-134, 142, 151
Real Numbers 153
Reciprocals 268-270
Roots 278-283
   of powers 281-284

Simplifying Polynomials 17-20, 22-23
Simplifying Radicals 116-128, 198-201
Slope
   and axes of different scales 102-103
   of a line 63-71, 92-99
   two point formula 77-78
Slope-Intercept Form 71-77, 94-97
Square Root 109, 278
   and calculators 110-111, 280
   principle 109
Squaring
   binomials 31-33, 146-149
   monomials 24-25
   radicals 129, 143-149
Squaring Principle
   for radical equations 169-170
   for radical formulas 184-192
Substitution Method for Solving Systems 402
Subtraction
   of complex numbers 157-158
   of polynomials 22-23
   of radicals 135-136, 143
Systems of Equations
   solution by Cramer's Rule 405-408, 427-430
   solution by the addition method 401, 411-416
   solution by the equivalence method 402, 416-419
   solution by the substitution method 402-403
   three equations 411-419, 427-430
   two equations 401-410

Trinomial(s) 11
   factoring of 40-51

Variables 215
   dependent 215-217, 238, 242, 246
   independent 215-217, 238, 242, 246
Variation 235-256
   direct 237-240
   inverse 240-243
   direct square 244-247
   inverse square 248-250
   joint 252-253
   combined 254-256
Vertex
   of a parabola 226-227
Vertical Intercept 60-61, 72, 90-91

Vertical Lines 85-87
Vertical Line Test 218-219

X-intercept 60-61

Y-intercept 60-61

Zero
   as an exponent 7, 303